E-Commerce Trends for Organizational Advancement:
New Applications and Methods

Mehdi Khosrow-Pour
Information Resources Management Association, USA

Information Science REFERENCE

INFORMATION SCIENCE REFERENCE
Hershey · New York

Director of Editorial Content: Kristin Klinger
Senior Managing Editor: Jamie Snavely
Assistant Managing Editor: Michael Brehm
Publishing Assistant: Sean Woznicki
Typesetter: Michael Brehm, Kurt Smith
Cover Design: Lisa Tosheff
Printed at: Yurchak Printing Inc.

Published in the United States of America by
Information Science Reference (an imprint of IGI Global)
701 E. Chocolate Avenue
Hershey PA 17033
Tel: 717-533-8845
Fax: 717-533-8661
E-mail: cust@igi-global.com
Web site: http://www.igi-global.com/reference

Library of Congress Cataloging-in-Publication Data

E-commerce trends for organizational advancement : new applications and methods / Mehdi Khosrow-Pour, editor.
p. cm.
Includes bibliographical references and index.
Summary: "This book gathers essential research on the changing face of commerce, investigating the development, delivery, and perception of e- and m-commerce systems and tools"--Provided by publisher.
ISBN 978-1-60566-964-9 (hardcover) -- ISBN 978-1-60566-965-6 (ebook) 1. Electronic commerce. 2. Mobile commerce. I. Khosrowpour, Mehdi, 1951-
HF5548.32.E18655 2010
658.8'72--dc22
2009042992

British Cataloguing in Publication Data
A Cataloguing in Publication record for this book is available from the British Library.

Advances in Electronic Commerce (AEC)

ISBN: 1935-2921

Editor-in-Chief: Mehdi Khosrow-Pour, D.B.A.

Consumer Behavior, Organizational Development, and Electronic Commerce: Emerging Issues for Advancing Modern Socioeconomies

Mehdi Khosrow-Pour, Information Resources Management Association, USA

Information Science Reference • copright 2008 • 354pp • H/C (ISBN: 978-1-60566-126-1)

Considering the global spread of e-commerce technologies and the rapid pace of organizational adoption of these technological advancements, there is a need for reliable research results on e-services, outsourcing applications, and consumer resources management. Consumer Behavior, Organizational Development, and Electronic Commerce: Emerging Issues for Advancing Modern Socioeconomies offers readers a one-stop resource for contemporary issues, developments, and influences in e-commerce. Through in-depth literature reviews and inventive methodologies, academics, students, and practitioners will find this publication to be a ready reference—suitable for university or corporate library collections and advanced coursework—for a complete depiction of technologies and their impact on modern global socioeconomics.

Outsourcing and Offshoring of Professional Services: Business Optimization in a Global Economy

Amar Gupta, University of Arizona, USA

Information Science Reference • copyright 2008 • 438pp • H/C (ISBN: 978-1-59904-972-4)

A growing number of companies are opting to perform increasing types of professional services in foreign countries, creating, for some companies, unprecedented opportunities to reduce costs and nucleate strategic relationships, while, for others, representing a major threat to current prosperity. Outsourcing and Offshoring of Professional Services: Business Optimization in a Global Economy discusses the considerations and implications surrounding the outsourcing and offshoring of professional services, such as software development computer-aided design, and healthcare, from multiple global perspectives. This Premier Reference Source, offering industry professionals, policymakers, students, and educators with a balance between a broad overview and detailed analysis of offshore outsourcing, is an invaluable addition to academic, research, and corporate libraries. This publication includes a foreward by Lester C. Thurow, Jerome and Dorothy Lemelson Professor of Management and Economics and Former Dean, MIT Sloan School of Management.

Utilizing and Managing Commerce and Services Online

Mehdi Khosrow-Pour, Information Resources Management Association, USA

CyberTech Publishing • copyright 2007 • 351pp • H/C (ISBN: 1-59140-932-2)

As businesses, researchers, and practitioners look to devise new and innovative technologies in the realm of e-commerce, the human side in contemporary organizations remains a test in the industry. Utilizing and Managing Commerce and Services Online broadens the overall body of knowledge regarding the human aspects of electronic commerce technologies and utilization in modern organizations. Utilizing and Managing Commerce and Services Online provides comprehensive coverage and understanding of the social, cultural, organizational, and cognitive impacts of e-commerce technologies and advances in organizations around the world. E-commerce strategic management, leadership, organizational behavior, development, and employee ethical issues are only a few of the challenges presented in this all-inclusive work.

The Advances in Electronic Commerce (AEC) Book Series is designed to provide comprehensive coverage and understanding of the social, cultural, organizational, and cognitive impacts of e-commerce technologies around the world. These accounts can be viewed from the impacts of electronic commerce on consumer behavior, as well as the influence of e-commerce on organizational behavior, development, and management in organizations. The secondary objective of this book series is to expand the overall body of knowledge regarding the human aspects of electronic commerce technologies and utilization in modern business organizations, assisting researchers and practitioners to devise more effective systems for managing the human side of e-commerce. It is through these objectives that the AEC Series seeks to make available literature on emerging research and methodologies in E-Commerce. All audiences, students, researchers, academicians and practitioners, will find useful the breaking accounts of E-Commerce principles and will benefit from the high-quality sources of reference this series intends to provide. By becoming a contributor to the Advances in Electronic Commerce (AEC) Book Series , you will be granted an opportunity that few ever gain. Your work will be showcased in a collection that finds wide acceptance by both libraries and international indexes, and it will be supported by a diverse international editorial advisory board. Given that this series is closely associated with the Journal of Electronic Commerce in Organizations, many students, practitioners, and researchers who use the journal as a resource will now also turn to the series to find new, salient content.

IGI GLOBAL
DISSEMINATOR OF KNOWLEDGE

Hershey • New York

Order online at www.igi-global.com or call 717-533-8845 x100 –
Mon-Fri 8:30 am - 5:00 pm (est) or fax 24 hours a day 717-5 33-7115

Editorial Advisory Board

Table of Contents

Detailed Table of Contents

The recent advances in wireless and mobile communication technologies enable users to conduct commercial activities anywhere and at any time. In this new environment, the designing of appropriate applications constitutes both a necessity and a challenge at the same time in order to support effectively the movable user. This chapter investigates the environment in which these applications operate, identifies possible categories of them and proposes guidelines for their development process.

The emergence of wireless and mobile networks has made possible the introduction of electronic commerce to a new application and research subject: mobile commerce. Mobile commerce is a promising trend of commerce because Internet-enabled smartphones such as iPhones are becoming very popular these days. People use smartphones to perform daily tasks like browsing the mobile Internet and making phone calls anytime and anywhere. However, understanding or constructing a mobile or an electronic commerce system is not easy because the system involves a wide variety of disciplines and technologies and the technologies are constantly changed. To facilitate understanding and constructing such a system, this chapter divides the system into six components: (i) applications, (ii) client devices or computers, (iii) mobile middleware, (iv) wireless networks, (v) wired networks, and (vi) host computers. Elements in these components specifically related to the subject are described in detail and lists of current technologies for component construction are discussed.

The combination of position fixing mechanisms with location-dependent, geographical information, can offer truly customized personal communication services through the mobile phone or other type of devices. Prompted by the avalanche of technology advances in the aforementioned areas, this chapter presents a generic framework for delivering Location Based Services (LBS). The framework is capable of providing the full functionality required for delivering LBS, starting from the specification of the service, covering issues like its deployment and maintenance of services, the service invocation and the final delivery of the produced results to the calling user. The main focus of the chapter is on the technical specification, the design and the functionality of the framework. However, with the purpose of assessing the proposed architecture, a prototype implementation based on the discussed specifications was built and its performance was evaluated using a series of pilot services.

This chapter deals with context-aware query processing in ad-hoc peer-to-peer networks. Each peer in such an environment has a database over which users execute queries. This database involves (a) relations which are locally stored and (b) virtual relations, all the tuples of which are collected from peers that are present in the network at the time when a query is posed. The objective of this work is to perform query processing in such an environment and, to this end, the authors start with a formal definition of the system model. Next, they formally define SQLP, an extension of SQL that covers the termination of queries, the failure of individual peers and the semantic characteristics of the peers of such a network. Moreover, they present a query execution algorithm as well as the formal definition of all the operators that take place in a query execution plan.

The service encounter occurs whenever a customer interacts with a company personally or through technology through, for example, the Internet, e-mail, or telephone. Nowadays, customers frequently initiate the encounter as, for example, inquiries, information searches, and complaints are conveniently performed online. This chapter explores the role of digital service encounters on customer perceptions

Based on customer cognitive, affective and conative experiences in Internet online shopping, this study, from customers' perspectives, develops a conceptual framework for e-CRM to explain the psychological process that customers maintain a long-term exchange relationship with specific online retailer. The conceptual framework proposes a series of causal linkages among the key variables affecting customer commitment to specific online retailer, such as perceived value (as cognitive belief), satisfaction (as affective experience) and trust (as conative relationship intention). Three key exogenous variables affecting Internet online shopping experiences, such as perceived service quality, perceived product quality, and perceived price fairness, are integrated into the framework. This study empirically tested and supported a large part of the proposed framework and the causal linkages within it. The empirical results highlight some managerial implications for successfully developing and implementing a strategy for e-CRM.

Advances in wireless communications and information technology have made the Mobile Web a reality. The Mobile Web is the response to the need for anytime, anywhere access to information and services. Many wireless applications have already been deployed and are available to customers via their mobile phones and wirelessly-connected PDAs. However, as communications and other IT usage becomes an integral part of many people's lives and the available products and services become more varied and capable, users expect to be able to personalize a service to meet their individual needs and preferences. The involved sectors have to meet these challenges by reengineering their front-end and back-end office. This chapter examines the interaction requirements regarding a friendlier, personalized and more effective multi-channel services environment. It presents the mobility challenges and constraints implemented into the business sector, investigating the current m-commerce situation and the extended user characteristics presenting a high level user-centric m-commerce architecture.

The growth of business-to-consumer (B2C) e-commerce has gained a lot of attention among SMEs. Most B2C firms are turning their attention on how to retain new customers and are left in a situation to compete with larger firms. This chapter aims to examine the evolution and impact of online relationships in B2C firms. Drawing upon the theories of customer relationship management, e-commerce, trust and loyalty, the author develops an integrated model to illustrate the stages customers go through when shopping online and its impact on trust, hard and soft factors that reflect the level of transactional and relational components of communication thereby impacting the customers shopping time lifecycle experiences. This chapter develop a number of hypotheses to facilitate testing of the model. The author then provides a discussion of the model leading to the theoretical and practical implications of this study and directions for future research.

This chapter presents the consumer view of an innovative m-service, that is, the e-newspaper, which is published for a mobile reading device equipped with an e-paper display. The research question of this chapter is: What are the implications of the consumer view of future m-service innovation on business models? In this chapter, the authors present empirical results from an online survey with 3,626 respondents representing the consumer view. The factor analysis revealed three aspects of consumer preferences, that is, Ubiquitous access, Prestige of news source and Local anchorage and advertising. These were then correlated with media behavior and e-newspaper preferences to indicate three possible market segments. The authors then discuss the implications that these consumer views could have on developing new business models, followed by a comparison to earlier research. Finally, they propose an integrated e-newspaper business model framework consisting of three models, ubiquitous, local, and prestige.

E-services have been recognized for their contribution to strengthening customer relationships and fostering customer loyalty. While prior research has focused on the role of e-services in business-to-consumer interactions, this chapter studies e-services in the business-to-business domain. More specifically, it explores how the preferences for e-services and electronic channels change over time as a result of developing technology and evolving customer requirements. Based on the case study of ETA SA, a Swiss manufacturer of watch movements and components, three stages of increasing electronic interaction with customers have been identified. The company pioneered online ordering of watch spare parts based on a first generation e-commerce solution and subsequently introduced a comprehensive customer service portal. This research suggests that moving from the e-commerce solution to the customer process portal involved a significant extension of the e-service portfolio (service innovation), whereas the subsequent stage introduced an additional electronic channel (channel innovation). Every stage in this evolution

Preface

While buying and selling goods and services once necessitated a face-to-face transaction, much of the commerce we now undertake is completely electronic. Recent advances in electronic and mobile commerce, precipitated by innovations in technology and user acceptance, have led to subsequent changes in individual and organizational behavior. *E-Commerce Trends for Organizational Advancement: New Applications and Methods* gathers essential research on the changing face of commerce, investigating the development, delivery, and perception of e- and m- commerce systems and tools.

Chapter 1, "*Developing Mobile Commerce Applications*" by P. Benou and V. Bitos discusses the fact that the recent advances in wireless and mobile communication technologies enable users to conduct commercial activities anywhere and at any time. In this new environment, the designing of appropriate applications constitutes both a necessity and a challenge at the same time in order to support effectively the movable user. This chapter investigates the environment in which these applications operate, identifies possible categories of them and proposes guidelines for their development process.

Chapter 2, "*A Technological Perspective of Mobile and Electronic Commerce Systems*" by Wen-Chen Hu, Yanjun Zuo, Naima Kaabouch, and Lei Chen divides the process of understanding or constructing a mobile or an electronic commerce system into six components: (i) applications, (ii) client devices or computers, (iii) mobile middleware, (iv) wireless networks, (v) wired networks, and (vi) host computers. Elements in these components specifically related to the subject are described in detail and lists of current technologies for component construction are discussed.

Chapter 3, "*A Component-Based Software Architecture for Delivering Location Based Services*" by Ioannis Priggouris, Dimitris Spanoudakis, Manos Spanoudakis, and Stathes Hadjiefthymiades, presents a generic framework for delivering Location Based Services (LBS). The framework is capable of providing the full functionality required for delivering LBS, starting from the specification of the service, covering issues like its deployment and maintenance of services, the service invocation and the final delivery of the produced results to the calling user. The main focus of the chapter is on the technical specification, the design and the functionality of the framework. However, with the purpose of assessing the proposed architecture, a prototype implementation based on the discussed specifications was built and its performance was evaluated using a series of pilot services.

Chapter 4, "*Context-Aware Query Processing in Ad-Hoc Environments of Peers*" by Nikolaos Folinas, Panos Vassiliadis, Evaggelia Pitoura, Evangelos Papapetrou, and Apostolos Zarras, explores with context-aware query processing in ad-hoc peer-to-peer networks. Each peer in such an environment has a database over which users execute queries. This database involves (a) relations which are locally stored and (b) virtual relations, all the tuples of which are collected from peers that are present in the network at the time when a query is posed. The objective of this work is to perform query processing in such an environment and, to this end, the authors start with a formal definition of the system model. Next, they formally define SQLP, an extension of SQL that covers the termination of queries, the failure of individual peers and the semantic characteristics of the peers of such a network. Moreover, they present a query execution algorithm as well as the formal definition of all the operators that take place in a query execution plan.

Chapter 5, "*The Role of Digital Service Encounters on Customers' Perceptions of Companies*" by Kristina Heinonen investigates the role of digital service encounters on customer perceptions of companies. Digital service encounters in this chapter denote remote customer-company interactions via the Internet or e-mail. The focus is on active customers initiating interactions and on customer perceptions of company responses to these interactions. A conceptual framework that captures customer perceived service encounter value on two dimensions (responsiveness and personalization) is proposed. An empirical study exploring the value of company responses to digital contacts indicated that many contacts are responded to promptly and satisfyingly. However, there are also significant differences in the value of the service encounter. Some service encounters are perceived as unpersonalized, and some are even left without response.

Chapter 6, "*Contract Negotiation in E-Marketplaces: A Model Based on Dependency Relations*" by Larbi Esmahi, presents an overview of protocols commonly used to implement negotiation in e-markets. An analysis of the interaction process within e-markets according to different situation of individual and joint profit/cost is presented. The authors also present a case study of a marketplace for e-services using dependency relations within the negotiation process. The experimental results of this negotiation model show that a combination of utility functions and dependency relations increase the number of contacts and reduce the differences between agents' individual profit. Finally, this chapter concludes with the introduction of some potential research problems related to e-markets, which will be explored within future extensions of this work.

Chapter 7, "*Designing Reputation and Trust Management Systems*" by Roman Beck and Jochen Franke analyzes the handling of customer complaints after shipping ordered goods by applying automated reputation and trust accounts as decision support. Customer complaints are cost intensive and difficult to standardize. A game theory based analysis of the process yields insights into unfavorable interactions between both business partners. Trust and reputation mechanisms have been found useful in addressing these types of interactions. A reputation and trust management system (RTMS) is proposed based on design theory guidelines as an IS artifact to prevent customers from issuing false complaints. A generic simulation setting for analysis of the mechanism is presented to evaluate the applicability of the RTMS. The findings suggest that the RTMS performs best in market environments where transaction frequency is high, individual complaint-handling costs are high compared to product revenues, and the market has a high fraction of potentially cheating customers.

Chapter 8, "*An Empirical Study on the Integrated Framework of e-CRM in Online Shopping: Evaluating the Relationships Among Perceived Value, Satisfaction, and Trust Based on Customers' Perspectives*" by Changsu Kim, Weihong Zhao, and Kyung Hoon Yang develops a conceptual framework for e-CRM to explain the psychological process that customers maintain a long-term exchange relationship with specific online retailer. The conceptual framework proposes a series of causal linkages among the key variables affecting customer commitment to specific online retailer, such as perceived value (as cognitive belief), satisfaction (as affective experience) and trust (as conative relationship intention). Three key exogenous variables affecting Internet online shopping experiences, such as perceived service quality, perceived product quality, and perceived price fairness, are integrated into the framework. This study empirically tested and supported a large part of the proposed framework and the causal linkages within it. The empirical results highlight some managerial implications for successfully developing and implementing a strategy for e-CRM.

Chapter 9, "*Improving M-Commerce Services Effectiveness with the Use of User-Centric Content Delivery*" by Panagiotis Germanakos, Nikos Tsianos, Zacharias Lekkas, Constantinos Mourlas, and George Samaras, examines the interaction requirements regarding a friendlier, personalized and more effective multi-channel services environment. It presents the mobility challenges and constraints implemented into the business sector, investigating the current m-commerce situation and the extended user characteristics presenting a high level user-centric m-commerce architecture.

Chapter 10, "*The Evolution of Online Relationships in Business to Consumer E-Commerce*" by Pauline Ratnasingam, aims to examine the evolution and impact of online relationships in B2C firms. Drawing upon the theories of customer relationship management, e-commerce, trust and loyalty, the author develops an integrated model to illustrate the stages customers go through when shopping online and its impact on trust, hard and soft factors that reflect the level of transactional and relational components of communication thereby impacting the customers shopping time lifecycle experiences. This chapter develops a number of hypotheses to facilitate testing of the model. The author then provides a discussion of the model leading to the theoretical and practical implications of this study and directions for future research.

Chapter 11, "*Business Models for M-Services: Exploring the E-Newspaper Case from a Consumer View*" by Carina Ihlström Eriksson, Thomas Kalling, Maria Åkesson, and Tobias Fredberg, presents the consumer view of an innovative m-service, that is, the e-newspaper, which is published for a mobile reading device equipped with an e-paper display. The research question of this chapter is: What are the implications of the consumer view of future m-service innovation on business models? In this chapter, the authors present empirical results from an online survey with 3,626 respondents representing the consumer view. The factor analysis revealed three aspects of consumer preferences, that is, Ubiquitous access, Prestige of news source and Local anchorage and advertising. These were then correlated with media behavior and e-newspaper preferences to indicate three possible market segments. The authors then discuss the implications that these consumer views could have on developing new business models, followed by a comparison to earlier research. Finally, they propose an integrated e-newspaper business model framework consisting of three models, ubiquitous, local, and prestige.

Chapter 12, "*The Evolving Portfolio of Business-to-Business E-Services: Service and Channel Innovation*" by Christine Legner, studies e-services in the business-to-business domain. More specifically, it explores how the preferences for e-services and electronic channels change over time as a result of developing technology and evolving customer requirements. Based on the case study of ETA SA, a Swiss manufacturer of watch movements and components, three stages of increasing electronic interaction with customers have been identified. The company pioneered online ordering of watch spare parts based on a first generation e-commerce solution and subsequently introduced a comprehensive customer service portal. This research suggests that moving from the e-commerce solution to the customer process portal involved a significant extension of the e-service portfolio (service innovation), whereas the subsequent stage introduced an additional electronic channel (channel innovation). Every stage in this evolution path forces companies to substantially re-architect their interorganizational process and system linkages. From the experiences of ETA SA, we conclude that a well-designed B2B architecture is required to cope with the growing number of e-services and the complexity of serving multiple electronic channels.

Chapter 13, "*Conceptualizing Competences in E-Services Adoption and Assimilation in SMEs*" by Ada Scupola, investigates the competences deemed necessary both at top managerial and individual levels for the successful adoption and assimilation of business-to-business e-services in small and medium size enterprises. To this end, an in-depth case study of a business-to-business e-service system, a Web-based travel reservation system, was conducted. The results show that three main competences, namely vision, value and control, are important at top management level for the primary adoption of e-services. For secondary adoption and assimilation, three categories of competences were identified as being important either to have or to develop at the individual level, namely technical, interpersonal and conceptual skills.

Chapter 14, "*Adding Value to SMEs in the Courier Industry by Adopting a Web-Based Service Delivery Model*" by Paul Darbyshire designs a framework for a Web system that is intended for linking small and medium transport companies with their customers. The unique aspects of the framework are two-fold. The framework utilizes Web services, which means that it can be applied to existing software and hardware environments. This reduces the need for specialized integration and development, the cost of which becomes a further barrier to SMEs in adding value to customers through existing systems. The framework is additionally designed to link both communities of SMEs and customers in a fledgling digital ecosystem arrangement. Such arrangements offer inherent added value to both types of participants.

Chapter 15, "*Factors Relating to the Adoption of Internet Technology by the Omani Banking Industry*" by Salim Al-Hajri and Arthur Tatnall addresses the question: What are the enablers and the inhibitors of Internet technology adoption in the Omani banking industry compared with those in the Australian banking industry? The chapter does not attempt a direct comparison of the banking industries in these two very different countries, but rather presents a discussion of Internet technology adoption in Oman, informed by the more mature Australian experience.

Chapter 16, "*E-Government Service Delivery Performance: An Adaptation of the Balanced Scorecard to the Public Sector*" by Assion Lawson-Body, Jared Keengwe, Laurence Mukankusi, Abdou Illia, and Glenn Miller, examines the effectiveness of Web site-supported Balanced Scorecard's four dimensions (innovation and learning, internal process, veteran value proposition, and financial) in improving e-government service delivery performance. The study used content analysis to analyze the data obtained from a sample of 19 county veteran service officers (CVSOs) to test the hypotheses. CVSOs use Web sites to serve veterans on a Government-to-Citizen (G2C) basis.

Chapter 17, "*Development and Testing of an E-Commerce Web Site Evaluation Model*" by Luiz Antonio Joia and Luiz Cláudio Barbosa de Oliveira, develops and tests a model for evaluating B2C e-commerce Web sites. Consequently, this study seeks to investigate the relationship between the Web site interface of B2C e-commerce and virtual customer behavior, concentrating on purchasing attitude and intention. The objective of this chapter is, therefore, to research which features of a virtual store effectively influence the user purchasing process via the Internet. In order to achieve this aim, the technology acceptance model was adapted for pertinent theoretical references about consumer behavior on the Web. The model developed was then tested and validated through a structural equation modeling approach. An empirical analysis of the CD e-retailing industry in Brazil was then conducted. The study concludes that, for the electronic commerce of CDs, ease of use, trust, pleasure, and attractiveness as perceived by visitors to CD e-retailing Web sites are key issues for understanding customer attitude and purchasing intention.

E-Commerce Trends for Organizational Advancement: New Applications and Methods provides current coverage of e-commerce issues, trends, and methodologies with a focus on how organizations can adapt to and benefit from new systems and tools in the field. Through discussions of specific cases and projects, this collection aims to inform researchers, practitioners, and individual users of the latest advancements and challenges in the continual evolution of modern-day commerce.

Chapter 1
Developing Mobile Commerce Applications

P. Benou
University of Peloponnese, Greece

V. Bitos
TIM HELLAS, Greece

ABSTRACT

The recent advances in wireless and mobile communication technologies enable users to conduct commercial activities anywhere and at any time. In this new environment, the designing of appropriate applications constitutes both a necessity and a challenge at the same time in order to support effectively the movable user. This article investigates the environment in which these applications operate, identifies possible categories of them and proposes guidelines for their development process.

INTRODUCTION

The exponential growth in wireless and mobile communication technologies, coupled with the impressive mass adoption of mobile phone and the increasing number of users that possess other handheld devices, such as personal digital assistants (PDAs) and smartphones, pave the way to the extension and evolution of e-commerce to m-commerce.

Mobile commerce or *m-commerce* is defined as any activity that is related with a commercial transaction (or a potential one)—an exchange of services or goods for money—and is conducted via wireless and mobile communication networks and uses wireless and mobile devices as user interface. A few of the existing e-commerce services can be successfully expanded to the new mobile environment and brand new services can be built by reason of the distinct attributes of the mobile setting (Durlacher, 2000; Turban, King, Lee, & Viehland, 2004). These special attributes include:

- **Ubiquity:** It refers to the advantage of a mobile device to be available at any time and anywhere, fulfilling the need of both real-time information and communication independent of the user's location.
- **Reachability:** Holding a mobile terminal, the user can be instantly reached at any time.
- **Convenience:** It has to do with the easiness that the user interacts in the mobile environment, just using a smartphone or a PDA, without booting up a PC or placing a call via a modem.
- **Instant Connectivity:** It relates to the ability of the mobile device to connect easily and quickly to the Internet, intranets, other devices and wireless networks.
- **Context-Sensitiveness:** It refers to the ability of the mobile applications to sense and exploit the context—the information that surrounds the human-computer interaction—in order to offer personalized, localized and generally purpose-suitable services.

In this article we discuss issues related to the development of mobile commerce applications. We describe the framework in which the m-commerce operates and the possible categories of m-commerce applications. We also address the important issues of context and context-awareness and propose guidelines for the development process.

More precisely, in the section entitled "*The Framework of Mobile Commerce*" we present the most important participating entities in m-commerce operations. Regarding the section "*Mobile Commerce Application's Categories,*" we describe some of the possible m-commerce application categories. In the section "*The Special Issue of Context and Context-Awareness,*" we discuss the concept of context and highlight an important dimension of it, the location. In this section we also define the context-aware applications and the requirements that these applications should meet. In the section "*Guidelines for M-Commerce Applications Development*" we propose some guidelines for the m-commerce application development process organized in the following categories: (i) management-related propositions, (ii) development process-related propositions, (iii) user-related propositions and (iv) context-related propositions. The section "*Conclusion*" concludes the article.

THE FRAMEWORK OF MOBILE COMMERCE

Mobile commerce is a complex process and involves a number of participating entities that constitute the m-commerce value chain (Giaglis, 2004). Although an analysis of the strategic goals, the key advantages and the values which these entities share is beyond the scope of this paper; we address them briefly in order to have a comprehensive view of the framework in which the m-commerce operates.

The value chain of m-commerce consists of the following:

- **Mobile network operators (MNOs):** The main role of MNOs is to implement and operate the countrywide mobile networks. They also act as mobile service providers by owning a portal, exploiting the already known billing relationship with the customer.
- **Network equipment providers:** They consist of companies that manufacture the infrastructure and equipment needed to operate a mobile or wireless network.
- **Device manufactures:** The manufactures of all kinds of mobile devices are found under this category.
- **Technology platform vendors:** They deliver the operating systems and microbrowsers for portable devices.

- **Wireless internet service providers (WISPs):** They either provide access to Internet through mobile networks or they operate public access wireless local area networks (WLANs) providing access to the Internet through WLAN hotspots.
- **Wireless application service providers (WASPs):** Their role is to develop and host wireless applications for companies that wish to conduct m-commerce activities but do not have the necessary resources.
- **Application developers:** They build applications that are used off-line, run over the wireless networks or operate as mobile Internet services.
- **Content providers:** They originate and provide content that is delivered to the end-user. The content provider who collects content from various content providers and delivers the aggregated content in a most appropriate package is called content aggregator. For example, a provider of this kind acquires real-time stock price content from a stock exchange, financial news content from a news agency and financial analysis content from a consulting firm and bundles them into a new package.
- **Mobile portal providers:** The mobile portals act as entry points to a wide spectrum of content and services. They can be characterized as *horizontals* if they contain content and services regarding to a wide spectrum of markets or *vertical* if they target to a particular market. When their primary concern is the information distribution they are called *information* or *content-oriented*; *communication-oriented* when they provide communication mechanisms such as e-mail, instant messaging and calendars, and *commerce-oriented* when they provide transaction services such as banking, shopping, auctions, and so forth.
- **Trading companies:** They are firms that conduct the m-commerce activities.
- **Consumers:** The target audience of the m-commerce is the consumers, which are individuals, companies or business customers.

The above mentioned m-commerce participating entities utilize the available technology that underpins the mobile commerce and leads it to growth and spreading. These technologies include standards such as WiFi, and Bluetooth for communication through the wireless networks. It also includes global system for mobile communications (GSM), general packet radio services (GPRS) and universal mobile telecommunications system (UMTS) for communication through the mobile networks. A multitude of diverse mobile devices is available including both cellular phones, smartphones, PDAs and hybrid handheld devices that combine voice, data processing and communication capabilities. Operating systems like Windows CE, Palm OS, EPOC, Linux and proprietary systems as well, such as RTOS and GEOS, enable the various mobile devices to operate. Java 2 Micro Edition (J2ME), a light version of Java, has been introduced to meet the limitations and peculiarities of mobile devices. The wireless application protocol (WAP), based on existing Internet and network technologies, optimizes the wireless Internet access. The connectivity gateways lie between the Web application servers and mobile devices in order to adapt the various device specific network protocols, such as WAP, to the common TCP/IP-based communication protocols applied on the server side. Standards like XML, WML and various kinds of HTML are used for content delivery; furthermore voice XML technologies offer convenient user interfaces (Hansmann, Merk, Nicklous, & Stober, 2003).

In the relatively new mobile environment, the application's designers and developers are invited to design and implement innovative mobile commerce solutions, taking into account the constraints that this new environment imposes, but also exploiting the distinct benefits, which this offers.

MOBILE COMMERCE APPLICATION CATEGORIES

Just like e-business is a wide concept that contains the e-commerce, m-business includes m-commerce; moreover it has a common section with e-business, as shown in Figure 1. M-business, excluding the m-commerce, includes other activities as well, such as mobile office, mobile medicine and mobile learning; but in this work we focus on the m-commerce applications.

Although there is an unlimited number of potential m-commerce applications, under both business to business (B2B) and business to consumer (B2C) models, here we are going to discuss only some important classes of them, including mobile financial applications, mobile advertising, mobile shopping and some others in more details.

Mobile Financial Applications

Mobile financial applications are one of the most important components of m-commerce and constitute a new service distribution channel for financial institutions. This new channel enables them to offer more value-added services and at the same time to further cut the operational costs. These applications include mobile banking, mobile brokerage services and mobile payments.

- **Mobile banking:** The simplest of the mobile banking solutions enable users to receive information about their account balance via an SMS request or via a voice call to an IVR (interactive voice response) system. More complex banking services can be offered through the wireless Internet (Mallat, Rossi, & Tuunainen, 2004). Customers can track their account and credit card transactions, transfer funds between accounts and pay bills using a menu-based interface (Herzberg, 2003). Additionally they can be informed about check exchange and interest rates and learn about new investment solutions. Furthermore they have access to information regarding housing, consumer and professional loans, and they can find out the location of automatic teller machines (ATMs) and branches of banks.
- **Mobile broking:** More and more brokerage firms are moving towards mobile broking as they realize the opportunity for service differentiation in order to attract and retain customers (Loney, Jessup, & Valacich, 2004). Investors—private or professional ones—are no longer constrained by wires and can complete their investment activities while out of the office or away from home. Mobile broking offers functionalities such as: receive alerts about price movements, buy and sell stocks, options, mutual funds and other investment products, browse and delete existing orders, receive a message when

Figure 1.

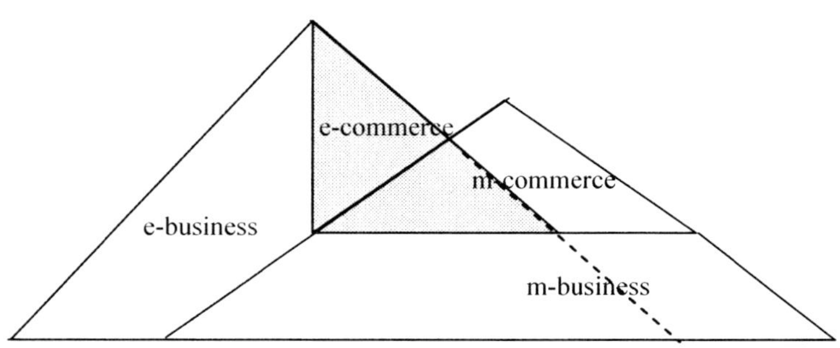

an order is executed and manage portfolio (Muntermann, Rossnagel, & Rannenberg, 2004).

- **Mobile payments:** One of the interesting applications of m-commerce is mobile payments. They are commonly categorized into small payments—for small purchase less than 10 euros—and large payments. They are further subdivided into proximity and remote payments, depending on whether the purchase takes place at the point of sale (POS) or at a distance via a mobile network (Herzberg, 2003). POS small payments include small purchases in shops and kiosks, such as of soft drinks, cigarettes, fast food and instant photos. An example of remote small payments is the purchase of mobile content such as ring tones, logos, games and information. Paradigms of POS large payments are car washing, parking services, taxi payments, retail shopping and paying at restaurants. Examples of remote large payments constitute receiving services and purchasing goods such as books, roses and gifts. A common way to charge mobile payments is to add them to the monthly mobile phone bill, providing the user with the advantage of no additional service enrollment. There is also the credit card billing scheme, where the mobile payments are included in the consumer's credit card bill (Durlacher, 2000). Another billing solution is the direct debit of the customer's bank account or a special account used only for mobile payments for security reasons. Additionally there is the possibility, for the POS purchases, the payment to be made via Bluetooth technology (McKitterick & Dowling, 2003). Electronic cash can be loaded on the mobile handheld and transferred in the vending machine through a personal area network.

Mobile Advertising

Mobile advertising applications deliver promotional information to consumers using mobile phones and PDAs. Advertisements can be sent via SMS or MMS for various on-going specials (shops, malls, restaurants) in the surrounding area of the user, using the mobile network (Bulander, Decker, & Schiefer, 2005). When WiFi or Bluetooth communication has been established, advertisements can be sent as notification messages (Aalto, Gothlin, Korhonen, & Ojala, 2004). The advertisements can be sent to all users in a specific area or to users with specific needs and interests regardless of their current location (Kurkovsky & Harihar, 2006). This is known as the "push" method of advertising. In addition to the "push" method for advertising, the "pull" method is available as well, enabling the user to request a page where information about the product or service concerned is embedded. MNOs lie in a very advantageous position as far as exploiting the mobile advertising is concerned, because they have the demographic information of the user, knowing his current location and his purchasing habits—if they own a portal—so that a more targeted advertising can be achieved.

Mobile Shopping

Mobile shopping includes retailing, ticketing, reservations and auctions.

- **Mobile retailing:** Many vendors allow customers to shop from wireless devices enabling them to perform quick searches, compare prices, order and view the status of their order using cell phones or PDAs (Turban et al., 2004). Books, CDs and groceries (Kourouthanassis, Spinellis, Roussos, & Giaglis, 2002) are everyday items, which the user knows very well and he just needs a tool to make the purchase. He can buy them in his spare time, regardless of the shop's opening hours and his current location (Newcomb, Pashley, & Stasko, 2003).

- **Mobile ticketing:** One of the most compelling mobile services is ticket purchasing. The user makes a phone call to a service number and the tickets are delivered as an SMS or MMS message (Mobile Electronic Transaction, 2003). The ticket can contain comprehensive information about the transaction—customer's name, number of persons travelling, means of payment used, time of departure, time of arrival and so on. Potential applications of mobile ticketing include: transportation (flights, trains, boats, ferries, buses, trams), events (concerts, theatres, fairs, museums, sports events) and facilities (gyms, solariums, spas).
- **Mobile reservations:** Mobile reservations are one of the most featured applications in mobile commerce. The customers are able, using their mobile devices, to access hotel and restaurant reservation systems to check prices or availability and to make or change a reservation.
- **Mobile auctions:** Mobile devices provide the ideal environment for auctions and many firms, such as eBay, which conduct online auctions, have moved towards the wireless environment. The main reason for this extension is that bidders want to continue to participate even when they are not in front of a PC. Additional capabilities, except the mobile biting that can be offered by the mobile auctions, are searching for the items, looking at pictures depicting the items and checking the auction lists.

Mobile Information Provisioning

As in the Web, the information that can be provided using the mobile devices is unlimited. It can take the shape of an MMS message or can be offered through the mobile information portals. It can contain general news, sports news, program information of cinemas/theatre, TV and radio, and travel information such as train, bus and flight schedules.

Mobile Entertainment

The available entertainment services and the increased communication bandwidth enable the consumers to buy games, music and video.

Mobile Inventory Management

This category of applications includes the tracking of goods, trucks, or even people. Tracking goods help the providers to determine the time of delivery to the customer, thus improving the customer service. Tracking trucks and people can be exploited for efficient fleet management. Another example is the just-in-time delivery of components in an assembly plant based on the rate of consumption of existing components.

Mobile Supply Chain Integration

The integration of business processes along the supply chain is of significant importance for efficient trading. Additionally accurate and on time information, for the on the move business person, is critical to business success. With the wireless technology there is the possibility—for a pharmaceutical sales representative on the road, for example—to check availability of a particular item in the warehouse, order particular products, or have security access to obtain financial data from the backend enterprise resource planning (ERP) systems.

Mobile CRM

Mobile access expands the reach of customer relationship management (CRM) systems outside the company to employees and partners. One of the attractive functions of the mobile CRM is the sales force automation. Key data that can be retrieved by the sales force on the road could include contact management information, order entry and product availability.

THE SPECIAL ISSUE OF CONTEXT AND CONTEXT-AWARENESS

The Notion of Context

Unlike desktop computing, where the applications are designed to operate in the relatively stable environment of office or home, mobile applications are intended to operate in the dynamic environment of everyday life. The users change their location and social roles during the day: they are professionals at work, consumers while shopping, tourists when traveling and investors when they conduct stock trading activities. Additionally the computing environment varies, as the users have the ability to choose between their cellular phones, PDAs or other handheld devices. All this information surrounding the human-computer interaction, which can be exploited by applications, constitute the *context.*

First Schilit & Theimer (1994) refer to context as location, identities of nearby people and objects, and changes to those objects. Brown, Bovey, & Chen (1997) define context as location, identities of the people around the user, the time of day, season, temperature, and so forth. Pascoe (1998) defines context as the subset of physical and conceptual state of interest to a particular entity. Schmidt & Laerhoven (2001) identify context as the knowledge about the user's and the device's state, including surroundings, situation and location.

Dey & Abowd (1999) propose a more generic definition that states that context is any information that can be used to characterize the situation of an entity. An entity is a person or object that is considered relevant to the interaction between a user and an application, including the user and the application themselves. For example, when a user wants to book a room in a hotel, the preferences of the user constitute valuable "context," which can be exploited to propose alternative solutions to him.

The contextual information can be classified, according to which entity it concerns, into the following categories (Schilit, Adams, & Want, 1994; Chen, & Kotz, 2000):

- **User context:** user identity, collection of nearby people, user's profile, and so forth.
- **Computing context:** hardware characteristics, software characteristics, network connectivity, communication bandwidth, nearby resources such as printers, displays and other devices and so forth.
- **Physical Context:** lighting, noise level, temperature, humidity, and so forth.
- **Spatio-temporal context:** location, place, time of the day, week, month, season of year, time zone, and so forth.
- **Task context:** the goals, tasks and actions of the user.

The types of the context according to the manner of its acquisition can be divided in three categories:

- **Sensed context:** this context is acquired from the environment by means of physical or software sensors (identity, location, temperature, time)
- **Derived context:** this kind of contextual information is computed (for example the name of the city from GPS coordinates through a relation mechanism).
- **Context explicitly provided:** the context that the user provides explicitly (for example the entries in the user profile).

During the last ten years of the context's exploration, many techniques have been proposed for its modeling and representation, including key-value pairs, object-oriented, markup schemes and ontology-based models (Strang & Linnhoff-Popien, 2004).

Positioning Techniques

The accurate knowledge of the position of mobile devices and portable objects can be beneficial for a lot of mobile applications, such as mobile advertising, fleet management, object tracking and location-based information services, for example, find the nearest facilities—restaurants, theaters, parking, and so forth. The significance of the location context element prompts us to discuss in detail the various positioning techniques that currently exist. These techniques include satellite-based, cellular network-based, WLANs-based and RFID-based systems, as they are described in the following paragraphs.

GPS and Assisted GPS (A-GPS): This is a handset-based mobile positioning technique, called so because the handset itself is the primary means for positioning the user and relies on the global positioning satellite system (GPS). The GPS consists of 24 satellites put into orbit, 20,000 kilometers above the surface of the earth, which transmit encoded signals. The GPS receiver, which is integrated in the mobile device, uses these signals to determine its position in 3D—in latitude, longitude and altitude—with accuracy of 10 meters or less. The GPS device needs to have a clear line-of-sight with at least four satellites before it can calculate its position. In the assisted GPS technique, the mobile network provider can, through an assistant server, aid the handset by directing it to look at specific satellites and by collecting data from the handset to perform location identification calculations that the handset is unable to perform due to its limited power.

Indoor GPS: The indoor global positioning system has been developed due to the fact that satellite GPS can not be applicable indoors, because the signal strength is too low to penetrate buildings. The signal of an indoors GPS is emitted by a number of pseudo-satellites and is similar to satellite GPS signal (Zeimpekis, Giaglis, & Lekakos, 2003).

Cell Identification or Cell of Origin (COO): This is referred to as a network-based method, because the mobile network is used to position the mobile device. This method relies on the fact that the mobile network operator is able to identify approximately the position of a mobile device by knowing the cell that the mobile device uses at a given time. Higher location precision can be achieved through the *angle of arrival (AOA)* and *time of arrival (TOA)* methods.

Wireless Local Area Networks (WLANs): In WLAN's environment, the position of the mobile device is determined by the location of the access point of the connection (Varshney, 2003).

Radio Frequency Identification (RFID): According to this method the location of a device, an object or a user, which carries an RFID card, can be determined from the location of the stationary reader that captures the signals emitted by the RFID card.

Context-Aware Applications

The context, which has been discussed in the previous paragraphs, could be exploited in order to provide more valuable services to the end-user. The ability of an application to have knowledge about the context that surrounds it, and to adapt its interface, content and behavior accordingly, is called *context-awareness*

Context-aware computing was first introduced by Schilit & Theimer (1994) to be software that adapts according to the location of use and the collection of nearby people and objects, as well as to changes to such things over time. Pascoe (1998) defines context-awareness as the ability of a device or program to sense, react or adapt to its environment of use. Brown, Bovey, & Chen (1997) define context-aware applications as applications that automatically provide information and/or take actions according to the user's present context as detected by the sensors. Fickas, Korteum, & Segall (1997) define context-aware applications (called environment-directed) to be applications that monitor changes in the environment and adapt the operation according to predefined or user-defined guidelines. Dey & Abowd (1999) characterizes a system as context-aware if it uses context to provide relevant information and/or services to the user, where relevancy depends on the user's task.

The functions that a context-aware application should implement (Schilit, Adams, & Want, 1994) are:

- **Proximate selection:** a user interface-level technique where the nearby-located objects are emphasized or otherwise made easier to choose.
- **Automatic contextual reconfiguration:** a process of adding new components, removing existing components, or altering the connections between components due to context changes.
- **Contextual information and command:** queries on contextual information can produce different results according to the context in which they are issued. Similarly, context can parameterize "contextual commands."
- **Context-triggered actions:** "If-then" rules used to specify how context-aware systems should be adapted.

Dey & Abowd (1999) propose that the features that context-aware applications may support are:

- **Presentation:** of information and services to the user or use context to propose appropriate selections
- **Automatic execution:** of a service according to context changes
- **Tagging of context:** to information for later retrieval.

GUIDELINES FOR M-COMMERCE APPLICATIONS DEVELOPMENT

Mobile commerce applications have a number of unique properties that differentiate them from traditional e-commerce applications and demand new development approaches. Although no comprehensive suggestions have yet been presented, many researchers have pointed out the necessity for new design methodologies and have put forward recommendations for each development phase. Kohler & Gruhn (2004) and Valiente & Heijden (2004) attempt to put forward methods for the identification of the business processes, which could be automated in the mobile environment. Stanoevska-Slabeva (2003), Gebauer, Gribbins & Shaw (2006) and Andreou et al. (2002) propose alternations of existing development approaches in brief. Holtzblatt (2005) has focused on a customer-center design while Rupnik & Krisper (2003) and Tarasewich (2003) point out the importance of context exploitation. Van de Kar (2005) reveals the importance of the interests of the participating entities; Krogstie et al. (2003) and Banavar et al. (2000) highlight the character of mobility, while Kemper & Wolf (2002) relies their proposal on the axis of *innovation*, *risk* and *speed of development*.

Moving to the direction of shaping a comprehensive m-commerce applications' development methodology, we present the unique properties of mobile applications and propose the following guidelines:

Management-Related Propositions

Property 1: *The usage of m-commerce applications is extended over time and space.*

Due to this characteristic, new innovative services can be devised, enabling the companies to remain in a competitive position in the future.

Guideline 1: *Managers should encourage the generation of novel ideas regarding the m-commerce solutions.*

This can be done through the examination of the e-commerce counterpart applications for opportunities to expand in the mobile environment, the observation of advances in technology, the investigation of international practices and the encouragement of brainstorming discussions.

Property 2: *Sometimes the return of investment—in terms of absolute numbers—doesn't justify the cost of the mobile undertaking.*

Guideline 2: *The undertaking of innovative m-commerce solutions does not only aim at generating revenues but also diffusing the reputation, acquiring knowledge and gaining competitive advantage.*

Property 3: *The development of m-commerce applications is a complex process and involves enough participating entities.*

The involving entities could be the trading company, the service provider, the mobile network operator, the content provider, the payment institution and so on.

Guideline 3: *The roles and the interests of the participating entities should be carefully determined, as well as the values that they will share in "a win to win situation."*

Property 4: *There is a relative lack of experience regarding the development of mobile information systems.*

The wireless and mobile technology is relatively new as well as the m-commerce undertakings.

Guideline 4: *Companies should form inter-organizational ties during the development of m-commerce solutions to complement their cognitions and competitive advantages.*

Development Process-Related Propositions

Property 1: *There is a plethora of available technologies that support the moving user.*

Guideline 1: *The alternative technologies, in each mobile case, should be carefully compared with each other and the suitable tools, platforms, protocols, devices and networks should be selected.*

Property 2: *There is a lack of long-time experience regarding the development of mobile information systems among the designers and developers.*

In the mobile environment, it is even harder than in the traditional case to identify non-anticipated difficulties arising from the new technology and new usage situations.

Guideline 2: *In the development of m-commerce information systems, prototyping development approaches should be adopted (Krogstie et al., 2003).*

Property 3: *Time to market for a mobile application is short (Kolos-Mazuryk, Poulisse, & Eck, 2005).*

The speed of development is crucial in the mobile environment since the technology is quickly bypassed and often strong competition factors press for quick solutions.

Guideline 3: *For the development of m-commerce information systems, rapid development approaches should be selected.*

Property 4: *The gathering and exploitation of the context in the m-commerce applications make the code more complicated.*

The mixing of context management and adaptivity code with the applications' core code renders the application code more complex, difficult to maintain and impossible to reuse.

Guideline 4: *During the m-commerce application's design, techniques of separation of context and adaptivity concerns from the core functionalities should be searched out.*

Property 5: *The mobile applications are intended to be used in many different devices with different interfaces, for example, voice interface.*

The main concern of m-commerce applications—in an interface independent design—is the tasks and subtasks that they implement (Banavaar et al., 2000).

Guideline 5a: *The structure of a mobile program should be described in terms of tasks and subtasks and their navigation order.*
Guideline 5b: *The logic of an application should be materialized through distinct services that can be bound into a presentation unit.*

Property 6: *The users don't have great experience of m-commerce applications; therefore it is difficult to identify their requirements.*

Guideline 6: *For the user requirements gathering, repetitive and scenario-driven approaches should be adopted.*

Property 7: *The spatio-temporal environment in which mobile applications operate make the evaluation of mobile services hard, since it is often not possible to create a realistic usage situation (Holmquist, Hook, Juhlin, & Persson, 2002).*

Guideline 7a: *Field test and evaluation planning should be carefully designed and executed.*
Guideline 7b: *The mobile applications should be refined and adapted to the working environment through an iterative process and repeated feedback loops.*

User-Related Propositions

Property 1: *The target audience of mobile applications is broader than desktop applications, with lack of homogeneity.*

Guideline 1 a: *During the mobile application design it is crucial to identify the potential target user groups, their age, cultural background, cognitive capabilities, as well as their information and communication needs.*
Guideline 1 b: *The widespread adoption of mobile applications is highly dependent on their usability in terms of ease to learn and ease to use.*

Property 2: *The user can only pay intermittent and limited attention to the mobile application over a long period of time.*

When the user interacts with a mobile application he is usually busy, as he simultaneously conducts other activities such as walking, meeting somebody or shopping.

Guideline 2 a: *The mobile application should offer clearly value-added services, such as saving time and increasing efficiency.*
Guideline 2 b: *Simple solutions will work better than complicated solutions.*
Guideline 2c: *Only the information that is absolutely critical should be displayed.*
Guideline 2 d: *During the mobile applications design the minimization of clicks should be sought.*

Guideline 2 e: *Convenient user interfaces, such as voice or sound interfaces, should be investigated.*

Property 3: *User activity occurs in spurts (Kolos-Mazuryk et al., 2005).*

Guideline 3: *There should be a matching between the mobile applications services and the tasks that the mobile user tries to complete.*

This can be done through the identification of the core tasks and subtasks that the user performs in order to achieve a certain goal.

Property 4: *In the mobile environment, the devices are usually personal.*

Although there is a need for generalization, the fact that the mobile devices are personal dictates the need for specialization.

Guideline 4: *Personalized services, tailored to users' preferences, could be provided.*

Context-Related Propositions

Property 1: *The mobile environment is dynamic and the context is rapidly changing.*

Guideline 1: *The context can be captured and exploited in order to provide adaptive mobile applications. These applications recommend content and services to the user or provide suitable information and services in each situation.*

Property 2: *Compared to desktop computers, mobile devices have small screens, limited interaction facilities, low processing power, limited memory and restricted network capabilities.*

Guideline 2: *During the mobile applications' design performance issues, regarding the computing devices and the network should be considered.*

Property 3: *The cellular networks have broadcasting capabilities.*

Guideline 3: *Mass information dissemination can be achieved through mobile networks to users with same interests or in the same location, for example, stock prices or advertising messages.*

Property 4: *The same application could operate on different devices.*

Guideline 4: *The mobile applications should provide consistent functionality and user interfaces through the different devices in which they run.*

CONCLUSION

In this article we discuss issues regarding the development of mobile commerce applications. We explore possible categories of them and address context-related topics. We strongly believe that the successful development of the m-commerce applications is highly dependent on the exact specification and efficient exploitation of the context. The collection and implementation of the context information should be done in a way so that it is reusable and easily extended to cover not yet predictable aspects of it. Moreover, new methods should be investigated in order to separate the core applications logic from the adaptive functionality.

Additionally, we realize that the distinct characteristics of the mobile setting reveal the necessity for new development methodologies. Although some researchers have pointed out the necessity for new development approaches—especially due to the device's characteristics and the changing context—there has not been suggested a complete proposition yet. While working on the shaping of a comprehensive methodology as such, we propose some guidelines for the development process.

Finally, although in this paper we do not address these matters, we believe that security and privacy factors will intensely influence the widespread adoption of m-commerce. The user's personal data, his potentially depicted behaviour and his preferences should be utilized in a clear way and with his consent. In addition, although current communication technologies incorporate strong authentication systems and cryptographic techniques, security issues should be taken into account in the nodes in which decryption and re-encryption occur.

REFERENCES

Aalto, L., Gothlin, N., Korhonen, J., & Ojala, T. (2004). Bluetooth and WAP push based location-aware mobile advertising system. In *Proceedings of the 2nd International Conference on Mobile Systems, Applications, and Services* (pp. 49-58). ACM.

Andreou, A., Chrysostomou, C., Leonidou, C., Mavromoustakos, S., Pitsillides, A., Samaras, G. (2002). *Mobile commerce applications and services: A design development approach.* The First International Conference on Mobile Business M-Business2002, July 8-9, Athens, Greece, 2002.

Banavar, G., Beck, J., Gluzberg, E., Munson, J., Sussman, J., & Zukowski, D. (2000). Challenges: An application model for pervasive computing. In *Proceedings of the 6th Annual International Conference on Mobile Computing and Networking* (pp. 266-274).

Brown, J., Bovey, D., & Chen, X. (1997). Context-aware applications: From the laboratory to the marketplace. *IEEE Personal Communications, 4*(5), 58-64.

Bulander, R., Decker, M., & Schiefer, G. (2005). Comparison of different approaches for mobile advertising. In *Proceedings of the Second IEEE International Workshop on Mobile Commerce and Services* (pp. 174-182).

Chen, G., & Kotz, D. (2000). *A survey of context-aware mobile computing research.* Technical Report TR2000-381. Department of Computer Science, Dartmouth College.

Dey, A., & Abowd, G. (1999). *Towards a better understanding of context and context-awareness.* Technical Report 99-22. Georgia Institute of Technology.

Durlacher Research Ltd. (2000). *Mobile commerce report.* Retrieved from http://www.durlacher.com/fr-research-reps.htm.

Fickas, S., Korteum, G., & Segall, Z., (1997). Software organization for dynamic and adaptable wearable systems. In *Proceedings of the 1st IEEE International Symposium on Wearable Computers* (pp. 56-63).

Gebauer, J., Gribbins, J., & Shaw, J. (2006). *Task-technology fit for mobile information systems.* Working Paper, 06-0107. University of Illinois at Urbana-Champaign, College of Business.

Giaglis, G. (2004). Κινητό και Ασύρματο Επιχειρείν. Εκδόσεις Ι. ΣΙΔΕΡΗΣ. ISBN 960-08-0331-5.

Hansmann, U., Merk, L., Nicklous, M., & Stober T. (2003). *Pervasive computing.* Berlin, Heidelberg, New York: Springer-Verlag.

Herzberg, A. (2003). Payments and banking with mobile personal devices. *Communications of the ACM, 46*(5), 53-58.

Holmquist, L., Hook, K., Juhlin, O., & Persson P. (2002). Challenges and opportunities for the design and evaluation of mobile applications. In *Proceedings of the Workshop on Main Issues in Designing Interactive Mobile Services.*

Holtzblatt, K. (2005). Customer-centered design for mobile applications. *Personal and Ubiquitous Computing, 9*(4), 227-237.

Kemper, H.G., & Wolf, E. (2002). Iterative process models for mobile application systems: A framework. In *Proceedings of the Twenty-Third International Conference on Information Systems* (pp. 401-413).

Kohler, A., & Gruhn, V. (2004). Analysis of mobile business processes for the design of mobile information systems. In *Proceedings of the 5th International Conference on Electronic Commerce and Web Technologies* (3182) (pp. 238–247).

Kolos-Mazuryk, L., Poulisse, G., & van Eck, P. (2005). Requirements engineering for pervasive services. In *Proceedings of the Second Workshop on Building Software for Pervasive Services* (pp. 18-22).

Kourouthanassis, P., Spinellis, D., Roussos, G., & Giaglis, G. (2002). Intelligent Cokes and diapers: MyGROCER ubiquitous computing environment. In *Proceedings of the 1st International Mobile Business Conference* (pp. 150-172).

Krogstie, J., Lyytinen, K., Ophahl, A., Pernici, B., Siau, K., & Smolander K. (2003). *Mobile information systems – Research challenges on the conceptual and logical level.* Lecture Notes in Computer Science – Advanced Conceptual Modeling Techniques (Vol. 2784) (pp. 124-135). Berlin, Heidelberg: Springer-Verlag, .

Kurkovsky, S., & Harihar, K. (2006). Using ubiquitous computing in interactive mobile marketing. *Personal and Ubiquitous Computing Journal*, 227-240.

Loney, C., Jessup, L., & Valacich, J. (2004). Emerging mobile business models for mobile brokerage services. *Communications of the ACM, 47*(6), 71-77.

Mallat, N., Rossi, M., & Tuunainen, V.K. (2004). Mobile banking services. *Communications of the ACM, 47*(5), 43-46.

McKitterick, D., & Dowling, J. (2003). *State of the art review of mobile payment technology.* Technical Report, TCD-CS-2003-24.

Mobile Electronic Transaction Ltd. (2003). MeT white paper on mobile ticketing. Retrieved from http://www.mobiletransaction.org/pdf/R200/white_papers/MeT_White_paper_on_mobile_ticketing_v1.pdf.

Muntermann, J., Rossnagel, H., & Rannenberg, K. (2004). Mobile financial information services, security and certification. In *Post-Proceedings of the CSES 2004 2nd International Workshop on Certification and Security in Inter-Organizational E-Services, IFIP 18th World Computer Congress 2004.*

Newcomb, E., Pashley, T., & Stasko, J. (2003). Mobile computing in the retail arena. In *Proceedings of the ACM Conference on Human Factors in Computing Systems* (pp. 337-244). New York: ACM Press.

Pascoe, J. (1998). Adding generic contextual capabilities to wearable computers. In Proceedings of the 2nd International Symposium on Wearable Computers (pp. 92-99).

Rupnik, R., & Krisper, M. (2003). Mobile applications: A new application model in information systems. In *Proceedings of the 2nd International Conference on Mobile Business.*

Schilit, B., & Theimer, M. (1994). Disseminating active map information to mobile hosts. *IEEE Network, 8*(5), 22-32.

Schilit, B., Adams, N., & Want, R. (1994). Context-aware computing applications. In *Proceedings of the 1st International Workshop on Mobile Computing Systems and Applications* (pp. 85-90).

Schmidt, A., & Laerhoven, K. (2001). How to build smart appliances. *IEEE Personal Communications, 8*(4), 66-71

Stanoevska-Slabeva, K. (2003). Towards a reference model for m-commerce applications. In Proceedings of the 11th European Conference on Information Systems.

Strang, T., & Linnhoff-Popien, C. (2004). A context modeling survey. In *Proceedings of the 1st International Workshop on Advanced Context Modelling, Reasoning And Management* (pp. 34-41).

Tarasewich, P. (2003). Designing mobile commerce applications. *Communications of the ACM, 46*(12), 57-60.

Turban, E., King, D., Lee, J., & Viehland, D. (2004). *Electronic commerce*. New Jersey: Prentice Hall.

Valiente, P., & Heijden H. (2004). A method to identify opportunities for mobile business processes. In *Proceedings of E-Commerce and Web Technologies* (EC-Web 2004). Lecture Notes in Computer Science (pp. 238-247). Berlin, Heidelberg: Springer-Verlag.

Van De Kar, E. (2005). Guidelines for designing mobile information service systems in a value network. In *Proceedings of the 5th ACM International Conference on Electronic Commerce* (Vol. 50) (pp. 413-421). New York: ACM Press.

Varshney, U. (2003). Location management for mobile commerce applications in wireless Internet environment. *ACM Transactions on Internet Technology, 3*(3), 236-255.

Zeimpekis, V., Giaglis, G., & Lekakos, G. (2003). A taxonomy of indoor and outdoor positioning techniques for mobile location services. *ACM SIGECOM Exchanges, 3*(4), 19-27.

This work was previously published in Journal of Electronic Commerce in Organizations, Vol. 6, Issue 1, edited by M. Khosrow-Pour, pp. 63-78, copyright 2008 by IGI Publishing (an imprint of IGI Global).

Chapter 2
A Technological Perspective of Mobile and Electronic Commerce Systems

Wen-Chen Hu
University of North Dakota, USA

Yanjun Zuo
University of North Dakota, USA

Naima Kaabouch
University of North Dakota, USA

Lei Chen
Sam Houston State University, USA

ABSTRACT

The emergence of wireless and mobile networks has made possible the introduction of electronic commerce to a new application and research subject: mobile commerce. Mobile commerce is a promising trend of commerce because Internet-enabled smartphones such as iPhones are becoming very popular these days. People use smartphones to perform daily tasks like browsing the mobile Internet and making phone calls anytime and anywhere. However, understanding or constructing a mobile or an electronic commerce system is not easy because the system involves a wide variety of disciplines and technologies and the technologies are constantly changed. To facilitate understanding and constructing such a system, this chapter divides the system into six components: (i) applications, (ii) client devices or computers, (iii) mobile middleware, (iv) wireless networks, (v) wired networks, and (vi) host computers. Elements in these components specifically related to the subject are described in detail and lists of current technologies for component construction are discussed.

DOI: 10.4018/978-1-60566-964-9.ch002

INTRODUCTION

The world has witnessed the blossom of electronic commerce in the past. Today, another form of electronic commerce, mobile commerce, is gradually replacing or complementing electronic commerce. Mobile commerce is defined as the exchange or buying and selling of commodities, services, or information on the Internet through the use of mobile handheld devices. The future of mobile commerce is bright according to the following studies:

- According to Gartner, Inc., a market research company, the numbers of units of PCs, smartphones, and cellular phones shipped in 2008 are:
  - 302.2 million PCs including desk-based PCs, mobile PCs, and X86 servers (Gartner, Inc., 2009a),
  - 139.3 million smartphones, which are mobile phones with advanced functions such as PC-like functions (Gartner, Inc., 2009b), and
  - 1.22 billion mobile phones (Gartner, Inc., 2009c).

  The number of smartphones shipped is increased fast in recent years and it is a little less than half of the number of PCs shipped. It is expected the number of smartphones shipped will surpass the number of PC shipped in the near future.
- Cumulative sales of smartphones will reach 1 billion units by 2010 (Symbian Limited, 2009).

Mobile commerce is an effective and convenient way of delivering electronic commerce to consumers from anywhere and at any time. Realizing the advantages to be gained from mobile commerce, companies have begun to offer mobile commerce options for their customers in addition to the electronic commerce they already provide. However, it requires a tremendous effort to understand or construct a mobile or an electronic commerce system because it involves such a wide range of disciplines and technologies. To lessen the difficulty, this article will divide the system into six components: (i) applications, (ii) client devices or computers, (iii) mobile middleware, (iv) wireless networks, (v) wired networks, and (vi) host computers. Since each component is large enough to be a research area by itself, only elements in components that are specifically related to mobile or electronic commerce are explained in detail. Lists of the technologies used for component construction are also discussed. Related research on mobile commerce systems can be found in the article by Varshney, Vetter, & Kalakota (2000).

Requirements of a Mobile Commerce System

A wide variety of technologies are used to build mobile commerce systems. No matter what kinds of technologies are used, the requirements for a mobile commerce system include:

- The system should be easy to deploy and adapt by content providers, telecommunication companies, and handheld device manufacturers.
- The system uses the state-of-the-art technologies.
- The applications can be accessed from a wide range of handheld devices.
- The applications should be easy to use as their counterparts of electronic commerce.
- It should allow end users to perform mobile commerce transactions easily, in a timely manner, and ubiquitously.
- It should allow products to be personalized or customized upon request.
- It should fully support a wide range of mobile commerce applications for content providers.

- Maximum interoperability is desirable because so many technologies are now available and new techniques are constantly being invented for the use of mobile commerce systems.
- Program/data independence is held, that is, changing the system components will not affect the existing programs/data.
- End-to-end security and user privacy are rigorously enforced.

SYSTEM STRUCTURES

This section illustrates the system structures of electronic and mobile commerce and explains the procedures of mobile commerce transactions.

An Electronic Commerce System Structure

Electronic commerce describes the manner in which transactions take place over networks, mostly the Internet. It is the process of electronically buying and selling goods, services, and information. An electronic commerce system is inherently interdisciplinary and there are many different ways to implement it. Figure 1 shows the structure of a traditional electronic commerce system and a typical example of such a system. The system structure includes four components, some of which are at least partly shared by mobile commerce systems: (i) electronic commerce applications, (ii) client computers, (iii) wired networks, and (iv) host computers.

A Mobile Commerce System Structure

Much like the concept of e-commerce, mobile commerce is a type of business conducted 100% electronically through computer networks; m-commerce is a concept of buying and selling goods and services via wireless networks with a mobile handheld device. Compared to an electronic commerce system, a mobile commerce system is much more complicated because components related to mobile computing have to be included. To facilitate understanding and constructing mobile commerce systems, this chapter will use a modular approach to analyze the system. Figure 3 shows the structure of a mobile commerce system, consisting of six components, and an example of such a system that is currently possible based on existing technology. The network infrastructure for mobile commerce systems consists of both wired and wireless networks. The wired networks component has the same structure and implementation as that needed by an electronic commerce system. A mobile com-

Figure 1. An electronic commerce system structure

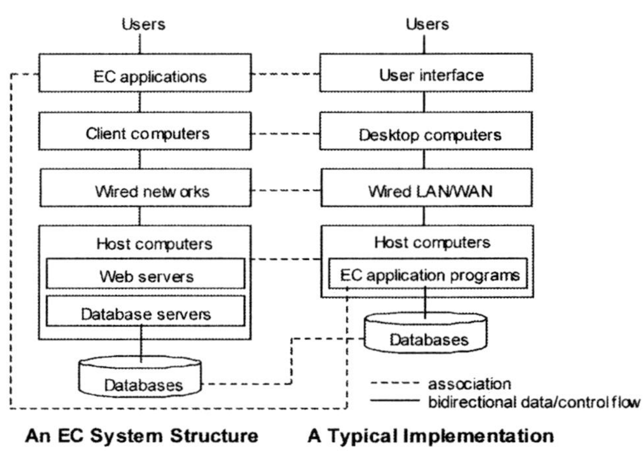

Figure 2. A mobile commerce system structure

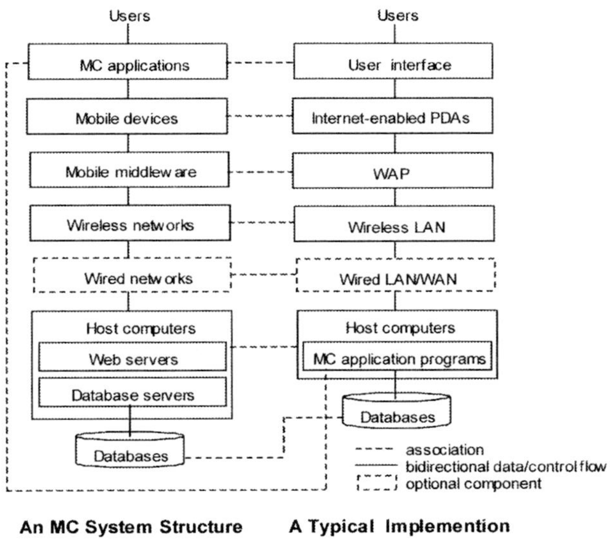

Figure 3. A flowchart of a user request processed in a mobile commerce system

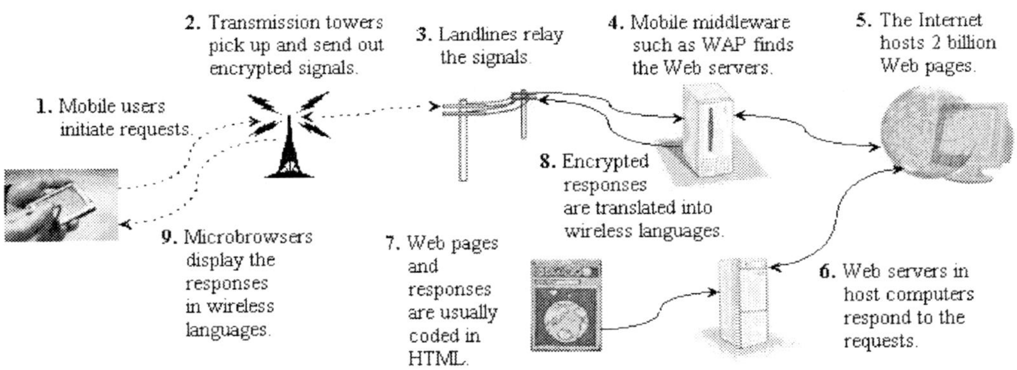

merce system is inherently interdisciplinary and could be implemented in various ways. Figure 2 shows the structure of a mobile commerce system and a typical example of such a system (Hu, Lee, & Yeh, 2004). The system structure includes six components: (i) mobile commerce applications, (ii) mobile handheld devices, (iii) mobile middleware, (iv) wireless networks, (v) wired networks, and (vi) host computers.

Mobile Commerce Transaction Processing

Mobile commerce transaction processing is complicated. To explain how the mobile commerce components work together for a transaction, Figure 3 shows a flowchart of how a user request is processed by the components in a mobile commerce system, along with brief descriptions of how each component processes the request:

1. **Mobile commerce applications***:* A content provider implements an application by providing two sets of programs: client-side programs, such as user interfaces on microbrowsers, and server-side programs, such as database access and updating.
2. **Mobile handheld devices***:* Handheld devices present user interfaces to the mobile end users, who specify their requests on the interfaces. The devices then relay the user requests to the other components and later display the processing results using the interfaces.
3. **Mobile middleware***:* The major purpose of mobile middleware is to seamlessly and transparently map Internet contents to mobile handheld devices that support a wide variety of operating systems, markup languages, microbrowsers, and protocols. Most mobile middleware also encrypts the communication in order to provide some level of security for transactions.
4. **Wireless and mobile networks***:* Mobile commerce is possible mainly because of the availability of wireless networks. User requests are delivered to either the closest wireless access point (in a wireless local area network environment) or a base station (in a cellular network environment).
5. **Wired networks***:* This component is optional for a mobile commerce system. However, most computers (servers) usually reside on wired networks such as the Internet, so user requests are routed to these servers using transport and/or security mechanisms provided by wired networks.
6. **Host computers***:* Host computers process and store all the information needed for mobile commerce applications, and most application programs can be found here. They include three major components: Web servers, database servers, and application programs and support software.

APPLICATIONS

The emergence of electronic and mobile commerce creates numerous business opportunities and applications. Electronic commerce, defined as the buying and selling of goods and services and the transfer of funds through digital communications, includes a wide variety of applications, such as auctions, banking, marketplaces and exchanges, recruiting, and retailing, to name but a few. Mobile commerce applications not only cover the electronic commerce applications, but also include new applications, e.g., mobile inventory tracking, which can be performed at any time and from anywhere by using mobile computing technology.

Electronic Commerce Application

One of the major characteristics of e-commerce is that it enables the creation of new business models. Other than the "buy-and-sell" model, the following list gives some other common models (Turban, et al, 2004):

- *Affiliate marketing*: Affiliate marketing is an arrangement where a marketing partner has an arrangement with a company to refer consumers to the company's Web site so that a consumer can purchase a service or product. The marketing partner receives a 3 to 15 percent commission on the purchase price when a customer they refer to the company's Web site makes a purchase there. For the example of the Amazon.com's Associates Program, the partners can receive up to 10% in referral fees on all qualifying revenue made through their links to Amazon products and services.
- *Comparing prices*: A consumer specifies his or her needs and then this model lists

prices from various companies for that service or product. mySimon.com is a comparison shopping site for apparel, computers, electronics, jewelry, video games, and more. It gathers prices on millions of products from thousands of stores, so customers can compare products and find the best price before he or she buys.

- *Customization and personalization*: Customization or personalization is to design and creation of content that meets a customer's specific needs. For example, Dell Inc. is based on a simple concept: by selling computer systems directly to customers. This direct business model eliminates retailers that add unnecessary time and cost.
- *Electronic marketplaces and exchanges*: Electronic marketplaces are Internet Websites acting as a meeting point between supply and demand and electronic exchanges are a central marketplace with established rules and regulations where buyers and sellers meet to trade futures and options contracts or securities. Electronic marketplaces and exchanges provide benefits to both buyers and sellers because they are more efficient than traditional ones.
- *Electronic tendering systems*: Tendering is potential suppliers bid competitively for a contract, quoting a price to the buyer. Large buyers, private or public, usually make their purchases through a tendering (bidding) system, which is more effective and efficient with the help of electronic commerce.
- *Group purchasing*: Discounts are usually available for quantity purchasing. Using the concept of electronic aggregation, a small business, or even an individual, can get a discount. Electronic commerce has spawned the concept of electronic aggregation, where a third party finds the individuals or small, medium enterprises (SMEs), aggregates orders, and then negotiates (or conducts a tender) for the best deal.
- *Name your price*: Pioneered by Priceline.com, this model allows a buyer to set the price that he or she is willing to pay for a specific service or product. Most Priceline.com services and products are travel-related (e.g., airline tickets, hotels). The following example shows how the "Name Your Price" of Priceline.com works. With Priceline.com's "Name Your Own Price" hotel reservation service, customers choose the star level of hotel they want, along with the desired neighborhood, dates and price they want to pay. Priceline.com then works to find a hotel room at the customer's desired price. There is no guarantee that any "Name Your Own Price" offer will be accepted due to the changeability of room availability and pricing. Customers learn the specific hotel name and location after the purchase is completed.
- *Online auctions*: eBay.com is the world's largest online auction site. It offers an online platform where millions of items are traded each day. Several hundred other companies, such as Amazon.com and Yahoo.com, also conduct online auction.

Mobile Commerce Applications

Mobile commerce applications cover almost everything in our daily lives such as traveling and foods. Table 1 lists some major mobile commerce applications along with explanations of three applications (Sadeh, 2002):

- *Map services*: Map services provide various useful functions to mobile users. Some of the functions include:
  - *Directions*, which are driving/walking directions from the starting location to destination,

Table 1. Major mobile commerce applications

Mobile Category	Major Applications	Clients
Advertising	Targeted ads, Location-based ads	Business
Commerce	Mobile transactions and payments	Business
Education	Mobile classrooms and labs	Schools and training centers
Enterprise resource planning	Resource management, Managing a mobile workforce	All
Entertainment	Games/images/music/video downloads and on-line gaming	Entertainment industry
Health care	Accessing and updating patient records	Hospitals and nursing homes
Inventory tracking and dispatching	Product tracking and dispatching	Delivery services and transportation
Traffic	Global positioning, Routing services, Toll paying, Traffic advisories	Transportation and auto industries
Travel and weather	Reservation services	Airlines, hotels, travel agencies

- *Maps*, which include traditional clear maps,
- *Local hangouts and businesses recommendations*, which provide suggestions for restaurant/gas-station/grocery-store/movie-theater, and
- *Satellite imagery*, which includes real images from satellites.

A few mobile map services are available. Google Maps for Mobile (n.d.) lets users find local hangouts and businesses across town or across the country—right from your phone, and Figure 4 shows three screen shots from the Google's map services where

(a) a clear map of the location with a postal code 58202,
(b) directions from the postal code 58201 to 58203, and
(c) a satellite map of (b) and a menu:
   - *Travel*: Travel expenses can be costly for a business or an individual. Mobile commerce could help reduce

Figure 4. Screen shots of the Google's map services showing (a) a clear map, (b) directions, and (c) a satellite image of (b) and a menu

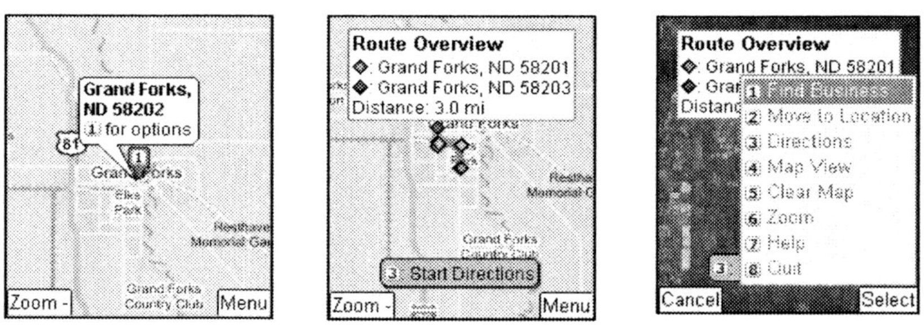

(a) (b) (c)

Figure 5. Screenshots of Yahoo! Mobile - Travel

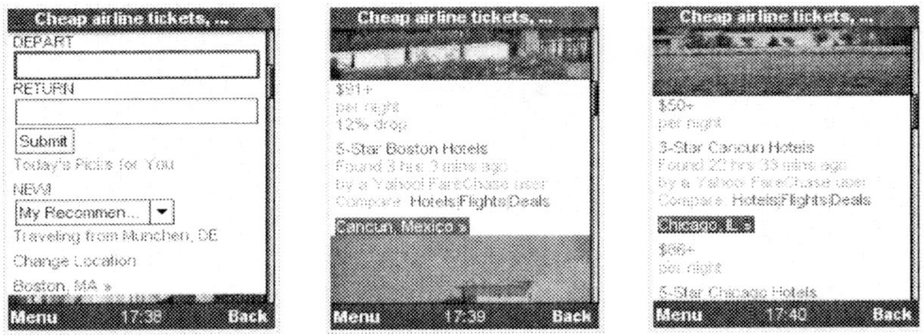

operational costs by providing mobile travel management services to travelers. It can be used to provide assistance to customers by using the mobile channels to locate a desired hotel nearby, purchase tickets, make transportation arrangements, and so on. The Travel Section of Yahoo! Mobile (n.d.) includes the following services:

- ◦ *Travel guides*: Allow mobile users to research 500,000 places to stay and things to do in over 40,000 cities worldwide with user reviews, photos, and maps, save favorite places into a custom trip plan, and get great travel deals.
- ◦ *Trip planner*: It is a tool that lets mobile users save hotels, attractions, restaurants, maps and more to a customized travel guide. You can add travel dates, your own comments, even bookmarks for other sites to your trip.
- ◦ *FareChase*: Yahoo! FareChase is a travel search engine that helps travelers scour the Web for the best flights and hotels that meet their budget and travel schedule.
- ◦ *Deals*: This service provides various top deals from hotels to car rentals.

Of course, mobile users can search travel-related information on Yahoo! Travel. Figure 5 shows screenshots of Yahoo! Travel.

- *Weather*: Weather reports are important for travelers, who may pack their bags or plan their trips based on the weather predictions. Most mobile service providers offer weather reports. For example, the Google provides the following local weather

Figure 6. Screenshots of Google weather report

information: (i) 3-day weather forecasts including lowest and highest temperatures, (i) current temperature, (iii) humidity, (iv) weather conditions with pictures, and (v) wind including direction and speed. Figure 6 shows screenshots of Google weather reports.

CLIENT COMPUTERS AND DEVICES

Desktop and laptop computers are on the client-side of electronic commerce systems, whereas mobile handheld devices are for mobile commerce systems. An Internet-enabled mobile handheld device is a small general-purpose, programmable, battery-powered computer that is capable of handling the front end of mobile commerce applications and can be operated comfortably while being held in one hand. It is the device via which mobile users interact directly with mobile commerce applications. The differences between these two client machines are given in Table 2. There are other kinds of computers such as tablet computers, which are a special kind of PCs.

Client-Side Programming

Electronic and mobile commerce programming involves a wide variety of technologies and languages. It consists of two kinds of programming:

- *Client-side programming*, which is to develop software running on client computers or devices. It is mostly related to Web interface construction. The popular languages for Web interface construction include CSS, DOM, (X)HTML, JavaScript, WML, WMLScript, XML, XSL(T), etc. Other than Web interface construction, client-side programming can build client-side applications such as address and schedule books. The tools and languages used for client-side application development are based on the client-side operating systems, e.g. Visual Studio for Windows and C/C++ for Palm OS.
- *Server-side programming*, which is to develop software running on servers. The software normally receives requests from browsers and sends the results from databases/files/programs back to the browsers for display. The popular server-side

Table 2. Differences between desktop & laptop computers and handheld devices

	Desktop and Laptop Computers	Mobile Handheld Devices
Browser	Desktop browsers	Microbrowsers
Functions	Full	Limited
Major Input Methods	Keyboards and mousse	Stylus and soft keyboards
Major Output Methods	Screens and printers	Screens
Mobility	Low	High
Networking	Wired	Wireless and mobile
Transmission Bandwidth	High	Low
Power Supply	Electrical outlets	Batteries
Screen	Normal	Small
Size	Desktop	Handheld
Weight	Normal	Light

languages include C/C++, Java, Perl, PHP, etc. Other than Web applications, it can be used to implement numerous applications such as instant messaging and telephony. However, this kind of applications is normally related to network programming such as TCP/IP programming and will not be covered in this chapter.

This sub-section discusses Web interface construction and the server-side programming will be covered in the Section of Host Computers. Other than building a Web system by using various languages and tools, some common software packages are available for Web development. Those packages can be divided into three categories: (i) multimedia editors, (ii) HTML editors, and (iii) integrated development environment (IDE):

- *Multimedia editors*, which are used to create, edit, and post animation, audio, images, and videos on Web pages. Three popular multimedia editors are
  - *Adobe Flash*, which is an authoring environment for creating animation, advertisements, various Web-page components, to integrate video into Web pages, and more recently, to develop rich Internet applications. Flash Professional is an IDE while Flash Player is a virtual machine used to run, or parse, the Flash files.
  - *Adobe Photoshop*, which is image-editing and graphics creation software.
  - *Microsoft Silverlight*, which is a programmable Web browser plugin that enables features such as animation, vector graphics and audio-video playback that characterize rich Internet applications.
- *HTML editors*, which are used to create static Web pages. Three popular HTML editors are
  - *Adobe Dreamweaver*: It is a WYSIWYG (What You See Is What You Got) authoring software that allows Web developers to generate HTML and JavaScript source code while viewing the site as they work.
  - *Microsoft Expression Web*, which is a design tool to create sophisticated standards-based Web sites. It combines both FrontPage and Visual Studio technologies in a new user interface for creating XHTML, CSS, XML, XSLT, and ASP.NET 2.0. Where appropriate, the user interface and features of Expression Web and Visual Studio are identical.
  - *Microsoft SharePoint Designer*, which will enable information workers to develop applications and solutions on top of the SharePoint platform to enable organizational agility, business process automation, and get the value of Microsoft Office applications on the SharePoint platform.

The category of integrated development environment (IDE) will be covered in the Host Computers Section.

MOBILE MIDDLEWARE AND WIRELESS NETWORKS

Mobile middleware and wireless networks are for mobile commerce systems only. The mobile middleware is optional, but the system will be greatly simplified with it. A mobile commerce system is already complicated enough. Without mobile middleware, the mobile system becomes even more complicated.

Mobile Middleware

The term middleware refers to the software layer between the operating system and the distributed applications that interact via the networks. The primary mission of a middleware layer is to hide the underlying networked environment's complexity by insulating applications from explicit protocols that handle disjoint memories, data replication, network faults, and parallelism (Geihs, 2001). The major task of mobile middleware is to seamlessly and transparently map Internet contents to mobile stations that support a wide variety of operating systems, markup languages, microbrowsers, and protocols. WAP and i-mode were the two major kinds of mobile middleware:

- *WAP (Wireless Application Protocol)*, which is a secure specification that allows users to access information instantly via mobile handheld devices such as smart phones and PDAs (Open Mobile Alliance Ltd., n.d.). WAP supports most wireless networks including CDPD, CDMA, GSM, PDC, PHS, TDMA, FLEX, ReFLEX, iDEN, TETRA, DECT, DataTAC, and Mobitex. WAP is supported by all operating systems. Ones specifically engineered for handheld devices include PalmOS, EPOC, Windows CE, FLEXOS, OS/9, and JavaOS. WAPs that use displays and access the Internet run what are called microbrowsers—browsers with small file sizes that can accommodate the low memory constraints of handheld devices and the low-bandwidth constraints of a wireless-handheld network. Although WAP supports HTML and XML, the WML language is specifically designed for small screens and one-hand navigation without a keyboard.
- *i-mode*, which is a mobile Internet service that has caused a revolution in both business and private lifestyles in Japan (NTT DoCoMo, Inc., 2007). 46 million subscribers have been attracted to this service since its start in February 1999 and currently more than 95,000 Internet sites are providing a variety of contents. The use of packet transmissions offers continuous access, while the use of a subset of HTML makes content creation easy and provides simple conversion of existing Websites.

Table 3 compares i-mode to WAP. Today, most telecommunication companies developed their own mobile middleware based on their own needs.

Wireless Networks

Wireless communication capability supports mobility for end users in mobile commerce systems. Wireless LAN, MAN, and WAN are the major components used to provide radio communication channels so that mobile service is possible. In the WLAN category, the Wi-Fi standard with 11 Mbps throughput dominates the current market. However, it is expected that standards with

Table 3. A comparison between the two major types of mobile middleware

	WAP	i-mode
Developer	Open Mobile Alliance	NTT DoCoMo
Implementation	A protocol	A complete mobile Internet service
Web Language	WML (Wireless Markup Language)	CHTML (Compact HTML)
Major Technology	WAP Gateway	TCP/IP development
Key Features	Widely adopted and flexible	Highest number of users and easy to use

much higher transmission speeds, such as IEEE 802.11a and 802.11g, will replace Wi-Fi in the near future. Compared to WLANs, cellular systems can provide longer transmission distances and greater radio coverage, but suffer from the drawback of much lower bandwidth (less than 1 Mbps). In the latest trend for cellular systems, 3G standards supporting wireless multimedia and high-bandwidth services are beginning to be deployed. The wireless telephone technology includes several generations as follows:

- *0G (1945 - 1973)*, which refers to mobile radio telephone systems.
- *1G (1980s)*, which is analog cellphone standards including NMT and AMPS.
- *2G (1990s)*, which is digital cellphone standards divided into TDMA-based and CDMA-based standards depending on the type of multiplexing used.
- *2.5G (late 1990s)*, which is implemented a packet switched domain in addition to the circuit switched domain.
- *3G (early 2000s)*, which includes wide-area wireless voice telephony and broadband wireless data, all in a mobile environment.
- *4G (2000s)*, which provides end-to-end IP solution where voice, data and multimedia streaming can be served at higher data rates with anytime-anywhere concept.

A wide variety of technologies and standards for wireless telephones are available. Some of the major ones include:

- *CDMA (Code Division Multiple Access)*, which is based on a spread spectrum method. The method transmits a signal by "spreading" it over a broad range of frequencies. This provides reduced interference and can increase the number of simultaneous users within a radio frequency band. With CDMA, each conversation is digitized and then tagged with a code.
- *GSM (Global System for Mobile communications)*, which is one of the most popular standards for mobile phones and is specifically developed to provide system compatibility across country boundaries, especially the Europe. It is based on TDMA (Time Division Multiple Access) technology, which works by dividing a radio frequency into time slots and then allocating slots to multiple calls. Therefore, GSM allows eight simultaneous calls on the same radio frequency.
- *IEEE 802.11*, includes an encryption method, the Wired Equivalent Privacy algorithm. WLAN (Wireless Local Area Network), based on 802.11, allows a mobile user connecting to a local area network (LAN) through a wireless (radio) connection. This wireless data transmission speed of WLAN is up to 54 Mbps.
- *IEEE 802.16*, which ensures compatibility and interoperability between broadband wireless access equipment. WiMAX (Worldwide Interoperability for Microwave Access), based on 802.16, provides wireless data over long distances, in a variety of different ways, from point to point links to full mobile cellular type access. In practical terms this enables a user, for example, to browse the Internet on a laptop computer without physically connecting the laptop to a wall jack.

Table 4 shows major technologies and standards used in the wireless telephone generations.

WIRED NETWORKS

Wired networks are used to transmit data for electronic/mobile commerce. This component is a requirement for electronic commerce, but not necessary for mobile commerce, though mobile commerce would be greatly benefited by apply-

Table 4. Wireless telephone technology evolution

	2G (10 Kbps – 40 Kbps)	2.5G (20 Kbps – 171 Kbps)	3G (60 KBps – 54 Mbps)	4G (50 Mbps – 1 Gbps)
CDMA track	IS-95	CDMA 2000	W-CDMA	UMTS Revision 8 (LTE)
GSM track	GSM	GPRS	EDGE	
IEEE 802.11 track			WLAN	
IEEE 802.16 track				WiMAX

ing wired networks to its data communication because data transmission using wireless networks is more expensive than using wired networks. Among several types of wired networks, three major types are

- *Local Area Network (LAN)*, which spans a relatively small space of only a few square kilometers or less such as an office building. It generally offers a throughput of 10 Mbps or 100 Mbps and is usually based on Ethernet technology, which is a network protocol using a bus topology and defining a specific implementation of the Physical and Data Link Layers in the OSI model (IEEE 802.3).
- *Metropolitan Area Network (MAN)*, which spans a geographical area greater than a LAN but less than a WAN such as few city blocks or a whole city. MAN typically uses wireless infrastructure or optical fiber connections to link its sites and it may connect multiple LANs together. Its maximum throughput is no less than 44 Mbps and it uses the Distributed Queue Dual Bus technology based on the IEEE 802.6 standard.
- *Wide Area Network (WAN)*, which spans a wide geographic area, such as state or country, and uses specialized computers to connect smaller networks, such as LANs. It generally offers a throughput of 1.5 Mbps or more. WANs typically use wide area network services from telecommunications carriers, whose technologies include standard phone lines, ISDN (Integrated Services Digital Network), or other high-speed services. Two examples of WAN are the Internet, the largest network in the world, and an airline using WAN to connect its offices around the world.

HOST COMPUTERS

This component is similar for both electronic and mobile commerce systems because host computers are usually not aware of the differences among the targets, browsers or microbrowsers, they serve. The application programs are responsible for apprehending their clients and responding to them accordingly. Most of the electronic/mobile commerce application programs reside in this component, except for client-side programs such as cookies or user interface using markup languages. A user request such as checking out or adding items to the shopping cart is actually processed at a host computer, which contains three major kinds of software specifically for e/m-commerce transactions: (i) Web servers, (ii) databases and database servers, and (iii) application programs and support software. Figure 7 shows a structure of three-tiered client-server Web systems. Before examining the three kinds of software in more detail, let us look at the LAMP stack, which is one of the popular Web technologies used these days.

Figure 7. Three-tiered client-server Web system structure

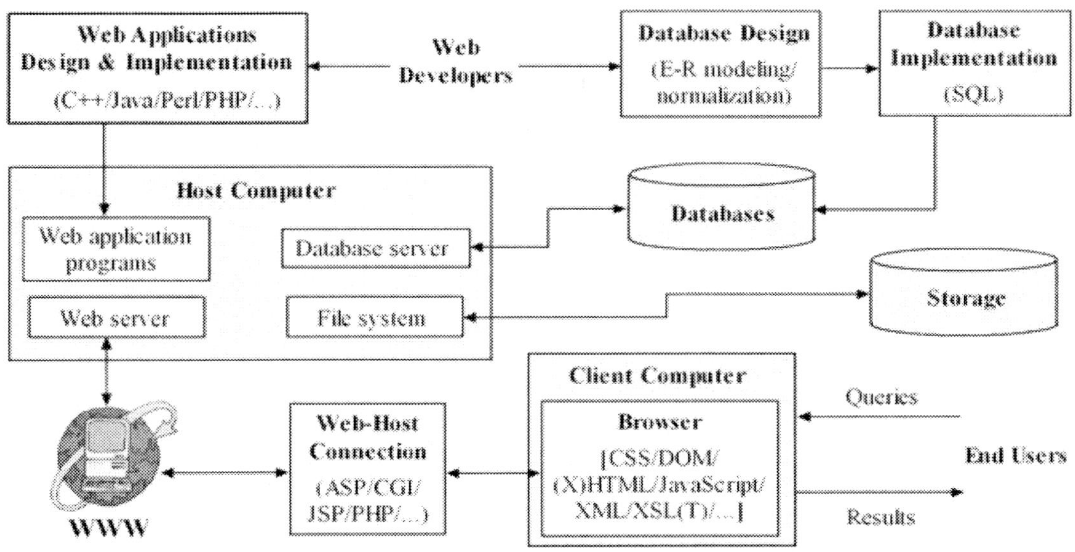

The LAMP Stack

One of the popular technologies for Web content implementation is the LAMP stack, which includes four components (Lawton, 2005):

1. *Linux*, which is a free open-source operating system based on Unix.
2. *Apache Web server*, which is an open-source HTTP server for modern operating systems including UNIX and Windows NT.
3. *MySQL*, which is an open source relational database management system (RDBMS) that uses Structured Query Language (SQL), the most popular language for adding, accessing, and processing data in a database.
4. *Script languages*, such as Perl, PHP, and Python:
   - *Perl (Practical Extraction and Report Language)*, which is often used for scanning text and printing formatted reports. It provides extensive support for regular expression matching, dynamically scoped variables and functions, extensible run-time libraries, exception handling and packages, provide/require. The use of Perl has grown significantly since its adoption as the language of choice of many World Wide Web developers. CGI interfaces and libraries for Perl exist for several platforms and Perl's speed and flexibility make it well suited for form processing and on-the-fly page creation. Perl5 is a major rewrite and enhancement of Perl version 4. It adds nested data structures and object-oriented features.
   - *PHP (HyperText Preprocessor)*, which is a general-purpose server-side scripting language that is especially suited for dynamic and interactive Websites and can be embedded into HTML. It is the widely-used, free, and efficient alternative to competitors such as Microsoft's ASP. The PHP syntax is very similar to Perl and C. PHP is often used together with MySQL (DBMS) and Apache (Web server) on various operating systems.

A PHP file may contain text, HTML tags, and scripts. Scripts in a PHP file are executed on the server.

- *Python*, which is a dynamic object-oriented programming language that can be used for many kinds of software development. It offers strong support for integration with other languages and tools, comes with extensive standard libraries, and can be learned in a few days. Python runs on Windows, Linux/Unix, Mac OS X, OS/2, Amiga, Palm Handhelds, and Nokia mobile phones. Python has also been ported to the Java and .NET virtual machines. Python is distributed under an OSI-approved open source license that makes it free to use, even for commercial products.

LAMP has several key advantages over proprietary software development approaches. Two of them are

- *Cost*: Open source software is either free or low cost compared to proprietary software.
- *Open source*: Anyone can review, modify, and work with open source software; developers can improve and fix the technology faster.

Web Servers

A Web server is a server-side application program that runs on a host computer and manages the Web pages stored on the Web site's databases/files. There are several Web server software including public domain software from NCSA and Apache, and commercial packages from Microsoft, Netscape, and others. Three popular Web servers are

- *Apache HTTP servers*, which are a collaborative software development effort aimed at creating a freely-available source code implementation of an HTTP (Web) server. They are jointly managed by a group of volunteers located around the world. Since April 1996, Apache has been the most popular HTTP server on the Internet. It was developed in early 1995 based on code and ideas found in the most popular HTTP server of the time, NCSA httpd 1.3. It has since evolved to rival almost any other Unix based HTTP server in terms of functionality and speed. It features highly configurable error messages, DBM-based authentication databases, and content negotiation (The Apache Software Foundation, n.d.).
- *Microsoft's Internet Information Services (IIS)*, which provide a Web application infrastructure for all versions of Windows servers (Microsoft, Corp., n.d.a). It is the world's second most popular Web server after Apache.
- *Sun Java System Web Servers*, which provide organizations with a single deployment platform for Web services, JavaServer Pages (JSP), Java Servlet technologies, NSAPI and CGI (Sun Microsystems, Inc., n.d.). They also offer built-in HTTP reverse-proxy capabilities to provide a highly scalable HTTP front-end to application servers or other HTTP origin servers. Its bundled FastCGI interface provides a stable and scalable platform for third party scripting technologies such as PHP, Ruby On Rails, Perl, Python, and more.

Database Servers

A database server manages database access functions, such as locating the actual record being requested or updating the data in databases. Some popular databases include:

- *IBM DB2*: DB2 9 is a hybrid data server with management of both XML and relational data. It includes the following major features:
    - XML data store,
    - integration with relational data,
    - Eclipse-base developer workbench, and
    - integration with leading application infrastructures like PHP, Java, and .NET
- *Microsoft*: Microsoft provides two kinds of databases:
    - *Access*: For the desktop computers, Microsoft offers Access. The Microsoft Access is a full-featured multiuser relational database management system that designed for the Microsoft Windows operating systems. Access is visually oriented and easy to use. It makes extensive use of drag-and-drop and visual design for queries, forms, and reports. Access comes with an integrated development environment (IDE), including incremental compilation, a fully interactive visual debugger, breakpoints, and single step-through. These capabilities combine to make Microsoft Access a powerful platform for developing client-server database solutions.
    - *SQL Server*: For the server engines in client-server solutions, Microsoft provides the SQL Server. SQL Server is a comprehensive database software platform providing enterprise-class data management and integrated business intelligence (BI) tools. The SQL Server data engine lies at the core of this enterprise data management solution. In addition to providing support for relational databases or XML, SQL Server combines the best in analysis, reporting, integration, and notification. This enables your team to build and deploy cost-effective BI solutions with which they can drive data into every corner of your business through scorecards, dashboards, Web services, and mobile devices. Close integration with Microsoft Visual Studio, the Microsoft Office System, and a suite of new development tools, including the Business Intelligence Development Studio, sets SQL Server apart.
    - *MySQL*: MySQL is an open-source, multithreaded, multi-user SQL relational database management system (Microsoft Corp., n.d.b). It is used in more than 11 million installations ranging from large corporations to specialized embedded applications. Not only is MySQL the world's most popular open source database, it is a key part of LAMP, a fast growing open source enterprise software stack. More and more companies are using LAMP as an alternative to expensive proprietary software stacks because of its lower cost and freedom from lock-in. MySQL is flexible and runs on more than 20 platforms including Linux, Windows, OS/X, HP-UX, AIX, and Netware.
- *Oracle databases*, whose newest version is Oracle11g. The following list shows the Oracle database migration (Oracle, n.d.):
    - *Oracle7.2*, which is a client/server based relational database management system. The query language is based on SQL. In order to access the system, an Oracle account must be created for a specific database.
    - *Oracle8i*, which is a client/server based relational database management system, with object capabilities

included. The query language is SQL. Java has been added to the database capabilities.
- *Oracle9i*, features full XML database functionality with the new Oracle XML DB feature, and other improvements.
- *Oracle 10g*, which is the first database designed for enterprise grid computing, which provides an environment in which individual users can access computers, databases, and experimental facilities simply and transparently, without having to consider where those facilities are located.
- *Oracle 11g*, which is able to work for Linux and Microsoft Windows.

Other than the server-side database servers, a growing trend is to provide a client-side mobile database or an embedded database to a handheld device with a wide range of data-processing functionality. The functionality is frequently very sophisticated, and the flat file system that comes with these devices may not be able to adequately handle and manipulate data. Embedded databases have very small footprints, and must be able to run without the services of a database administrator and accommodate the low-bandwidth constraints of a wireless-handheld network. Some leading embedded-databases are Progress Software databases, Sybase's Anywhere products, and Ardent Software's DataStage (Ortiz, 2000).

Application Programs and Support Software

Application programs and support software are responsible for handling server-side processing. Three generations of programming languages and environments are used for server-side Web application development:

1. *1st generation*: Traditionally, conventional programming languages such as C/C++ and Java are used for Web development.
2. *2nd generation*: Dynamic programming languages such as Perl and PHP gradually replace conventional languages for Web development. A dynamic language enables programs that can change their code and logical structures at runtime, adding variable types, module names, classes, and functions as they are running. These languages frequently are interpreted and generally check typing at runtime.
3. *3rd generation*: Recently, a couple of IDEs (Integrated Development Environments) are used for Web development:
   - *Adobe ColdFusion*, which is an application server and software development framework used for the development of computer software in general, and dynamic Web sites in particular.
   - *Microsoft ASP.NET*, which is part of Microsoft's .NET platform and is the successor to ASP technology. ASP.NET is a free technology that allows programmers to create dynamic Web applications.
   - *Microsoft Visual Studio*, which is Microsoft's flagship software development product for computer programmers. It lets programmers create standalone applications, Web sites, Web applications, and Web services that run on any platforms supported by Microsoft's .NET Framework.
   - *NetBeans IDE*, which is an open-source Integrated Development Environment for software developers. It is used to create professional cross-platform desktop, enterprise, Web, and mobile applications.
   - *Ruby On Rails (ROR)*, which is a full-stack framework for developing

database-backed Web applications according to the Model-View-Control pattern.

- o *Sun Java Studio IDE*, which is a development platform with features such as UML modeling, instant collaboration, and application profiling. It is used to develop, debug, tune, and deploy enterprise applications, Web services, and portal components based on the Java EE platform.
- o *Zend Core*, which is the production PHP 5 stack that provides the certified, enhanced capabilities with support and services that professionals need for PHP development and production.

SUMMARY

The emerging wireless and mobile networks have extended electronic commerce to another research and application subject: mobile commerce. A mobile or an electronic commerce system involves a wide range of disciplines and technologies. This level of complexity makes understanding and constructing such a system an arduous task. To facilitate this process, this article divided a mobile or an electronic commerce system into six components, which can be summarized as follows:

1. *Applications*: Electronic commerce applications are already broad. Mobile commerce applications not only cover the existing applications, but also include new applications, which can be performed at any time and from anywhere by using mobile computing technology.
2. *Client computers or devices*: Mobile handheld devices are limited by their small screens, limited memory, limited processing power, and low battery power, and suffer from wireless network transmission problems. Numerous mobile devices, such as PDAs or Web-enabled cellular phones, are available in the market, but most use one of three major operating systems: Palm OS, Microsoft Pocket PC, and Symbian OS. At this moment, Palm OS leads the market, although it faces a serious challenge from Pocket PC.
3. *Mobile middleware (mobile commerce only)*: Mobile middleware is a set of software that intends to simplify the high complexity of interactions between mobile applications and wireless networks. WAP and i-mode were the two major kinds of mobile middleware few years ago. WAP was widely adopted and flexible; while i-mode had the highest number of users and is easy to use. However, most telecommunication companies develop their own mobile middleware that fits their needs these days.
4. *Wireless networks (mobile commerce only)*: Wireless communication capability supports mobility for end users in mobile commerce systems. Wireless LAN, MAN, and WAN are major components used to provide radio communication channels so that mobile service is possible. In the WLAN category, the Wi-Fi standard with 11 Mbps throughput dominates the current market. It is expected that standards with much higher transmission speeds, such as IEEE 802.11a and 802.11g, will replace Wi-Fi in the near future. Compared to WLANs, cellular systems can provide longer transmission distances and greater radio coverage, but suffer from the drawback of much lower bandwidth (less than 1 Mbps). In the latest trend for cellular systems, 3G standards supporting wireless multimedia and high-bandwidth services are beginning to be deployed. WCDMA and CDMA2000 are likely to dominate the market in the future.

5. *Wired networks*: This component is a requirement for electronic commerce, but not necessary for mobile commerce, though mobile commerce would be greatly benefited by applying wired networks to its data communication because data transmission using wireless networks is more expensive than using wired networks. Among several types of wired networks, three major types are (i) LAN (Local Area Network), (ii) MAN (Metropolitan Area Network), and (iii) WAN (Wide Area Network).
6. *Host computers*: Host computers process and store all the information needed for mobile commerce applications, and most application programs can be found here. They include three major components: (i) Web servers, (ii) database servers, and (iii) application programs and support software.

Electronic and mobile commerce programming involves a wide variety of technologies and languages. It consists of two kinds of programming:

- *Client-side programming*, which is to develop software running on client computers or devices. It is mostly related to Web interface construction. The popular languages for Web interface construction include CSS, DOM, (X)HTML, JavaScript, WML, WMLScript, XML, XSL(T), etc.
- *Server-side programming*, which is to develop software running on servers. The software normally receives requests from browsers and sends the results from databases/files/programs back to the browsers for display. The popular server-side languages include C/C++, Java, Perl, PHP, etc.

REFERENCES

Apache Software Foundation. (n.d.). *Apache HTTP Server Project*. Retrieved June 21, 2008, from http://httpd.apache.org/

Gartner, Inc. (2009a). *Gartner Says in the Fourth Quarter of 2008 the PC Industry Suffered Its Worst Shipment Growth Rate Since 2002*. Retrieved March 15, 2009, from http://www.gartner.com/it/page.jsp?id=856712

Gartner, Inc. (2009b). *Gartner Says Worldwide Smartphone Sales Reached Its Lowest Growth Rate With 3.7 Per Cent Increase in Fourth Quarter of 2008*. Retrieved March 18, 2009, from http://www.gartner.com/it/page.jsp?id=910112

Gartner, Inc. (2009c). *Gartner Says Worldwide Mobile Phone Sales Grew 6 Per Cent in 2008, But Sales Declined 5 Per Cent in the Fourth Quarter*. Retrieved March 19, 2009, from http://www.gartner.com/it/page.jsp?id=904729

Geihs, K. (2001). Middleware challenges ahead. *IEEE computer, 34*(6), 24-31.

Google. (n.d.). *Google Maps for Mobile*. Retrieved March 12, 2009, from http://www.google.com/gmm/

Hu, W.-C., Lee, C.-w., & Yeh, J.-h. (2004). Mobile commerce systems. In S. Nansi (Ed.), *Mobile Commerce Applications* (pp. 1-23). Hershey, PA: Idea Group Publishing.

Lawton, G. (2005). LAMP lights enterprise development efforts. *IEEE Computers, 38*(9), 18–20.

Microsoft, C. (n.d.b). *SQL Server2005*. Retrieved May 6, 2008, from http://www.microsoft.com/sql/default.mspx

Microsoft, C. (n.d.a). *Internet Information Services*. Retrieved June 15, 2008, from http://www.microsoft.com/WindowsServer2003/iis/default.mspx

Microsystems, S. Inc. (n.d.). *Sun Java Sysem Web Server*. Retrieved June 19, 2007, from http://www.sun.com/software/products/Web_srvr/home_Web_srvr.xml

NTT DoCoMo, Inc. (2007). *i-mode*. Retrieved June 12, 2007, from http://www.nttdocomo.com/services/imode/index.html

Open Mobile Alliance Ltd. (n.d.). *WAP Forum*. Retrieved from June 13, 2008, from http://www.openmobilealliance.org/tech/affiliates/wap/wapindex.html

Oracle. (n.d.). *Oracle Databases*. Retrieved August 25, 2008, from http://www.oracle.com/database/index.html

Ortiz, S. Jr. (2000). Embedded databases come out of hiding. *IEEE Computer*, *33*(3), 16–19.

Sadeh, N. (2002). *M-commerce: Technologies, services, and business models* (pp. 177-179). New York: John Wiley & Sons.

Silicon Valley Daily. (2006). *HP Regains Lead in Global PC Sales*. Retrieved June 11, 2007, from http://www.svdaily.com/gartner1.html

Symbian Limited. (2009). *Fast Facts*. Retrieved March 10, 2009, from http://www.symbian.com/about/fastfacts/fastfacts.html

The Yankee Group. (2001). *Over 50% of large U.S. enterprises plan to implement a wireless/mobile solution by 2003*. Retrieved December 10, 2002 from http://www.yankeegroup.com/public/news_releases/news_release_detail.jsp?ID=PressReleases/news_09102002_wmec.htm

Turban, E., King, D., Lee, J., & Viehland, D. (2004). *Electronic Commerce 2004: A Managerial Perspective*. Prentice Hall.

Varshney, U., Vetter, R. J., & Kalakota, R. (2000). Mobile commerce: A new frontier. *IEEE Computer*, *33*(10), 32–38.

Yahoo! (n.d.). *Yahoo! Mobile*. Retrieved February 21, 2009, from http://mobile.yahoo.com/

Chapter 3
A Component-Based Software Architecture for Delivering Location Based Services

Ioannis Priggouris
University of Athens, Greece

Dimitris Spanoudakis
University of Athens, Greece

Manos Spanoudakis
Credit Suisse, Switzerland

Stathes Hadjiefthymiades
University of Athens, Greece

ABSTRACT

The combination of position fixing mechanisms with location-dependent, geographical information, can offer truly customized personal communication services through the mobile phone or other type of devices. Prompted by the avalanche of technology advances in the aforementioned areas, this chapter presents a generic framework for delivering Location Based Services (LBS). The framework is capable of providing the full functionality required for delivering LBS, starting from the specification of the service, covering issues like its deployment and maintenance of services, the service invocation and the final delivery of the produced results to the calling user. The main focus of the chapter is on the technical specification, the design and the functionality of the framework. However, with the purpose of assessing the proposed architecture, a prototype implementation based on the discussed specifications was built and its performance was evaluated using a series of pilot services.

INTRODUCTION

The mobile communications market experiences an unprecedented boom in the recent years. New handheld devices with increased capabilities are introduced, while mobile operators are striving to gain a significant portion of the market by delivering new state-of-the art value added **network services**, that can fully utilize the given technology. **Location**

DOI: 10.4018/978-1-60566-964-9.ch003

Based Services (LBS) is just one such category of services, on which both manufacturers and mobile operators have invested a lot (Kaasinen, 2003). However delivering new services requires developing means and tools that will assist in their creation, provision and maintenance.

This paper presents an extensible cross-platform framework, which facilitates the provision of LBS services with minimum effort from all involved sides (e.g. service provider, mobile operator etc.) and with no change to the wireless Internet infrastructure. Moreover, the framework is based on open standards so that it can accommodate future evolved technologies and be fitted in future telecommunication infrastructures.

LBS PROVISIONING PLATFORMS REQUIREMENTS

The development of a provisioning platform, which would cover all aspects of LBS provisioning is a primary goal of all vendors involved in the world of LBS software systems. The set of requirements and desired characteristics for such platform includes:

- Means for supporting the service creation process without the need for special programming skills.
- Support for service deployment and operation through automatic procedures, which do not require special IT personnel or programming skills.
- Service provisioning through a variety of access protocols so that the service is accessible from different networks.
- Portability over different operating systems and hardware platforms, so that integration to different infrastructures is possible.
- Reusability: The same platform may host a number of different services, with different requirements and functionality. The introduction of new services should not require changes to the platform and changes to the platform should not affect the execution of existing or future services.
- Access to the system should be possible through a variety of devices, using different transport protocols.
- Independence from underlying **positioning** and GIS technologies: The platform should not bind to any specific networking technologies. GSM/GPRS and WLAN interfaces should be treated in a unified way, while the introduction of UMTS and 3G networks will not require changes to either services or the platform. This also applies to the GIS interface which should be open enough so that integration with different implementations to be possible.
- Security. Interfaces with external entities have to be secure and privacy should be maintained. These characteristics will make the user feel safer and as a result more receptive to using the platform.
- Roaming across different infrastructures. Both indoor and outdoor environments should be supported.
- Scalability. Possibility of hosting large number of services, each capable of serving numerous concurrent requests, is sure to enhance the marketing potential of the platform.

Currently, to the best of our knowledge, no integrated platform, which covers all these aspects, exists. Most of the existing platforms facilitate LBS operations by providing very simple functionality, which fulfils only a subset of these requirements. During the design of the discussed framework the goal was to cover all the aforementioned requirements and provide all these features that could maximize the benefit from the adoption of the provided system.

Figure 1. Framework architecture

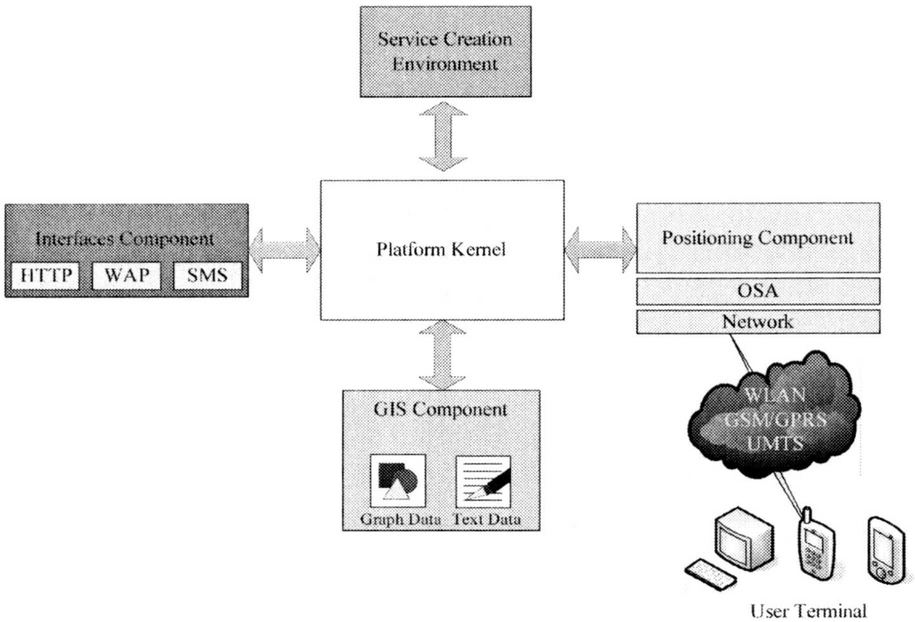

FRAMEWORK OVERVIEW

The proposed framework builds upon technical advances in the area of Info-mobility, distributed/mobile computing, and **Geographical Information Systems** (GIS) and aims to provide all the functionality needed for delivering location-based services in a single platform and at the same time respect all requirements set in the previous section. The platform focuses on the development, deployment and execution of services and consolidates several technologies.

A modular approach has been followed during the design of the platform resulting to an architecture, which defines a central processing unit: the kernel and independent functional entities on the boundary, each covering a certain operational area in the chain of LBS provisioning (Figure 1). The autonomy granted to each component through this architectural approach makes possible the extension of the platform by adding new components or modifying the existing ones, in order to enhance the overall functionality, which was designed to facilitate the creation of location-based services.

In order to understand the philosophy behind the proposed framework we will provide in the next section an overview of the technologies, involved in delivering location-dependent services. These technologies, more or less, dictate the functional areas that need to cover a platform aiming to provide Location-based services.

Positioning Technology

Positioning technology (Hightower & Borriello, 2001) is a key point in the LBS context. Research in this area during the last years is impressive. A plethora of solutions that provide accurate estimations of user's location have emerged, each with distinct characteristics and capabilities and hence applicable in different circumstances. Two main categories of positioning solutions are available today:

- Satellite Based such as the Global Positioning System (GPS) (Leick, 1995), which provide high accuracy with the drawback of needing additional equipment on the terminal side. In GPS, the determination of the user's location is based on the timed difference of arrival of different satellite signals (at least 3 satellite signals are required). A newcomer in this category is ESA's Galileo (Benedicto, Dinwiddy, Gatti, Lucas, & Lugert, 2000), which promises accuracy similar if not superior to GPS. The major drawbacks of GPS technology are its inability to operate in indoor environments as well as the need for specific hardware.
- Terrestrial Infrastructure-Based Solutions which are less accurate than the satellite based ones; however despite this drawback they are widely used, as they don't require expensive add-ons to the terminal devices (i.e., GPS receivers).

A certain shortcoming of the most popular positioning systems is their inability to produce accurate estimates in inside-building environments. However, lately, significant progress has been achieved in this area through WLAN based positioning solutions (Castro, Chiu, Kremenek, & Muntz, 2001), (Bahl & Padmanabhan, 2000).

GIS Technology

The Geographic Information Systems (GIS) technology, although not strictly required in order to provide location-dependent services, comprises a key technology in this area as its presence can greatly enhance the quality of the delivered service, by incorporating detailed visual representations (e.g., maps) or specific info that is otherwise difficult to be acquired (e.g., points of interests etc.).

The GIS represents a comprehensive database related to positional information with an integrated set of tools for querying, analysing, and displaying information. A standard RDBMS (e.g. Oracle or MS Access) is usually used for storing the spatial information. Numerous data formats have been defined for storing GIS data in these databases (geotiffs, shapefiles, RDFs etc.). Apparently each data format needs specific software in order to be read and further processed, resulting to a GIS market flooded with proprietary GIS software packages. However, products that perform translation between the most popular formats do exist today; thereupon binding to a specific format does not impose serious problems. A coarse categorization of the requests submitted towards a GIS system includes:

- Navigation requests for routing between two distinct locations.
- Proximity requests, for queries regarding points of interest or other information relative to user's current position.
- Find location requests, where a information about a specific location is requested.
- Geocoding requests, where specific knowledge of a certain area can be used to deduce the corresponding coordinates.

As delay is a crucial factor for such requests, existing GIS software implements algorithms for handling all abovementioned queries with the minimum possible delay.

Interfaces Technologies

The term "interfaces technologies" refers to the transport technologies used for communicating and transferring data between two end-points. As location-based services are strongly coupled with the mobile communications domain bearers that are used in the latter are of prime importance for transferring data between the platform server and the corresponding client. Until recently the exchange of data between 2 actors of the mobile infrastructure was limited to the SMS and the WAP technology. The former offers a basic transport protocol for textual data while the latter a much more advanced protocol, which could be considered analogous to the HTTP protocol but for the mobile domain. SMS is widely used today but WAP failed to gain wide acceptance mainly due to unavailability of bandwidth and cost-efficiency reasons.

Evolution towards 3G networks benefited greatly both technologies as the EMS and MMS technologies enhanced the SMS protocol with advanced multimedia capabilities while the GPRS technology not only revived the WAP technology, but together with the evolution of handheld devices made possible the use of the HTTP protocol in the mobile domain. In the WLAN environments the HTTP protocol remains the simplest and most promising technology for interfacing with servers and other remote systems.

PLATFORM ARCHITECTURE

After this short review of enabling technologies in this section we proceed with the analysis of the architecture of the LBS provisioning system. A brief overview of the main components of the platform (Figure 1) is presented in the following section. Four main components have been identified, namely: a) the Kernel, b) the Interfaces Component, c) the Positioning Component and d) the GIS Component,

The Kernel

The Kernel is the heart of the platform. The role of the Kernel within the proposed architecture is to serve as the runtime environment for the services offered to users as well as to coordinate the operation of the peripheral components according to the directives of the service logic.

Moreover the kernel provides all the functionality required for deploying new services through automatic procedures. Interfaces for injecting new services in the Kernel have been provided. These interfaces can be invoked either through command line or they can be used to create advanced graphical IDEs that will assist the service creation process.

Figure 2. Kernel architecture

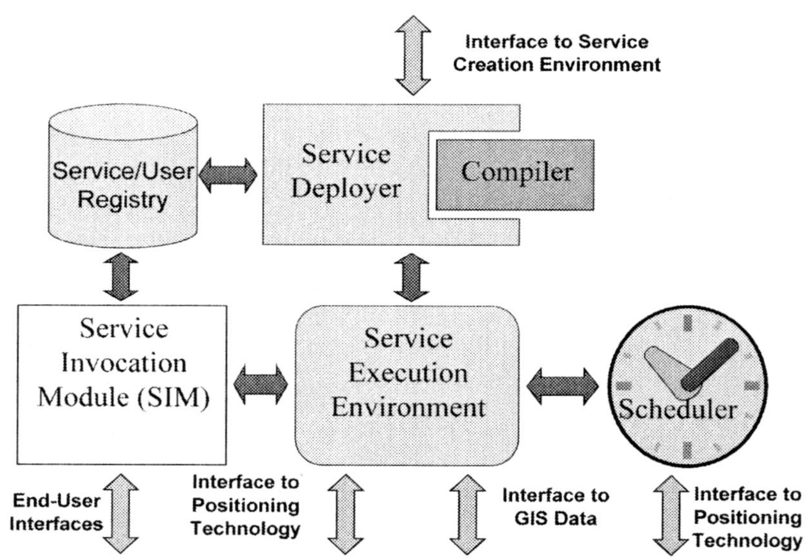

The internal architecture of the Kernel, which is illustrated in Figure 2 consists of a set of co-operating sub-modules designated to provide the full functionality needed during the life cycle of a service. An overview of these modules and their functionality is presented in the following sections.

The Service Deployer (with the Compiler)

It is the entry point for all new services entering the Kernel. The deployer consists of two distinct parts, the client and the server side. The client side resides externally to the Kernel (usually at the service creator's premises) and uses Web services to communicate with the server side. The Deployer provides methods needed for the deployment of new services on the platform. The same interface allows external clients to query the status of already deployed services, or to get information about a specific service. The Service Deployer receives the service specification and produces an executable form of the service which can run inside the Service Execution Environment. The architecture of the service deployer is illustrated in Figure 3.

In the beginning, the service specification document is checked for the integrity of received data, and after verification, a three-phase procedure is followed. The initial form of the service is an XML-based language. The initial phase consists of locating and extracting the service-control logic part that is contained within the service specification document. In the second phase, the compiler is invoked to perform lexical and syntactical analysis in order to produce the executable form of the service. The service parameters are extracted from the XML document and saved in a service-related repository so that they can be available during service invocation. This repository is located within the SEE. In the third

Figure 3. The internal architecture of service deployer

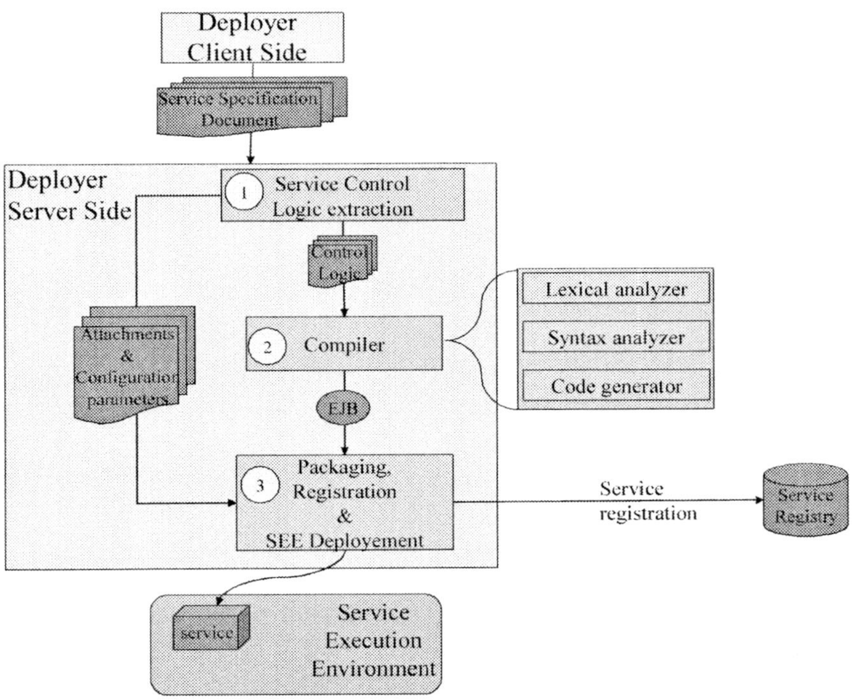

phase, the compiled service, along with its attachments, is packaged and a request for deployment inside the SEE of the kernel is performed. If the request completes successfully, the new service is registered within the service registry.

The Service Invocation Module (SIM)

The Service Invocation Module (SIM) receives all incoming requests and provides a standard interface towards the Interfaces component. An XML protocol that enables this communication has been designed. This protocol was designed taking into account the efficiency requirements imposed by the real-time communication needs, therefore it is simple and compact, but at the same time flexible enough in order to accommodate all information for service execution and handling of responses. The DTD of the protocol is presented in Figure 4.

In order to cope with large number of concurrent requests the SIM has been designed as a module, which incorporates a generic, transparent mechanism for invoking all possible services. The general functionality provided by the SIM has as follows:

a) Reception of incoming requests in the form of an XML message
b) Processing of the message to deduce the service to be invoked and its parameters
c) Consultation of the service registry
d) Invocation of the service
e) Packaging of service results according to the XML protocol

Figure 4. SIM-Interfaces communication protocol (DTD)

```
<!ELEMENT      interf-krn-message  (request-to-kernel | response-from-kernel)  >
<!ELEMENT      request-to-kernel   (user-data, service-invoc-data, security-data?)  >
<!ATTLIST      request-to-kernel
               msg-id     CDATA       #REQUIRED
               date-time  CDATA       #REQUIRED>
<!ELEMENT      user-data (term-cap?)>
<!ATTLIST      user-data
               user-id                CDATA              #REQUIRED
               user-address                     CDATA               #REQUIRED
               second-response-type             (sms | wap | http)  #IMPLIED>
<!ELEMENT      term-cap  EMPTY>
<!ELEMENT      service-invoc-data  (service-data, cookies*)>
<!ATTLIST      service-invoc-data
               service-id             CDATA              #REQUIRED
               service-type           (sms | wap | http) #REQUIRED
               rqst-method            CDATA              #REQUIRED>
<!ELEMENT      service-data           CDATA>
<!ELEMENT      cookies   EMPTY>
<!ATTLIST      cookies
               session-id             CDATA     #IMPLIED
               name                   CDATA     #REQUIRED
               value                  CDATA     #REQUIRED
               description            CDATA     #REQUIRED>
<!ELEMENT      security-data          EMPTY  >
<!ATTLIST      security-data
               url-proxy CDATA        #IMPLIED
               url-referer            CDATA     #IMPLIED>
<!ELEMENT      response-from-kernel             (xslt-data, service-invoc-data, user-data)
<!ATTLIST      response-from-kernel
               msg-id                 CDATA              #REQUIRED            ,
               cservlet-name          CDATA              #IMPLIED
               date-time              CDATA              #REQUIRED
               priority               (high | medium | low)        "medium">
<!ELEMENT      xslt-data EMPTY  >
<!ATTLIST      xslt-data
               type      (generic | specific)            "generic"
               name      CDATA                           #IMPLIED>
```

f) Dispatching the results back to the Interfaces Component in the form of an XML message

The Service Execution Environment (SEE)

The Service Execution Environment (SEE) acts as the runtime context for the deployed services. It is also responsible for coordinating communication of the deployed service with components residing externally to the SEE either inside the Kernel (i.e., the Scheduler) or outside (i.e., the Positioning and GIS peripheral components). Specific client modules inside the SEE namely the Component handlers undertake the task of translating the service control logic directives to synchronous method invocations on the aforementioned components. Concept-mechanisms such as configuration variables, variables shared between different instances of the same service, variables that persist between successive invocations of the same service for a specific user are also supported by the SEE through the definition of specific repository areas. These repositories are created per service and can directly be accessed inside the service logic.

The Scheduler

The role of the Scheduler (Figure 5) is to support the automatic execution of services within the platform. It features built-in support for both event and time-triggered service execution and a variety of scheduling paradigms are catered for (one time scheduling, infinite periodic scheduling, periodic scheduling with fixed deadlines, etc.). The Scheduler also provides means for delivering location events coming from the Positioning component to the service (e.g., the user entered a mall or a building block), with the minimum possible delay (Tsetsos et al., 2006). This is part of the functionality provided by the OSA/Parlay mobility specification (Parlay – ETSI, 2001)

Figure 5. Scheduler's internal architecture

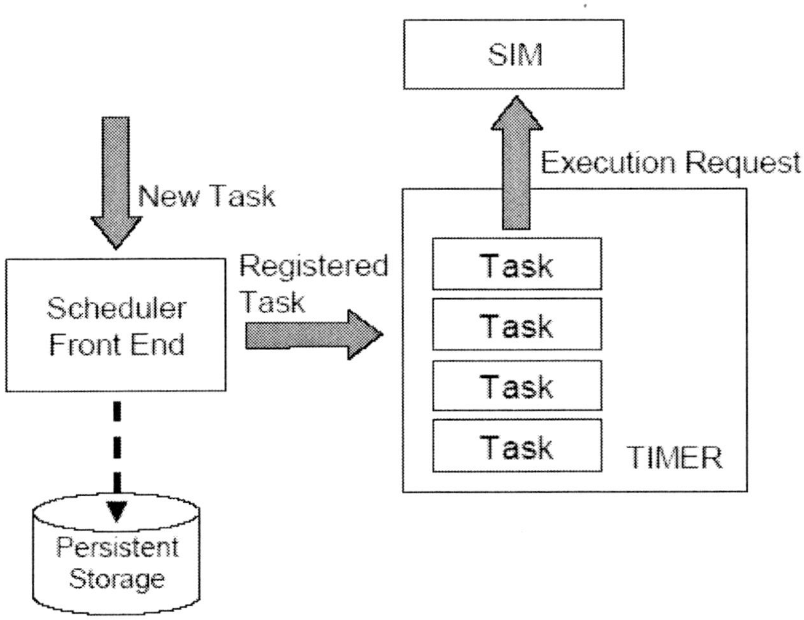

and is also particularly interesting in scenarios of personalized Location Based advertising. No priorities have been integrated in its architecture, which means that all events are treated equally. However, this approach does not impose any restrictions as the framework is multithreaded which guarantees, that even concurrent events are executed with the minimum possible delay. Particular effort during the design of the Scheduler was focused on the persistency of the scheduled tasks. Hence, aiming to maximum reliability the Scheduler features internal mechanisms for storing information pertaining to each scheduled task in order to cope with potentials crushes of the platform.

The Registry

The Registry is an integral part of the Kernel's architecture. It is implemented using Directory Services, thus allowing for the swift retrieval of the stored data. The Registry consists of two smaller registries (sub-registries): the Service Registry which contains service-specific data and the User Registry, each modelled as a different branch under the root node of the Registry.

When a new service is successfully deployed on the platform, the Service Deployer updates the Service Registry with the respective information. This information is used for the invocation, undeployment and redeployment of a service. Each registered service has numerous attributes, some of which are significant for its execution, while others are for informational purposes only.

Registered services have three possible states (Figure 6): Started, Stopped and Stopping. Stopped services are services which have been successfully deployed but are not available to end-users and do not have running instances within the SEE. This is the default state for all newly deployed services. In order for the service to be accessible by end-users the administrator must explicitly start it (using the administration console), thus setting its state to Started. Finally, Stopping services are services that are in the process of being stopped. They do not accept requests but possibly have running instances inside the SEE. After the completion of those instances the service moves to the Stopped state.

The registry contains, also, information pertaining to registered users (User Registry). The information in this registry is used for implementing, role-based policies controlling user access to specific services (Priggouris et al., 2004). In this perspective services deployed to the system can be configured so that only users with specific roles can access them. These are the restricted services while a service can be deployed without any ac-

Figure 6. Registered service state diagram

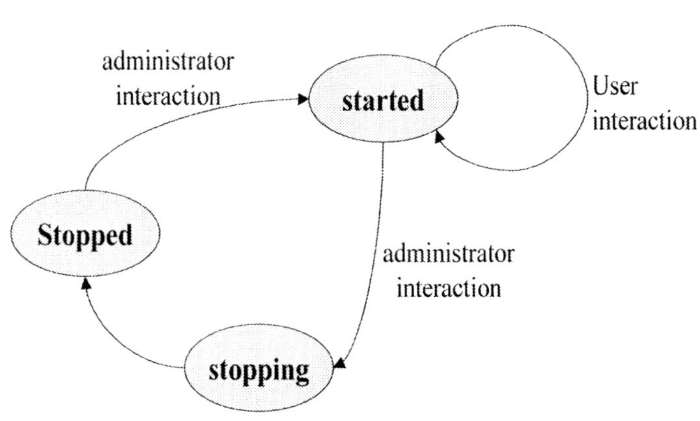

cess control policy, as well (free service). Every restricted service defines a number of roles, each having a different set of execution privileges on the service. Roles for each service are defined in the service script and are generated, upon service deployment, according to the information contained in this script. Registered users can be assigned to one or more of the abovementioned roles for each service they want access to. This is a task handled by the security administrator of the system, and administration tools for performing such associations are available. When a user logs in the system his roles are retrieved from the registry and are thereafter used for controlling access to every service instance, upon its invocation

The Peripheral Components

The peripheral components take their name from their relative position with respect to the Kernel. The functionality of each component features a well-defined API for interfacing with the Kernel. This modular architecture results in autonomous and independent components, which can be plugged out and replaced at any moment with others implementing the same APIs without the Kernel noticing any difference. As shown in Figure 1, three peripheral components have been incorporated, i.e. the Positioning component, the GIS component and the Interfaces component.

The Positioning Component

The positioning component (Figure 7) is responsible for providing location information to the Kernel. It integrates the underlying positioning technologies, providing a unified interface through a well-defined API. It abstracts the differences that are introduced by the diversity of the underlying location tracking techniques. This interface is fully configurable through a variety of parameters intended to cover all possible cases and operation paradigms.

The positioning component supports four different types of location requests. These types are:

Figure 7. Positioning Component internal architecture

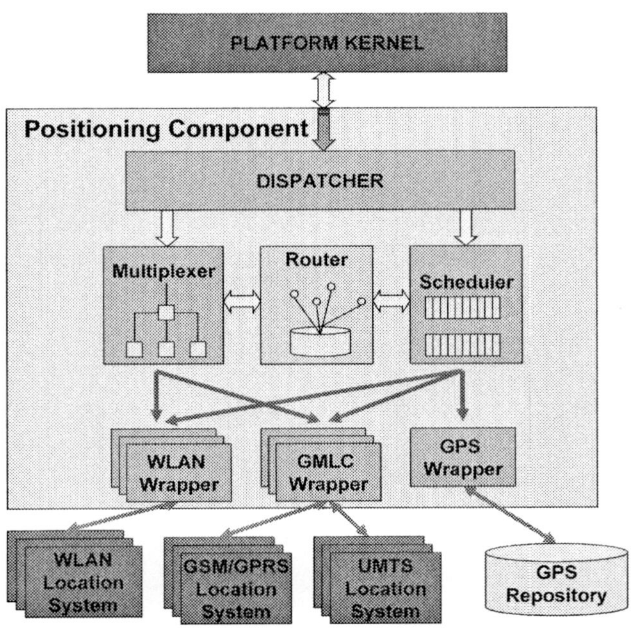

- Request-Response (RR), where an explicit location request is issued towards a specific network (GSM, UMTS, etc.)
- Periodic Request (PR): This request will generate a periodic location report for the designated user.
- Event-Driven Request (ER), which generates network-driven alerts, in case the user's location coincides with a predefined area.
- Generic Request (GR), which is a similar to the RR type, but the location request is forwarded to all supported networks where the user may have registered.

Communication with the underlying network infrastructure is based on open interfaces, such as OSA/Parlay (Parlay-ETSI, 2001), guaranteeing both the independence from the underlying location technology as well as the openness and extensibility of the component. The proposed implementation of the Positioning component includes support for location retrieval from 2G/3G networks and WLAN environments (Papazafeiropoulos, Prigouris, Marias, Hadjiefthymiades & Merakos, 2003). GPS support also exists through the use of specific software client-modules running on the terminal side for pushing location data towards the corresponding server module residing inside the positioning component. The framework support for various positioning technologies allows the seamless provision of location across different infrastructures and facilitates user roaming.

The main design orientations of the Positioning component were extensibility and openness. Therefore a number of core modules, implementing basic functions have been incorporated. Functions, like multiplexing, routing or scheduling are mandatory steps that every positioning request has to go through, depending on its type. Functionality specific to the positioning techniques has been placed in separate modules, which are called wrappers, as their task is to wrap the peculiarities of the different positioning system in order to provide a single interface towards the Kernel.

The GIS Component

The GIS Component (Figure 8) is responsible for the interaction of the Kernel with external GIS repositories. It covers visual and textual information retrieval and preparation in order to create representations (e.g., maps) for a certain geographical area. The GIS component implements algorithms like navigation routing and geocoding, which render it capable of answering both simple and sophisticated requests. It actually consists of two modules: a client module, which communicates with the Kernel and a server module, which interfaces with the GIS repository. All needed functionality is incorporated inside the latter, while the former is merely charged with the task of propagating the requests coming from the SEE (Service Execution Environment) and receiving the answers of the server module.

The GIS server component supports various services for both indoor and outdoor environments. For the outdoor domain four types of services have been developed as follows:

- Find Location. This GIS service generates a map with the defined locality pinpointed. The locality is specified upon the service invocation.
- **Navigation**. Takes as input two localities and generates a map with the path for reaching one from the other.
- **Geocoding**. The service generates a map with a locality pinpointed. The input of the service is a descriptive address string (e.g., Fifth Avenue 35).
- **Proximity**. Generates a map with all nearby Points of Interests (PoIs) pinpointed. The input of the service is a locality, a PoI type (e.g., churches) and the range in meters within which the search will take place,

Figure 8. The GIS component architecture

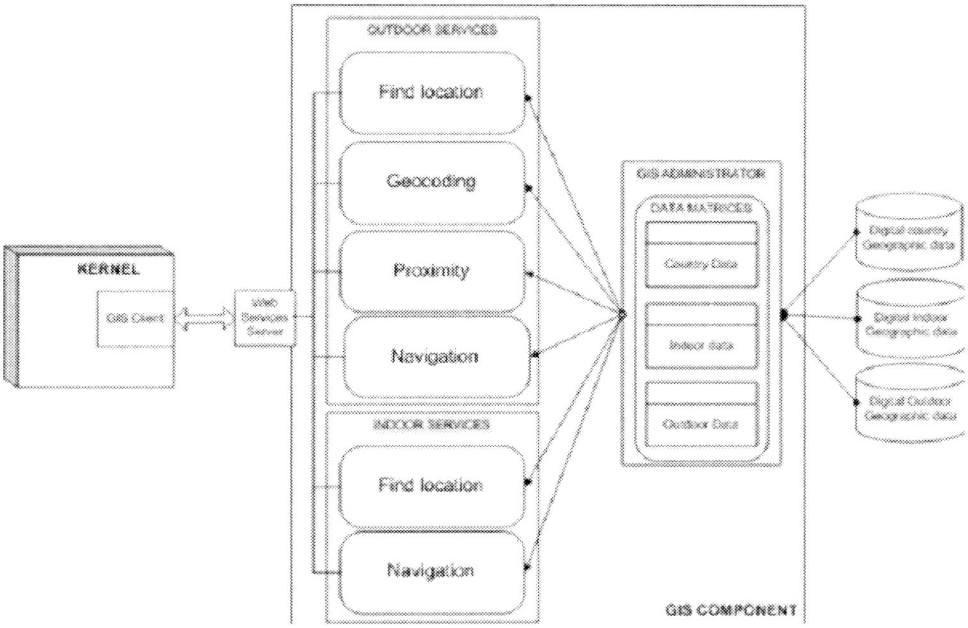

For the indoor domain only the first two services are supported. Each service beside the map can produce a text response as well.

The Interfaces Component

The Interfaces component is responsible for performing the communication of the Kernel with external entities (i.e., the user's equipment) irrespective of the underlying transport technology, which is used as bearer for the user's request. Its current specification supports three established and commonly used protocols (protocol suites), namely SMS, WAP and HTTP. However its design is modular enough so that support for additional protocols can be easily integrated in the future. The Interfaces Component features multiple interfaces with Proxy Servers (e.g. HTTP Server, WAP Gateway or SMS Center (SMSC)) in order to interact with different types of terminal devices. Proxy Servers can be part of the framework infrastructure; although their external placement is not precluded (e.g. the SMSC would normally belong to a mobile network operator).

The Service Creation Environment (SCE)

The Service Creation Environment (SCE) comprises an Integrated Development Environment (IDE) supported by a GUI, whose goal is to facilitate the creation and deployment of new services. The SCE by featuring a graphical interface, and customized support for the LBS-oriented language (SCL), can provide significant support to inexperienced users of the platform for creating, deploying and debugging new services. For more adept users a set of command-line tools is provided. These tools, interface with the web-services interface exported by the Service Deployer and the SEE inside the Kernel, allowing queries about deployed and debugged services or deployment requests to be posted. The IDE part of the SCE is based on the Eclipse software platform (http://www.eclipse.org), an open source, extensible graphical framework, provided by IBM, which enables customized GUIs to be developed. Eclipse supports, also, the Graphical Editor Framework (GEF) (Moore, Dean, Gerber, Wagenknecht & Vanderheyden, 2004), for developing graphical (or visual) editors, where service logic is presented to the users as an execution flow diagram. Within this diagram it is possible to add and remove commands corresponding to SCL directives, using a simple drag'n'drop mechanism, and to define the flow of execution between the various commands. Commands in the visual editor are represented as boxes. Each box has a set of attributes that need to be defined, resulting to one-to-one correspondence with an SCL directive. The standard text editor is also available for a conventional development. Switching between the two environments during the creation of the services is also possible.

Figure 9. Textual (top) and Visual (bottom) editors in action

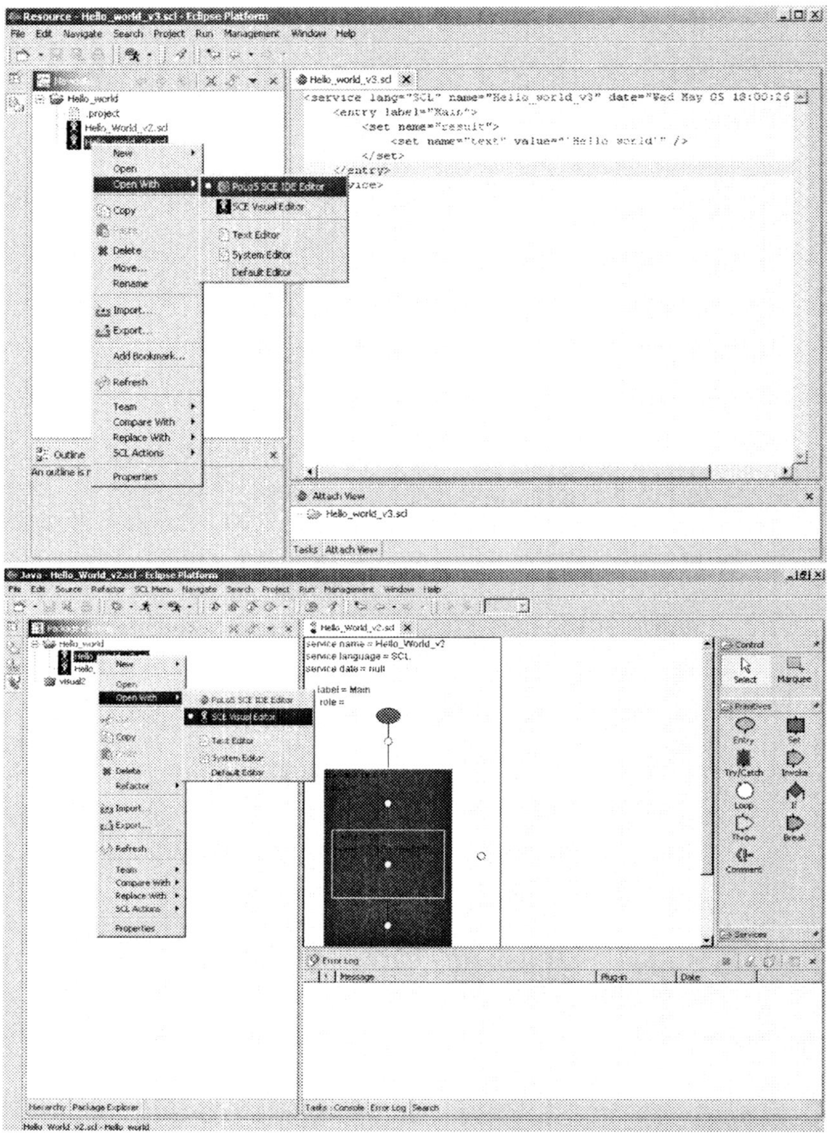

An example of a simple Hello World service as displayed in both editors can be seen in Figure 9. It is evident that the SCE is an add-on to the platform **middleware**, and its absence does not restrict the capabilities of the platform. Moreover, the open web-services interface provided by the Kernel allows potential users to develop their own proprietary SCEs for facilitating the service creation process, thus proving the openness of the system.

The Service Control Language (SCL)

The Service logic is specified through a new language, SCL (Service Control Language) developed specifically for supporting location-based services (Ioannidis et al., 2004). SCL is an XML-based language, built to facilitate implementations with simple functionality but also flexible enough to have the capabilities of a general-purpose language, similar to other scripting languages. We will provide a brief overview of its grammar and syntax through a simple example.

Figure 10 presents the SCL code of a simple service, named "GetMyLocation". This service defines a single "Main" method, with certain security constraints. Specifically, the method is accessible only to registered users with the specified role ("super") enabled. The "posdata" variable is defined using the "set" directive. The variable takes the value returned by a method invocation performed on the Positioning component. Component invocations in the SCL are defined with the "invoke" command, followed by a sequence of variables. The first variable in this sequence defines the service/method to be invoked on the

Figure 10. Service Control Language example

```
<service date="adate" name="GetMyLocation" lang="SCL">
  <entry label="Main" roles="super">

    <set name="posdata">
        <invoke component="POS">
            <set name="getLocation">
                <set name="transactionId" value="0"/>
                <set name="requestMethod" value="'RR'"/>
                <set name="priorityLevel" value="1"/>
                <set name="targetUserAddress" value="'0049932910729'"/>
            </set>
        </invoke>
    </set>
    <set name="gisdata">
        <invoke component="GIS">
            <set name="getLocationInfo">
                <set name="f" value="posdata.longtitude"/>
                <set name="l" value="posdata.latitude"/>
                <set name="userid" value="'10302020210'"/>
            </set>
        </invoke>
    </set>
    <set name="result">
        <set name="image" value="gisdata.imagedatal.image"/>
        <set name="text" value="gisdata.textdatal.text"/>
    </set>

  </entry>
</service>
```

component, whereas the rest define the parameters to be passed to it. In our example, the result of the invocation is returned in the variable "posdata" in the expected format. Note that the SCL, similar to most scripting languages, does not provide strong type checking and variables are cast to appropriate types dynamically during execution.

The next invocation is towards the GIS component, where the "getLocationInfo" service is requested. The contents of the "posdata" variable, which contains the result of the previous invocation, are used to define the parameters of the new one. The expected result is assigned in the variable "gisdata" and is next used for defining a new variable with the special name "result". The "result" variable is reserved by the language for specifying the return value of a method. Hence, the content of the "result" variable will be returned upon invoking the "Main" method of the "GetMyLocation" service.

SERVICES

A set of location-based services were developed and provided through the platform. These services were designed to support both textual and visual representation (in the form of maps). They also supported different means of access. In this perspective, their invocation was possible either through a standard web browser, as well as WAP browsers and SMSs. The developed set of services included standard LBS paradigms and more sophisticated ones, as follows:

1) Location Service
2) Navigation Service
3) Find PoI service
4) Locate Friend Service
5) Nearby Friend Service

A further elaboration of each service, along with screenshots from their invocation is cited in the following paragraphs.

Location Services

Indoor Location Service

This is the simplest indoor service, which locates and returns the room where the user is located. The output of the service is a map, where a special mark indicates the position where the user was detected, along with a brief text description of the specific location (Figure 11)

Outdoor Location Service

The service locates the user and returns his current position. The output on the user's device is a map with a spot on the location, together with a textual description of the location (i.e. the address of that specific point). The output of this service is illustrated in Figure 12.

Navigation Services

Indoor Navigation Service

The scope of this service is to successfully navigate the user from his current position to the requested place, via the most convenient path. The user may have the option either to select his destination from a drop-down menu (Figure 13) or to type the exact description of his destination.

When the request is submitted, the platform locates the user, checks for the destination point and calculates the route. The displayed result on the user's screen is a short text with directions on how to reach the destination point, along with a map illustrating the described path(Figure 14).

Outdoor Navigation Service

The outdoor Navigation Service is considered to be the most valuable service and absolutely necessary in order to consider a LBS system complete. The navigation service implemented provides the user with details in order to reach his destination.

Figure 11. Indoor Location Service output

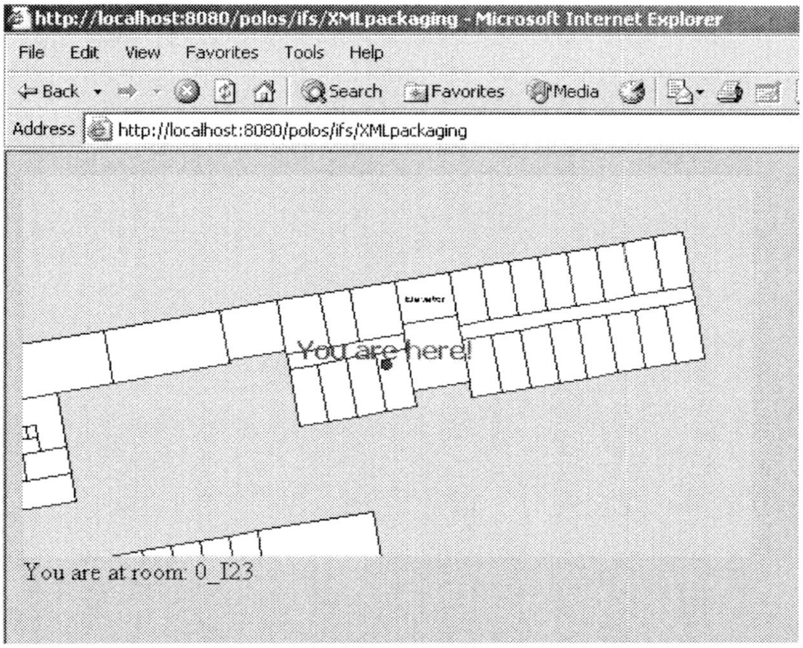

Figure 12. Outdoor Location Service output

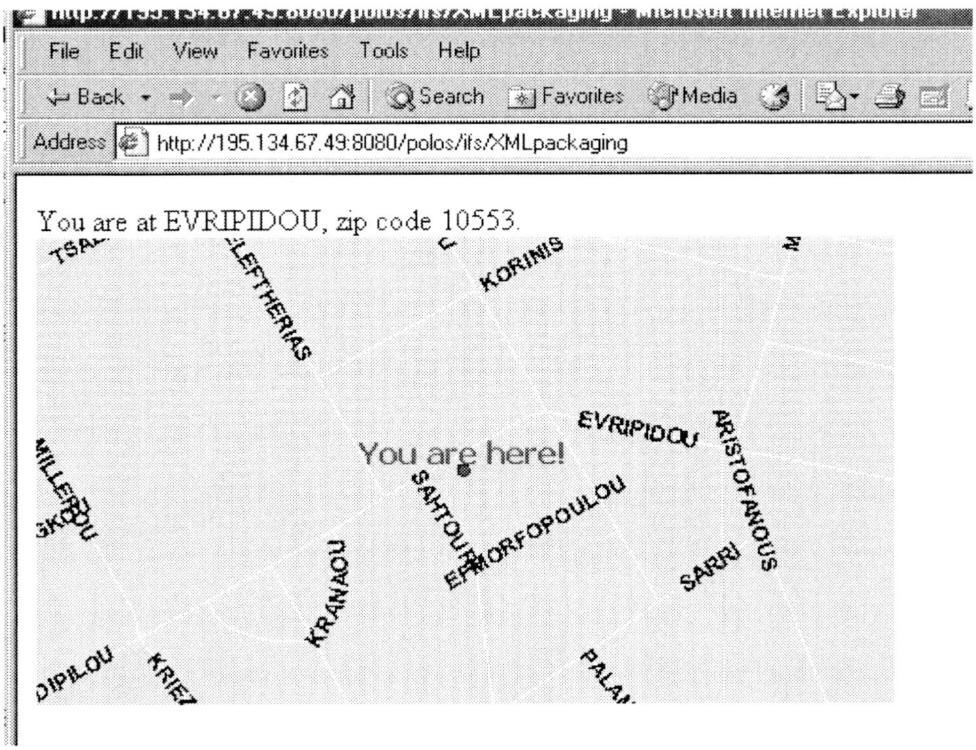

Figure 13. Indoor Navigation Service invocation interface

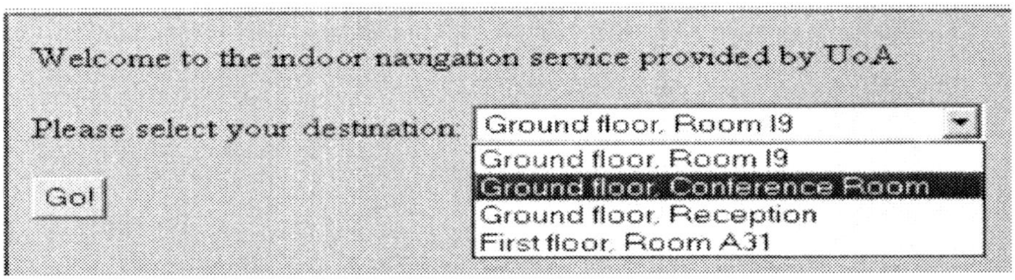

Figure 14. Indoor Navigation Service output

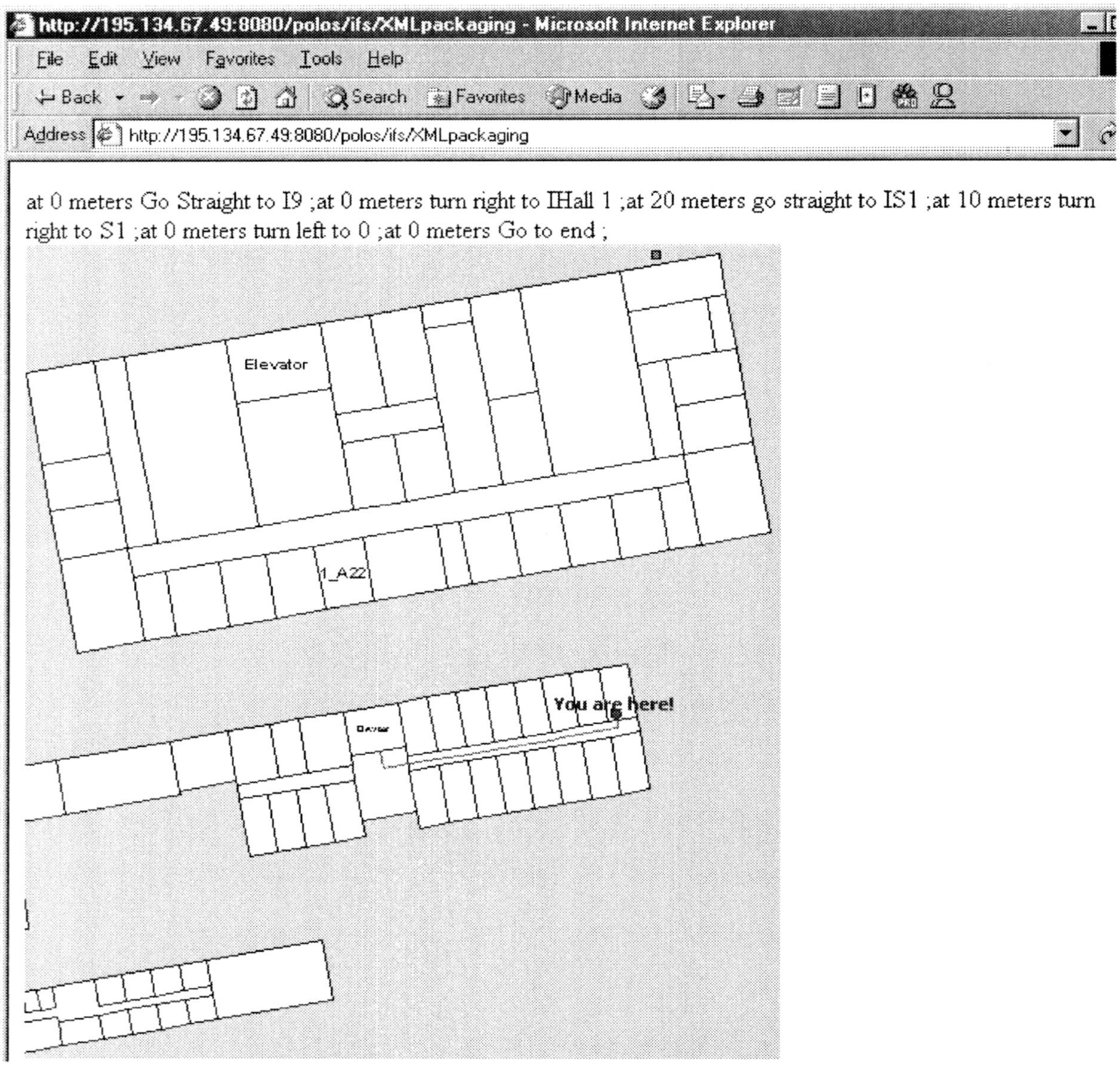

The destination is entered by the user via a simple interface, where the user enters street name and number, and the system, after locating the user's current position, creates the optimal path towards his desired destination. The path is then drawn on a map, which is returned to the user, together with a detailed description of each step the user has to follow in order to reach the final destination. Figure 15 illustrates the two different ways of visualization that have been developed

Find Point-Of-Interest Services

Indoor Find POI Service

A very common issue is the need to locate a resource, in a large complex building. This resource could be a fax machine, a network printer, a specific access point etc. Therefore, a *"Find Point-Of-Interest"* service for indoor environments was developed. In this service, the user asks for the specific resource, and the service, locates the user, scans for the required resource, and returns the path towards the resource displayed on a map along with some brief directions (Figure 16). This service resembles actually a navigation service towards the chosen POI.

Outdoor Find POI Service

One of the most useful and widely used services, is to provide the user with proximity information, for any Point of Interest, he might be looking for, in a specific range from his current location. Such points of interest could be banks, post offices, churches, bus or railway stops, etc. The *Outdoor Find Point-Of-Interest* service provides an interface for users to ask for a specific point, along with the preferred maximum distance from current position. The LBS platform locates the subscribed user, and after identifying his position, checks for the specified type of information, within the range requested. As soon as all possible answers are detected, their distances from current position are compared, and the smallest distance is selected. The nearest POI is marked on the map, together with the user's current position, and the result is visualized with a map together with the exact description of the POI illustrated, as shown in Figure 17. In the case of an actual invocation, the information displayed along with

Figure 15. The two variants of the Outdoor Navigation Service output

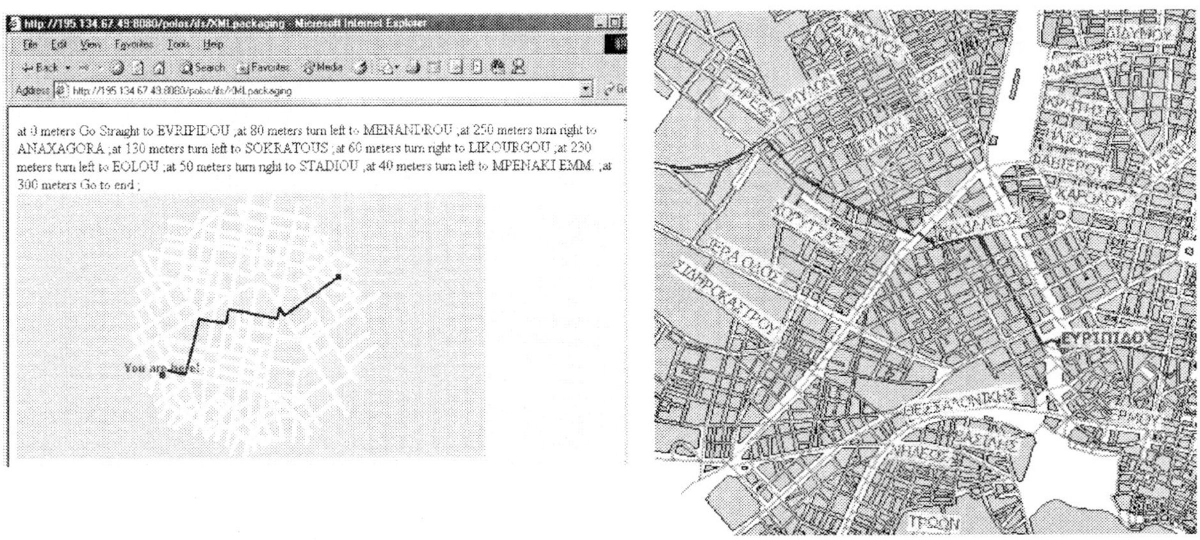

Figure 16. Indoor Find POI Service output

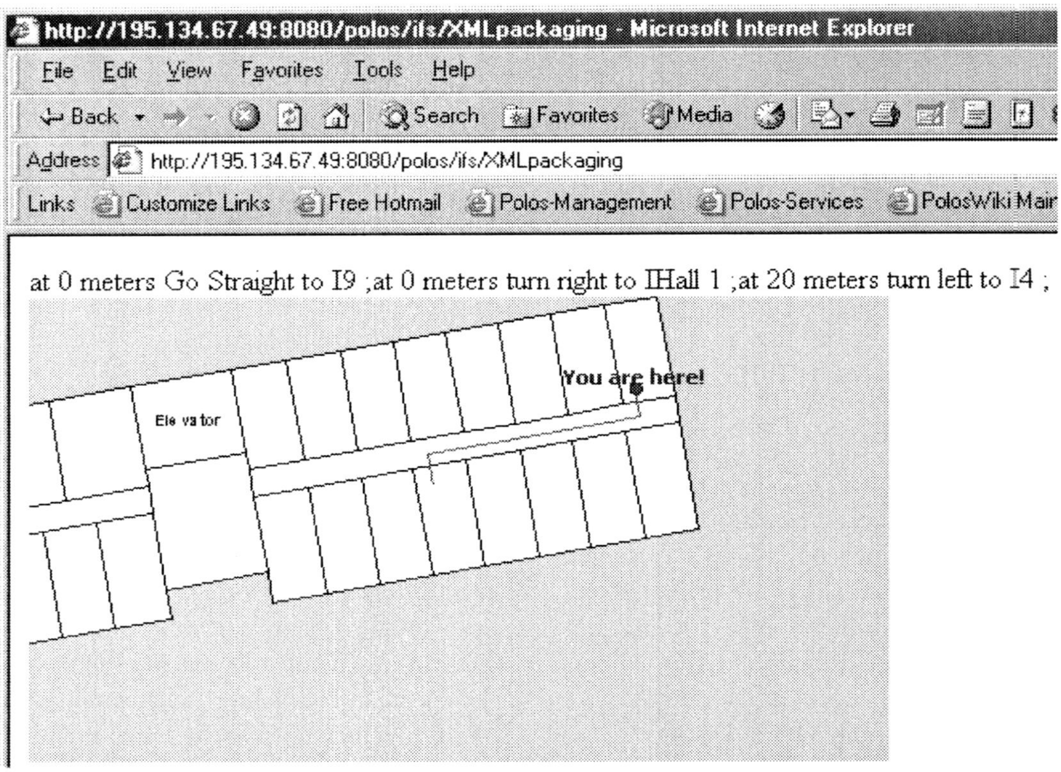

a map would include telephone numbers, open hours, or any other valuable information, apart from the exact address.

Locate "Friend" Service

The Locate "Friend" Service comprises a non typical outdoor LBS service that was developed and offered through the discussed LBS middleware. The service is designed for access through standard mobile devices and it is triggered through SMS messaging. The idea behind this service, is that a user (the service invoker), wants to find the exact location of one of his "friends". The "friend" is identified through his mobile number. Of course, in order for the service to provide accurate results the "friend's" terminal needs either to be equipped with positioning capabilities (e.g., a GPS) or be serviced through a location aware mobile home network.

The service invoker (User A) needs only to send an SMS with the service id #1043#, which corresponds to the Locate Friend service, followed by the mobile number of his "friend" (User B), whom he wants to locate. Respecting user's B privacy, the service informs him of the location request issued by User A, through an appropriate SMS. At the same time, User A is informed through another SMS, about the pending authorization (see Figure 18).

The service sets a 5 min timer, waiting for an authorization message from user B. The service flow hereafter depends on the actions performed by the "friend" (user B). He can either ignore the message or he can answer it. Ignoring the message corresponds to a non authorization, in which case position of user B is not disclosed. User A

Figure 17. Outdoor Find POI Service output

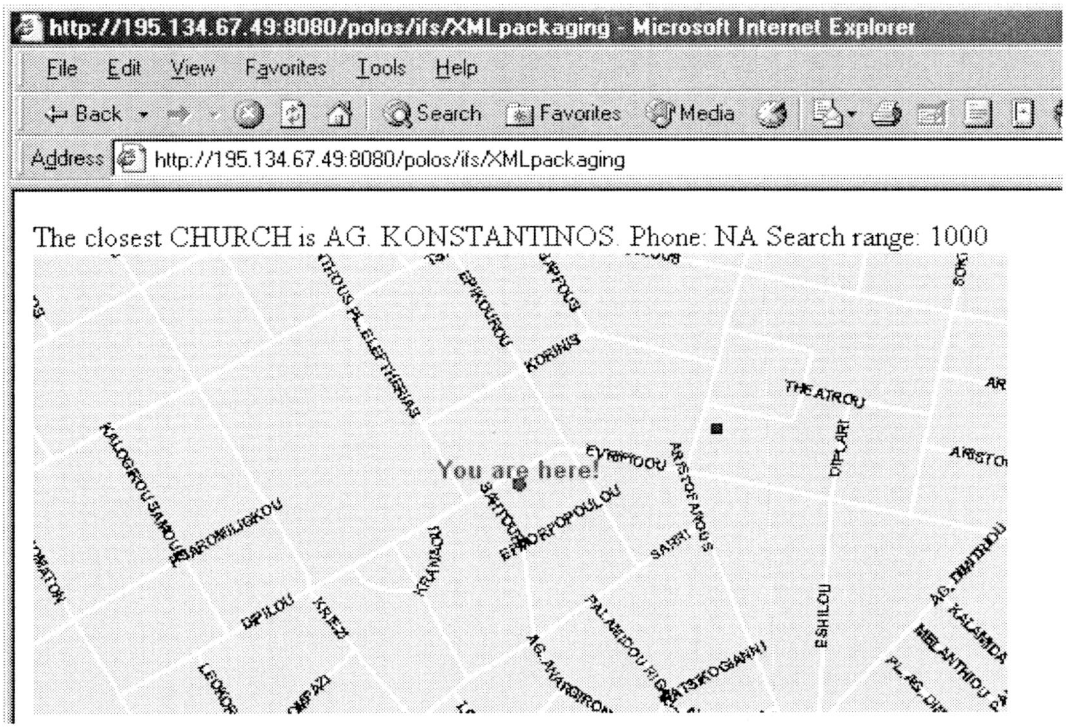

Figure 18. Locate "Friend" Service invocation

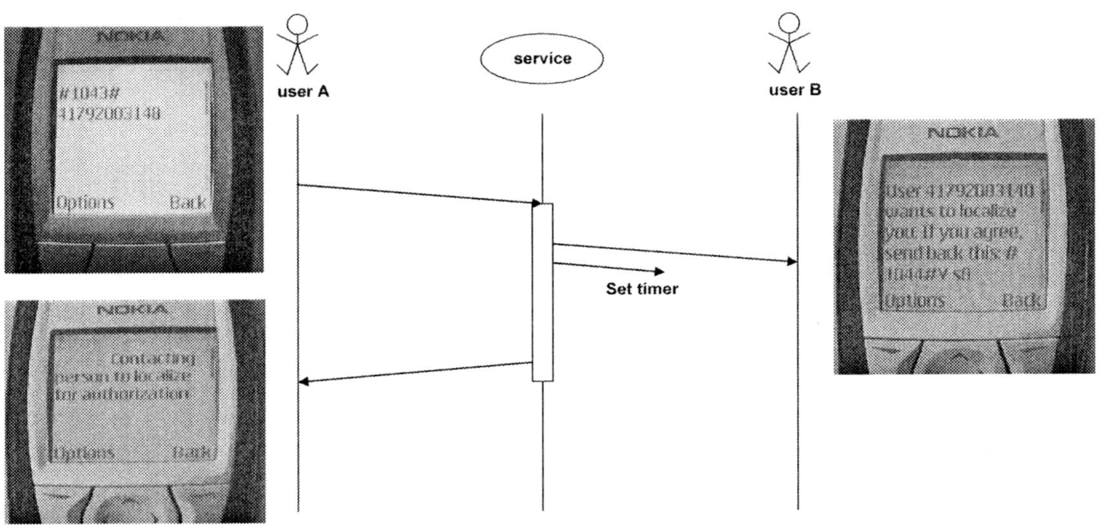

Figure 19. Locate "Friend" Service - service timeout

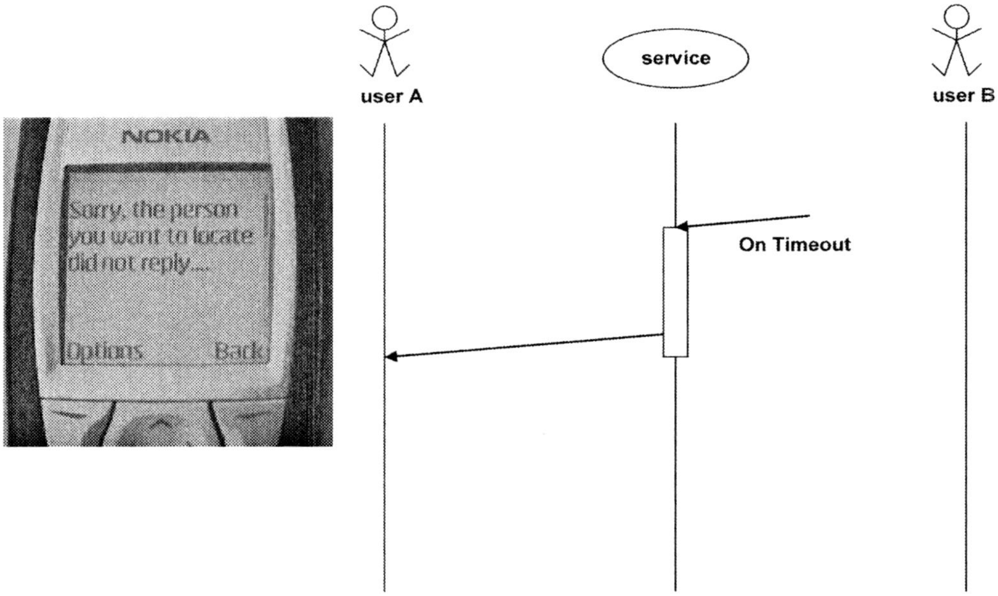

receives a respective informational message when the timer expires (Figure 19).

On the other hand, if user B answers the request, it means that he has accepted his position disclosure, to user A. In this case, the service proceeds to acquire user's B location, through the middleware's positioning component by issuing an RR type positioning request and forwards this location to user A, through the appropriate SMS (see Figure 20).

The Locate Friend Service, is one of the most sophisticated service paradigms developed using the suggested middleware. It exploits almost all components offered by the latter, from positioning and GIS to the SMS interface. Moreover, it uses core mechanisms provided by the kernel, such as the scheduler.

Nearby "Friend" Service

This is another service, targeted for mobile users. The service is an extension of the Locate "Friend" which was presented in previous paragraphs. This time the requestor has to register to the service. Upon his registration, he specifies the MSISDN number of his friends. This task is performed through an appropriate web application, provided by the service for configuring the service behavior (Figure 21).

Upon adding a phone number in the list of Nearby friends, an authorization procedure, similar with that of the Locate "Friend" service, is followed. However, this time, on user's B authorization, no position message is dispatched to the service requester. The authorization is kept by the service internally, and used to register an Event-based Location Request towards the positioning component of the system. The parameters of the registered event include a radius defined at 1km, around requester's user position, which defines the area of interest, which will be checked for nearby friends. If any of the specified MSISDNs enters this area, a corresponding location event will be dispatched to the service. The service will then send an appropriate SMS to the service requestor to inform him about his "friend"'s proximity (see Figure 22).

Figure 20. Locate Friend Service - "Friend"'s location retrieved

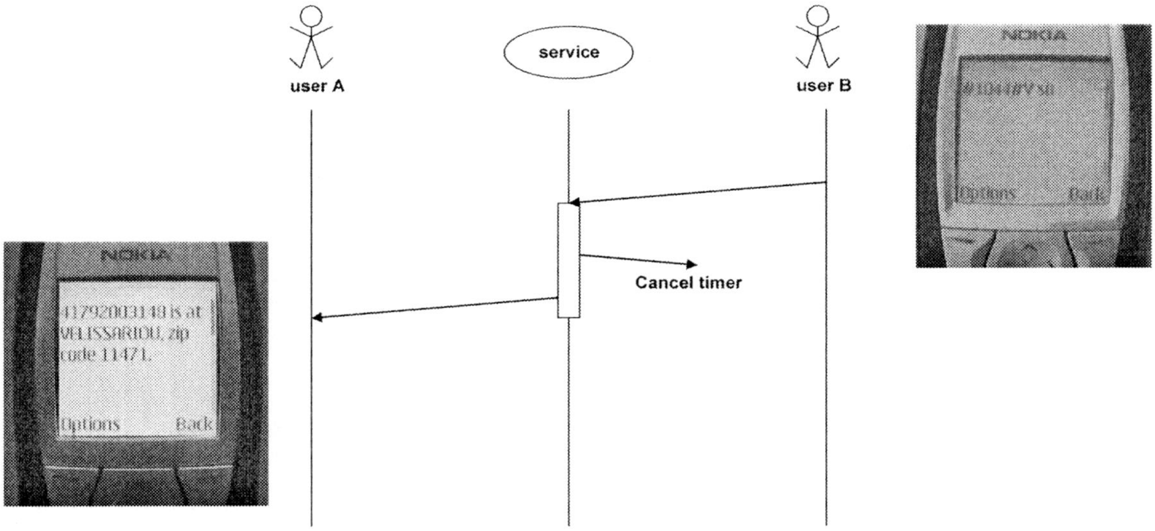

Figure 21. Nearby "Friend" Service configuration application

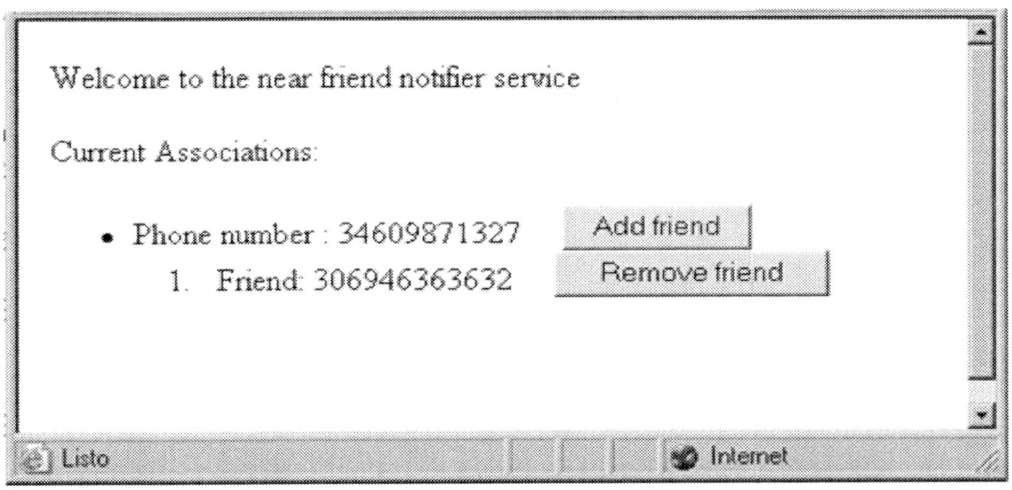

PERFORMANCE EVALUATION

Evaluation Setup

The prototype system that was developed and used for delivering the Location-Based services discussed in previous sections, undertook an extensive performance testing with the aim of evaluating its behavior under real conditions. This evaluation tried to assess every aspect of the architecture as a whole and as separate components as well. In order to perform this component evaluation, the Jprofiler (Ej Technologies, 2009) profiling software was used.

Figure 22. Nearby "Friend" Service - service output

Trial Setup

The trial setup involved the network topology displayed in Figure 23. The LBS middleware runs on an AMD Athlon XP 2000+ PC (768MB RAM), whereas the Positioning Gateway and the GIS server run on another AMD Athlon XP 1800+ (512 RAM) PC. Both PCs are located in the same local area network and run Microsoft Windows 2000 Server. Service invocations are performed from a remote computer (Pentium 4 1800 MHz with 256MB RAM, running Microsoft Windows XP Professional Edition), which accesses the network via an ADSL 384Kbps line. The aforementioned topology was selected because it is closer to a real invocation scenario, where high data rates are experienced in the server domain, while the link between the end user and the invoker is significantly slower.

One indoor and three outdoor services were used for the tests. Both services performed an invocation to the positioning gateway to obtain the user's position. A second outdoor service which performed an additional request towards the GIS to retrieve the map of the surrounding area was also created and tested.

Evaluation Metrics

In order to study the system's performance, a series of trial scenarios were developed. Each scenario used a different type of service and aimed to assess the performance of our prototype, under different load conditions. In this perspective, each scenario consisted of a series of experiments, with an increasing arrival rate for incoming requests. The general evaluation scenario is as follows: A total number of N users is assumed. Each user performs 1 to 3 services invocations. The user arrival rate (L) follows the exponential distribution. 60% of the users perform exactly one request, while the rest remain connected and perform a maximum of 3 requests (the number is randomly chosen for each individual user) before they leave the system. Time between two subsequent user requests follows the basic Pareto distribution with the shape parameter set to 0.9 (α=0.9), thus modeling the thinking time of the user. Starting with a small number of users (N=10) and a very low inter-arrival rate L (6 arrivals/min), the values of both N and L are increased until the platform fails to respond to a significant number of incoming requests. In order to be statistically correct, the Monte Carlo methodology (Fishman, 1996), consisting of 10

Figure 23. Trial network setup

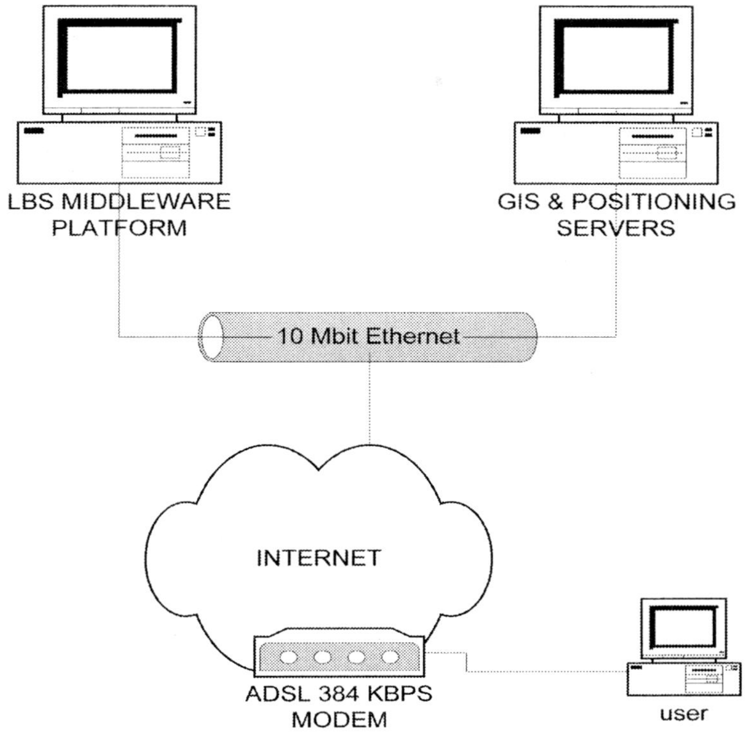

repetitions of each experiment, was followed. Final results are the statistical average of all 10 experiments.

Evaluation metrics that were used include: a) the success rate (% of total requests that were served successfully), and b) the average service execution time. An additional evaluation metric was the CPU usage per component.

Evaluation Results

Indoor Services

Indoor Location Service

In this section, results obtained from indoor services are presented. Figure 24 depicts the success rate for the indoor location service as a function of the number of users in the system (N) and the user arrival rate (L users/min). For N=10 the success rate decreases when the users arrival rate (L) increases to over 1200 users per minute, while for greater values of N (20, 50, 100, 200 and 500 users), the performance of the system degrades faster (for values of L between 120 and 240 users/min).

The overall behavior of the system for the indoor location service scenario is uniform, progressive and predictable. Degradation of performance is gradual and depends strongly on the system load. The statistical analysis of the results proved the existence of very small confidence intervals, which imply predictable behaviour as well as high reliability and precision for our measurements.

As presented in Figure 25, the system's response time (the time from the moment the request is received until the moment the user receives the response) is increased gradually when the rate of incoming requests is increased.

The CPU usage per component for the aforementioned service is depicted in Figure 26. The

Figure 24. Indoor location system success rate

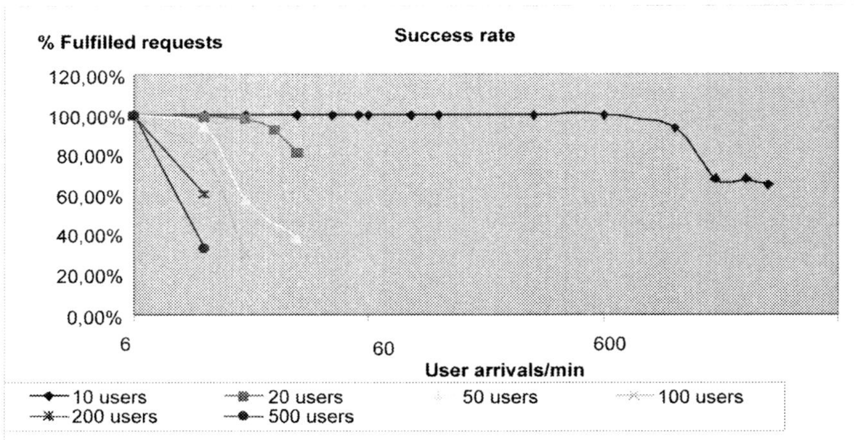

diagram shows that most CPU time is consumed inside the positioning component (47%). The GWLC component consumes about 28% of CPU time, while each one of the rest components consumes less than 1%.

Outdoor Services

Outdoor Location Service (with GIS Invocation)

The service retrieves the user's position and displays it on a map. As shown in Figure 27, the service performs reasonably well, and the performance decreases gradually as the user arrival rate increases. It is remarkable though that the system remains stable even when the user arrival rate exceeds the 100 users/min threshold.

The average service execution time, for this service, remains in acceptably low levels and increases slowly as the arrival rate increases. As shown in Figure 28, when the user arrival rate is low, the system responds in a few seconds (maximum response time in the order of 2-3 sec), which is an excellent performance. The response time goes over 20 seconds only for very high user arrival rates (>120 users/min) and a large pool of users (N=1000).

Outdoor Location service (without GIS Invocation)

In order to determine the possible cause of the delays experienced in the service invocation, we performed a series of profiling sessions. The results showed that the GIS component imposed a bottleneck for the system. To prove this, we developed a simplified service that did not perform a GIS invocation. We tested the performance of this service and the results are displayed in the following figures. This service performance is significantly enhanced, and as shown in Figure 29, the success rate is more stable, and degrades slowly while the user arrival rate increases.

A more obvious improvement was observed in the service execution time, which remained below 20 seconds even for very high arrival rates (Figure 30).

Outdoor Navigation Service

Figure 31 depicts the CPU usage per component for the outdoor navigation service. According to the diagram 61% of the CPU usage is consumed by the Service Execution Environment component, while 21% is consumed by the Positioning component and the remaining 18% by the rest of the components. A more detailed analysis of the results, proves that the high CPU usage of the SEE

Figure 25. Average service execution time

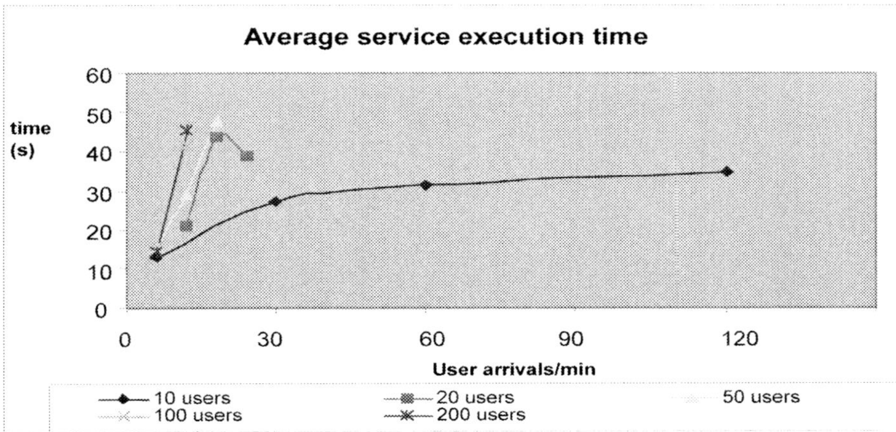

Figure 26. CPU usage per component

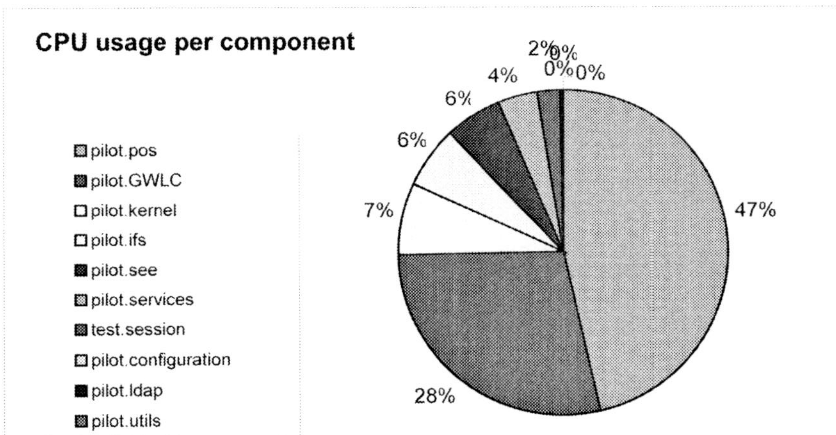

Figure 27. Success rate for Outdoor Location Service

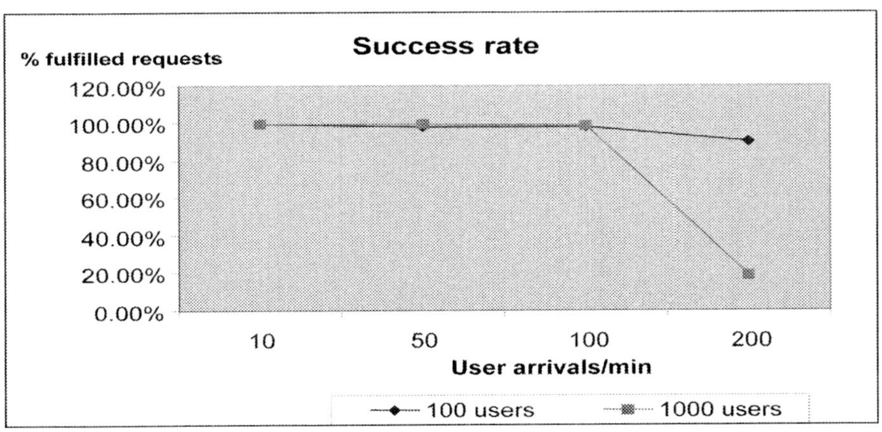

Figure 28. Average Outdoor Location Service execution time

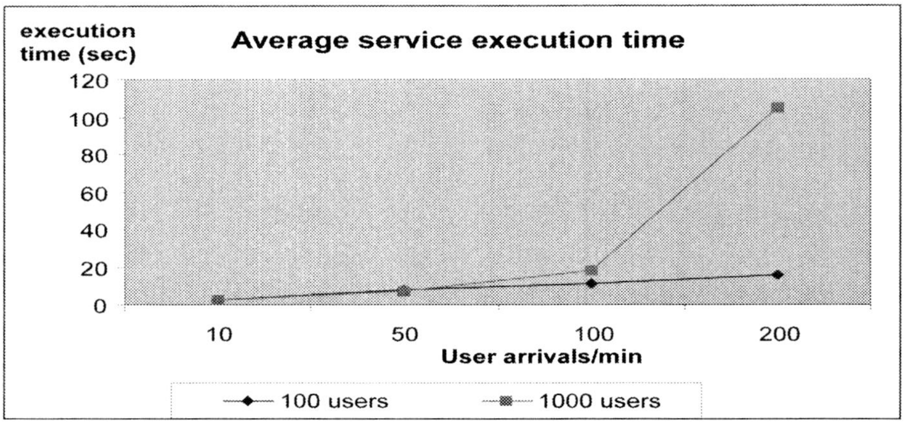

Figure 29. Success rate for the simplified outdoor test service

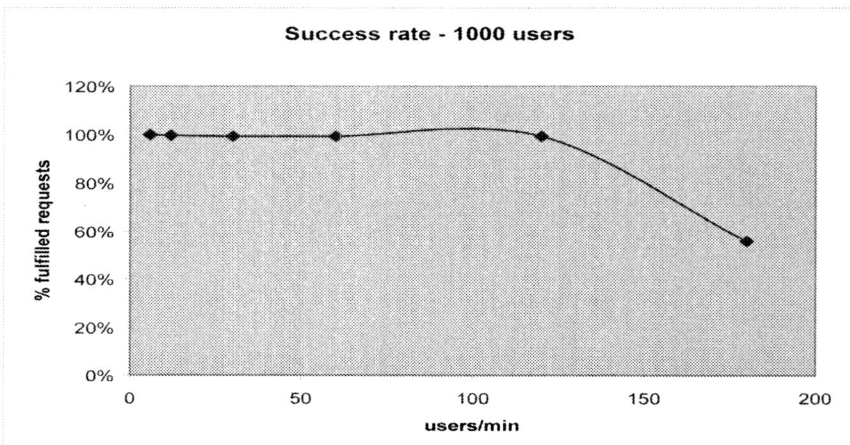

component is caused due to the complexity of the service and functionalities like address translation and path computation.

Evaluation Conclusions

In general, all tested services exhibited similar behavior, with slight variations. The more complex services which require larger processing power, in general exhibit slightly worse behavior. This is normal since in such services, apart from locating the user the system needs a remarkable amount of computations, in order to identify the destination point, find possible routes, compare, calculate and create the final output to the user. This is more obvious when the CPU usage statistics are studied carefully, but is also obvious from the remarkably larger delays exhibited, and the lower stability limits that were calculated on those services.

The exhibited user rates and delay times are considered reasonable, especially when considering the evaluation setup and its available resources, which were limited to simple Personal Computers. In case the same evaluation was performed on

Figure 30. Average response time for the simplified outdoor test service

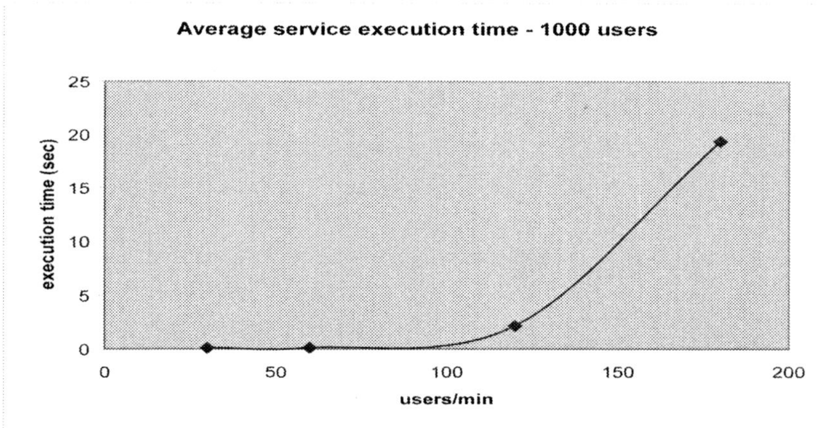

Figure 31. CPU usage per component

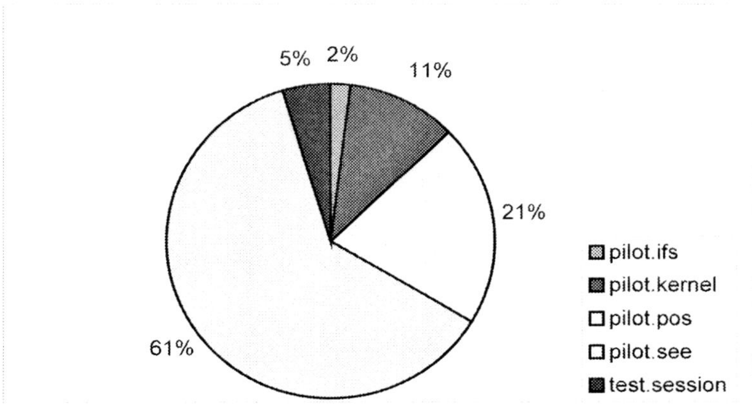

specially designed high speed server computers, with all necessary resources, the results are likely to improve significantly.

An issue that should be stressed here is that a lot of time in both indoor and outdoor services is spent in the GIS calls. This results from the fact, that most of the processing that is required for providing the results of the service is performed by the GIS server (e.g. routing algorithms etc.) Consequently in certain cases the LBS architecture has to wait for the answer of the GIS server, thus, resulting to significant delays in the service execution. Removing the GIS calls from the tested services showed significant improvement in both the success rate and the mean service execution time.

Another component of the system that can be considered to be relatively slow, and impose delay to the system, is the Positioning Component (POS). This component is usually on the top ranks on CPU usage. This is partially explained for the following reasons. This component is always invoked in every service and, additionally, the procedure of locating a user retrieving and processing his current location is a demanding task that requires a considerable amount of calculations, which in turn implies some time delay.

Some overall evaluation results, which can not be illustrated in figures and in plain numbers, concern the general behavior of the architecture. Although the platform did not exhibit very high user arrival rates, it maintained a remarkable stability, even when a heavy load of concurrent requests occurred. In such cases, the platform maintained its stability, managed to enqueue all requests and gradually responds to all, without the need to drop a number of them, and with no indications of a potential crash.

CONCLUSION

This chapter presented issues concerning the design of a generic software architecture platform for delivering location-based services. Within the document we focused both on the principles that drove the design of the proposed framework, as well as its internal architectural design. The main advantage of the discussed framework is its potential, since it caters for almost every aspect of the location-based service life-cycle. It can fully manage the LBS from its creation and testing until its deployment and provision to the end user. This is achieved through specific middleware services (positioning, gis, etc), which come as integral parts of the framework, and allow the easy interfacing with external systems and platforms, needed by such services.

A prototypical implementation of the LBS framework was developed and used for running a series of services. Extensive testing was performed to the prototype, with the purpose of assessing its performance and capability to serve large volumes of user requests for specific services. The results demonstrated that the behavior of the system is predictable, although its performance seems to degrade when high user arrival rates are experienced. However we believe that the conclusions produced by the evaluation of the system will further assist our research in order to improve its performance. The architecture presented in this chapter is only an experimental prototype, and as such, can not be compared to similar commercial systems; though the supported features and its potential surely supersede many of them. Our future work includes the development of a more robust version of this prototype, which can directly compare to other similar systems, in terms of performance as well.

ACKNOWLEDGMENT

This work is supported by the PYTHAGORAS programme of the Greek Ministry of National Education and Religious Affairs (University of Athens Research Project No 70/3/7411)

REFERENCES

Bahl, P., & Padmanabhan, V. N. (2000). *RADAR: Proceedings of IEEE INFOCOM: An in-building RF based user location and tracking system.*

Benedicto, J., Dinwiddy, S. E., Gatti, G., Lucas, R., & Lugert, M. (2000) *GALILEO: Satellite System Design and Technology Developments*. European Space Agency.

Castro, P., Chiu, P., Kremenek, T., & Muntz, R. (Eds. Abowd G. D., Brumitt B., & Shafer S.A.N.). (2001). *Proceedings of Ubicomp-2001: A Probabilistic Room Location Service for Wireless Networked Environments*. Springer-Verlag.

Ej-Techonologies. (2009). Jprofiler. Retrieved from http://www.ej-technologies.com/products/jprofiler/overview.html

Fishman, G. S. (1996). *Monte Carlo Concepts, Algorithms and Applications*. Springer Verlag.

Hightower, J., & Borriello, G. (2001). Location Systems for Ubiquitous Computing. *IEEE Computer Magazine*, *34*(8), 57–66.

Ioannidis, A., Spanoudakis, M., Sianas, P., Priggouris, I., Hadjiefthymiades, S., & Merakos, L. (2004). Using XML and related standards to support Location Based Services. In *Proceedings of SAC'2004, 19th ACM Symposium on Applied Computing, Web Technologies and Applications Special Track, Nicosia, Cyprus*.

Kaasinen, E. (2003). User needs for location-aware mobile services. In *Personal and Ubiquitous Computing* (Vol. 7).

Leick, A. (1995). GPS *Satellite Surveying*, Second Edition, John Wiley & Sons.

Moore, W., Dean, D., Gerber, A., Wagenknecht, G., & Vanderheyden, P. (2004). Eclipse Development using the Graphical Editing Framework and the Eclipse Modeling Framework. Retrieved March 2005, from http://www.redbooks.ibm.com/abstracts/sg246302.html.

Papazafeiropoulos, G., Prigouris, N., Marias, I., Hadjiefthymiades, S., & Merakos, L. (2003). *Proceedings of the IST Mobile & Wireless Communications Summit: Retrieving Position From Indoor WLANs Through GWLC, Portugal.*

Parlay – ETSI., (2001) *Open Service Access API. Part 6: Mobility SCF. ES 210 915-6 v.1.1.1.*

Priggouris, G., Hadjiefthymiades, S., & Merakos, L. (2004). *An XML framework for multi-level access control in the enterprise domain, Proceedings of the 6th International Conference on Enterprise Information Systems, Porto, Portugal, April.*

Sun Microsystems Inc. (2003). *Enterprise JavaBeans 2.1 Specification* (Public Draft).

Tsetsos, V., Sekkas, O., Priggouris, I., & Hadjiefthymiades, S. (2006). A Scheduling Framework for Enterprise Services. [JSS]. *Journal of Systems and Software*, *79*(2), 259–272. doi:10.1016/j.jss.2005.05.014

Chapter 4
Context-Aware Query Processing in Ad-Hoc Environments of Peers

Nikolaos Folinas
University of Ioannina, Greece

Panos Vassiliadis
University of Ioannina, Greece

Evaggelia Pitoura
University of Ioannina, Greece

Evangelos Papapetrou
University of Ioannina, Greece

Apostolos Zarras
University of Ioannina, Greece

ABSTRACT

In this article, we deal with context-aware query processing in ad-hoc peer-to-peer networks. Each peer in such an environment has a database over which users execute queries. This database involves (a) relations which are locally stored and (b) virtual relations, all the tuples of which are collected from peers that are present in the network at the time when a query is posed. The objective of our work is to perform query processing in such an environment and, to this end, we start with a formal definition of the system model. Next, we formally define SQLP, an extension of SQL that covers the termination of queries, the failure of individual peers and the semantic characteristics of the peers of such a network. Moreover, we present a query execution algorithm as well as the formal definition of all the operators that take place in a query execution plan.

INTRODUCTION

Nowadays, the synergy between network and database management systems provides opportunities for the integration and querying of various heterogeneous sources of information, spread over an ad hoc network of peers. *The fundamental topic of this article is the context-aware processing of queries in ad hoc networks of peers through Web services.* We assume the existence of a set of peers who communicate with each other, thus forming a time varying ad hoc network of peers. For reasons of interoperability, we also assume that these peers use Web services for their interactions. Each peer has a database where (a) data can be locally stored, or (b) descriptions of data are present, in a form that allows their collection from the appropriate peers and their subsequent querying with traditional database mechanisms. The querying and/or collection of these data is dependent on the state of the peer network and on the knowledge that the peer has about this state; therefore, each time a query is posed, its processing must be adapted to this state. *In other words, the state of the peer posing a query and, most importantly, the state of its surrounding network constitutes the context under which the query is processed.*

Assume the case where several kinds of vehicles are driving in a highway. Each vehicle is a part of a global pervasive computing environment where computations can be performed, data can be exchanged between computing devices of the environment and information is interactively requested and presented to the users. Cars interact with each other through Web services, providing dynamically changing information regarding the vehicle's location, velocity and fuel deposit. Moreover, each vehicle comprises services that offer static information concerning its type and technical characteristics. On the highway, there exist exits to parking areas, which may include facilities such as gas stations, fast food restaurants, medical help, and shopping centers. Each one of these facilities also comprises Web services, which range from simple ones, reporting the existence of the facility, to more complex ones providing information regarding, for instance, the price lists, the availability of certain goods or the number of patients waiting for medical help. The users of the facilities of the pervasive environment, for example, the drivers of the vehicles, can obtain information by posing queries to global information space of the environment. For instance, they may be interested in obtaining information like the closest gas station with a price of gasoline under 2€/gallon, the closest Italian restaurant, or notifications for the average speed of all the cars ahead.

To facilitate the smooth operation of peers within the aforementioned environment, specific technical challenges must be addressed. A significant problem is the fact that traditional query processing must be reconsidered to adapt to the particularities of our computing environment. In this article, we are specifically interested in the problem of formally defining a declarative query language that enables the posing of queries over an ad hoc network of peers as well as the introduction of a mechanism for the transformation of declarative database queries to query execution plans.

First, we start with the theoretic formulation of the problem. We construct a directed graph of peers, where each node corresponds to a peer and each edge to the physical connection among two peers. The graph of peers is time varying, since nodes and edges are added or invalidated as time passes. Apart from the possibility of communication, which dictates the structure of the graph, peers are further organized in *communities*, based on their semantic similarity, or *classes*, based on the interface of Web services they support. All our deliberations are based on the *principle of local scope*, that dictates that no peer has a global knowledge of the entire graph, and therefore, all its decisions must be made depending solely on the knowledge that this node has at a given time point. Specifically, the *viewpoint of a peer* is the subset of the graph known to this peer at a given time point and the communities of the peer are sets of peers whose publicized characteristics fulfill a logical condition that classifies them into the appropriate community. The only classification that is not local is the class of each peer: we assume that a set of finite classes exists, each with an interface comprising a set of public Web service operations that all class instances support. Every peer is created as an instance of one of the globally known classes.

With respect to the relationships among peers, each peer plays both the role of the server and the role of the client in this environment. As a server, the peer implements and exports the interface of Web service operations prescribed by its class. The other peers of the system can invoke these Web services at runtime. At the same time, the peer is responsible for answering queries posed by its users. In our framework, we discuss traditional database queries and, therefore, the peer hosts a relational database where query processing takes place. The database includes different categories of relations. First, the database includes relations that obey the traditional assumption that a database hosts locally stored relations, whose extents are finite sets of locally stored tuples. In this article, we extend this implicit assumption and assume that the extent of a relation can be spread among the peers forming the context of a peer. Therefore, only the description of the schema (or intention) of such a *virtual* relation is locally available, along with the description of the necessary Web services that must be invoked in order to locally collect the relation's extent before continuing query processing as usual. This collection procedure practically dictates that a workflow of Web services has to be executed for each peer of the viewpoint of the querying peer. Finally, a third category of relations involves hybrid relations, whose extent is partly locally stored and partly needs to be collected from the other peers.

The processing of queries in such an environment is inherently different to the traditional one. We have already mentioned the context-aware aspect of data collection for the population of virtual relations. Moreover, due to the volatile character of the state of the peer's graph, it is quite probable that the viewpoint of a peer is an inaccurate reflection of the state of the peer graph. In other words, it is quite possible that the graph has changed since the last refreshment of the viewpoint of a peer. In fact, the graph can possibly change also during the execution of a query; therefore, the processing of a query must be inherently designed to tolerate failures (i.e., Web service invocations that do not respond) and continue operating regularly. Also, due to the possible vastness of the graph, it is necessary to be able to stop collecting answers after a certain, satisfactory amount of information has been collected. Based on these fundamental differences with traditional query processing, we introduce an extension of SQL, SQL^P that allows the user to exploit the context-dependent nature of the environment by specifying the peers of interest though abstract criteria that involve their location in the graph, their community, their class, or QoS characteristics, like, for example, their availability. The usage of virtual tables is transparent in SQL^P. We exploit the previously introduced model to formally specify the semantics of SQL^P.

The processing of the queries in this extended version of SQL requires also an extension of the mechanism of query execution. Traditional relational database management systems translate the declarative SQL queries to procedural, executable plans that are expressed in the form of left-deep trees of relational operators. Therefore, we introduce novel operators, specifically tailored for the support of Web service invocation and composition, in order to populate the virtual tables. Then, query processing can continue as usual. We have also implemented a mechanism that allows us to determine the necessary set of peers that are supposed to participate in a query and to visually display the produced plans to the user.

This article is organized as follows: in the second section, we propose SQL^P, an extension of SQL for ad hoc P2P systems. To this end, we define a system model; we investigate language requirements and propose the syntax and semantics of SQL^P. In the third section, we extend the relational algebra with novel operators and algorithms in order to map SQL^P queries to query plans. In the fourth section, we discuss implementation issues. Finally, in the fifth section, we discuss related work and in the sixth section, we conclude our results and discuss topics for future work.

SQL FOR PEERS: SYSTEM MODEL, REQUIREMENTS, SYNTAX AND SEMANTICS

In this section, we formally define the system model. Then, we move on to formally define SQL^P, an extension of SQL for ad hoc P2P systems.

System Model

A bird's eye view of the system infrastructure is modeled by a graph *G(V,E)*, comprising a set of nodes *V* and a set of edges *E* (*Figure 1*). Each node in our system graph is a peer and each edge *e* = <*u,v*> stands for the fact that node *u* can communicate with node *v*. The notion *'can communicate'* means that peer *u* can send data or make a request for data to *v* - in other words, the edge <*u,v*> implies that peer *u* assumes (a) knowledge of existence and (b) network connectivity with node *v*. The edges are directed in the sense that although node *u* can communicate with *v*, the inverse does not hold (an edge <*v,u*> would be required to demonstrate such a fact). This is quite frequent in modern ad hoc networks and deeply affects the design of efficient routing protocols (Abolhasan, Wysocki, & Dutkiewicz, 2004). In the sequel, we will also refer to an edge between two nodes as a *direct link*. To discriminate between different nodes, each node is characterized by a globally unique identifier, *peer id*.

As usually, a *path* between two nodes, say u_1 and u_2, is an acyclic sequence of consecutive edges belonging to *E* that connects these two nodes. The distance of two nodes, say u_1 and u_2, is the cardinality of the minimum set of edges required to reach node u_2 through a path starting at u_1. In other words, the distance of two nodes is defined by the number of hops involved in the connecting path, which is a typical assumption in ad hoc networks research. We will denote the distance of two nodes as $distance(u_1, u_2)$.

It is quite important here to stress the following properties of the system's graph:

Figure 1. A system's graph G(V,E)

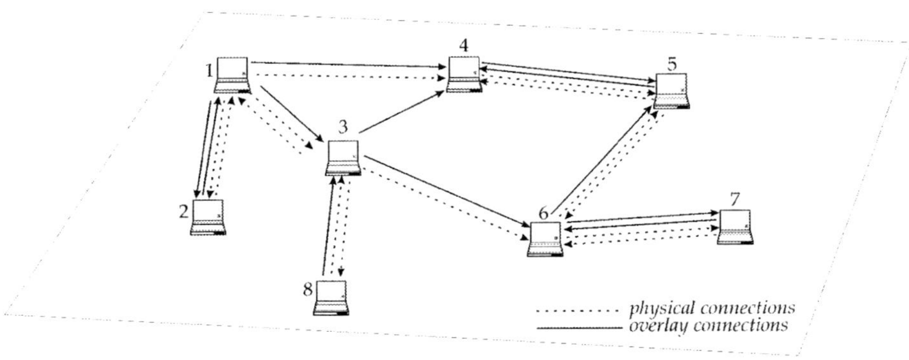

- The graph is time varying. In other words, nodes leave or enter the system as time passes. Furthermore, nodes move randomly, causing the destruction of existent links and the establishment of new ones.
- No node has a full knowledge of the system's graph at any time point. On the contrary, it is important to design a system where each node has only a personal, restricted viewpoint of the graph. A fundamental principle in our deliberations is the *locality of peer scope*: Each peer must be designed to operate by exploiting its own knowledge of a subset of the system, without counting on some higher-level authority to provide a global viewpoint of the system.
- It is also important that each node is designed to operate under the assumption that its knowledge of the graph is both incomplete and (possibly) inaccurate. This is a disadvantage related to the current networking technology for ad hoc networks (Chlamtac, Conti, & Liu, 2003).
- The overall graph is not fully connected. In other words, it is not always possible to reach any node v of V, starting from another node u.

Context = Viewpoint of a Node

At every time instant T, a node u is aware of a subset of the system's graph, as it was configured at a previous time point $T' \leq T$. This subset of the graph is called viewpoint of node v at time T and denoted by *viewpoint*(v,T). The subgraph *viewpoint*(v,T) is connected. This property is recursively defined as follows:

1. $v \in$ *viewpoint*(v,T)
2. All nodes u that are connected to a node x, $x \in$ *viewpoint*(v,T) through an edge (x,u) belong to *viewpoint*(v,T). In other words, first, all nodes u that are connected to v through an edge (v,u) belong to *viewpoint*(v,T). Then, the nodes that can be reached from these ones are also added. This is recursively continued.

Inaccuracy is inherent in this definition. Firstly, all the knowledge about direct links refers to a time point T' in the past. This means that whatever changes have happened between T and T' are obscure to v. The exact determination of time T' depends on the implemented routing protocol. Second, it is obvious that even if the overall set of nodes is finite (which is not an assumption that we have made so far), it is clear that it is impractical or even impossible to maintain all the knowledge for the graph for each node v. In fact this is the approach taken a large category of routing protocols known as on-demand routing protocols (Abolhasan et al., 2004).

Community

Apart from the physical connectivity among nodes, we can devise logical schemes for the connectivity of peers. In P2P terminology, the network of peers that share similar semantical properties is called an overlay network (Androutsellis-Theotokis & Spinellis, 2004). In our setting, a *community* of nodes is a subset of V who shares the same semantical properties. Each peer defines its own communities. Formally, semantical proximity is captured by a formula in a first-order predicate calculus. The principle of locality of a peer's scope imposes a design where each peer comprises a local set of communities, each defined as a subset of its viewpoint, upon fulfillment of the appropriate formula. Therefore, a community *comm_name* of a peer u is defined as:

$$community_{comm\ name}(u)=\{\ v \mid v \in viewpoint(u,T)\ and\ \varphi_{comm\ name}(v)=true\}$$

with φ being a formula in a first-order predicate calculus that returns true or false given the properties of a node v.

Clearly, a node u can have many communities and each node v in the viewpoint of u can belong to more than one communities of u. Moreover, assuming a simple community *Unclassified* that comprises all nodes that do not belong to any other community, the union of all communities of node u returns *viewpoint*(u,T), at a time point T. An interesting observation here is that if two or more nodes agree for a correspondence of communities, a P2P overlay is formed.

Web Services

Each node is equipped with a set of Web service operations that it publishes, therefore giving the possibility to the rest of the nodes to invoke them. Formally, each node $u \in V$ possesses a finite set of Web service operations $WS^u=\{ws^{u1}, ws^{u2},.....,ws^{um}\}$ that are made public to the rest of the peers. In the sequel, we will not discriminate between the terms Web service operations and Web services.

Peer Classes

In the context of the integration of peers at a large scale, each peer has to resolve the problem of mapping the external interface of the other peers to its internal state. In other words, if a peer u is to invoke a Web service operation of another peer v, how does u decide the mapping of the operation's parameters or the operation's result to its internal state? Typically, there are two well-known extremes from the database community to handle this problem, as well as intermediate solutions.

- In the first extreme, a *global schema* is assumed. In distributed database systems (Ozsu & Valduriez, 1991), a global schema is assumed for the whole environment and each local database comprises a subset of the global schema. This approach requires a universal common agreement over a global schema (and the implicit semantics hidden behind it). We find this requirement too restrictive for a large-scale P2P environment that needs to be dynamically readjusted to novel peers that appear.
- An intermediate approach would be to hard-code all mappings among all peers. Still, this approach is too labor-intensive and clearly unable to scale up to the full extent of a P2P environment.
- In the second extreme, semi-automated techniques for schema matching have recently appeared in the literature. In the context of the schema-mapping problem, where the mapping among two schemata must be discovered, semi-automated techniques have been proposed (Madhavan, Bernstein, Doan, & Halevy, 2005). Nevertheless, a certain degree of training and supervision is required for a mapping to be derived and --to the best of our knowledge-- there is no fully automated, fast method for this purpose. Therefore, although this technology would resolve the scalability problem and the ad hoc nature of the P2P environment, we cannot rely on its effectiveness for the moment.

To resolve the aforementioned problems of (a) scalability, (b) ad hoc nature of the environment and (c) schema mapping discovery, we resort to an intermediate solution that provides a reasonable balance to all the aforementioned issues. We classify peers to *peer classes*, with the members of each class exporting the same Web service operations. In other words, we assume a factory for each class, specifying the interface for each deployed instance.

We assume a traditional tree-based hierarchy of classes. Each subclass has a single superclass, whose interface it extends. All instances of the subclass are also instances of the superclass. Each node (a) *directly* belongs to exactly one class and (b) *indirectly* belongs to all the classes of the path that starts in the root and ends in its containing class in the tree of the class hierarchy.

We call the set of nodes that directly belong to a class *immediate extent* and the set of nodes that indirectly belong to a class (due to its subclasses) the extended extent. Classes that do not have any descendants are called *base*, or *leaf classes*. We denote the interface of a class C by *interface(C)* and its immediate and extended extents as $extent_i(C)$ and $extent_e(C)$.

In *Figure 2* we can see the base classes *VW*, *BMW*, *TOYOTA*, *SHELL*, *BP*, *HOTEL*, and *RESTAURANT* with their respective nodes. In *Figure 3* we can also observe the superclass *CARS* on top of the classes *VW*, *BMW* and *TOYOTA* and a class *GAS STATION* as a superclass of *SHELL* and *BP*.

The aforementioned problems of integration are resolved in a balanced fashion. With respect to the scale-up of the environment, the integration problem is only dependent on the number of peer classes and not on the number of their instances. Although we anticipate a reasonably small number of peer classes, still the problem of integration is present. We assume a hard-coded, intermediate solution between pairs of classes. This does not necessarily require that all classes be mapped to each other; the only effect of the absence of a mapping would be that two instances belonging to non-reconciled classes could not query each other without a total failure of the system. Moreover, it is straightforward to devise mechanisms for incremental updates of class mappings for the deployed instances, so that, as new classes are added and the interfaces of old classes are updated, the deployed instances are informed on the new situation. With respect to the ad hoc nature of the P2P environment, the problem of class integration is orthogonal and not affected. The last problem, discovery of schema mappings is resolved at the

Figure 2. Base classes with their corresponding nodes

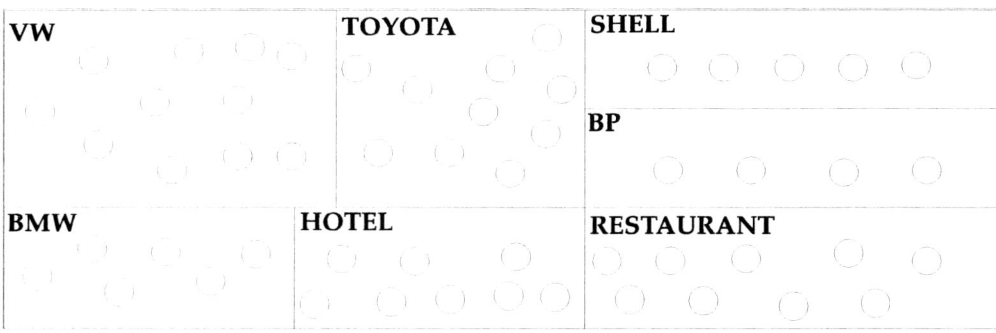

Figure 3. A hierarchy of classes with their corresponding nodes

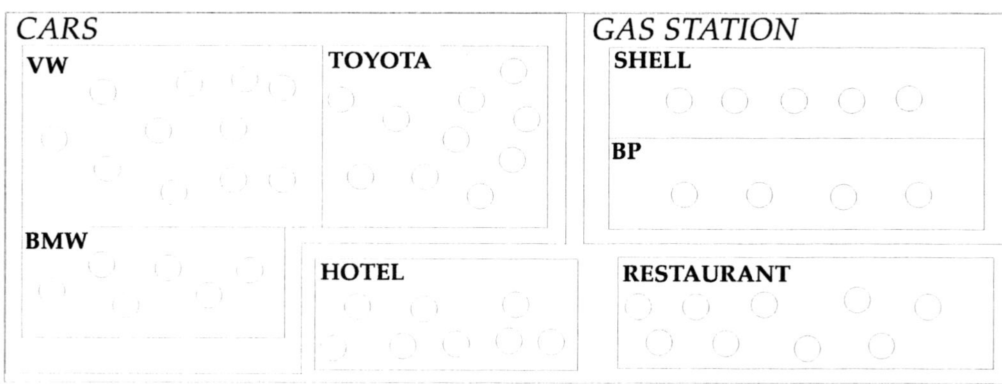

factory level (although we recognize that we still need the same amount of coding effort as in traditional mediator-wrapper environments).

Difference between Classes and Communities

The class of a node is an inherent property of the node, determined once and for all at the creation of the node, mainly for integration purposes, whereas the community (or communities) to which it belongs is a potentially time varying property that is determined individually by the other peers and is mainly used for querying purposes.

Clock

Each peer has its own clock. The clocks of the peers are not necessarily synchronized.

Peer Database

Each peer has a database, which comprises a set of relations. Each relation has a *schema* or *intention* comprised of a finite set of distinct attribute names. Also, each relation has an extension, which is a finite subset of the Cartesian product of the domains of the attributes of the relation's schema. The relations of a peer's database are classified in the following categories:

- **Locally stored (or local) relations:** Local relations are relations whose extension involves tuples that are locally stored at the peer that carries the relations' database. In other words, local relations are exactly the same as in traditional relational databases.
- *Virtual relations*: Virtual relations are relations whose schema is fixed and locally known, but whose extension is not locally stored in the database of the peer. On the contrary, the extension of a virtual relation is collected from the appropriate peers at query time. Practically, this means that each time a user poses a query involving a virtual relation, the peer determines the set of peers who are to be contacted (along with the appropriate sequence of Web service operations of these peers that are to be invoked), collects the respective tuples, transforms them to the schema of the virtual relations, and, finally, stores (or "materializes") them. Then, query processing can be performed as usual.
- *Hybrid relations*: Hybrid relations are variants whose extension includes both locally stored tuples and tuples to be collected from other peers.

Each tuple collected for a relation belonging to the last two categories is tagged with a *timestamp*, produced by the clock of the node that receives the incoming tuple. The timestamp corresponds to the transaction time of the tuple, that is, the exact time point of its entrance to the receiver's database. A tuple's timestamp will be used for caching purposes.

Peer Characteristics

Each peer is characterized by several properties that can either be determined by the peer itself or by the class to which it belongs. Specifically, the characteristics that we adopt are:

- **(Average) Availability**, that is, the probability that the peer is operational at a given time instant.
- **(Average) Response Time,** that is, the average time needed for a Web service operation of the peer to execute.

Peer's System Catalog

Each node u needs a system catalog for its proper operation. The catalog includes useful information about the nodes known to *u*. Specifically, this information refers to:

- Class of the other nodes.
- Communities of the other nodes.
- Distance from other nodes.
- Node characteristics, like availability, and response time.

Results Collection from Other Peers

In this subsection we discuss issues of tuple collection for the virtual and hybrid relations. First, we formally introduce workflows of Web service operations. Next we discuss how the mapping of the workflow's result to a peer's relation is performed and finally, we formalize issues of result materialization.

Workflow $wf^{u.R}(u_i)$

Assume a peer u that poses a query and invokes Web service operations from a set of peers u_1, u_2,....., u_z in order to collect their tuples. In principle, it is quite possible that the requested information from a certain peer can only be obtained after the invocation of a workflow of Web service operations (rather than a single operation). For example, assume that a peer using the European metric system collects the velocities of other peers of class *CAR*, and a certain class of cars returns miles instead of kilometers. The conversion can be performed through a simple BPEL workflow. We denote each of these workflows as $wf^{u.R}(u_i)$, with $1 \leq i \leq z$. Each such workflow w is an acyclic directed graph $G_w(V_w,E_w)$, with operations being modeled as nodes and edges being the representatives of control passing. Edges are tagged with the conditions under which they are fired at runtime. Each workflow has also a flat relational schema that includes a set of attributes that result from the possible un-nesting of the XML elements of the final message delivered by the workflow. Finally, the workflow has an extension, dynamically created at runtime, that instantiates the aforementioned schema.

Mapping of Other Peers' Web Services to Virtual Relations

In this paragraph, we formally discuss the mechanism that allows peers to collect tuples from the peers of their viewpoint. Assume a peer u that poses a query and invokes Web service operations from a set of peers u_1, u_2,....,u_z in order to collect their tuples. The application of the workflow $wf^{u.R}(u_i)$ results to a set of tuples under the schema $(B_1, B_2, ..., B_m)$, possibly after a set of XML un-nesting operations. Assume $R(A_1, A_2,...., A_n)$ to be the schema of R, the mapping between the two schemata is a function f_{map}, with f_{map}: $(A_1, A_2,...., A_n) \times (B_1, B_2, ..., B_m) \rightarrow \{true, false\}$. We impose the constraint that for each A_i, $1 \leq i \leq n$, there exists at most one B_j, $1 \leq j \leq m$, to which A_i is mapped. As usual, all attributes of the workflow schema that are not mapped to the schema of the target relation are projected out, whereas all the relation's attributes that are not populated by the workflow are filled with NULL values. The following example clarifies the aforementioned process. Assume the relation *R(E_ID, E_SALARY, E_AGE)* in the database of node u and let the workflow that is mapped to R for node v have the schema *(ID,AGE,NAME)*. The workflow provides no information on salaries and the database does not store any data on names. Therefore, our mappings resulting to true are:

$f_{map}(E\_ID,ID)=true$,
$f_{map}(E\_AGE,AGE)=true$,

with the rest of all the other possible mappings of the Cartesian product of the two schema being evaluated to false. The transformation at an instance level is simple: (a) we project out all unnecessary workflow attributes, (b) we introduce NULL-valued attributes for the relation's attributes for which no workflow attribute exists, (c) we appropriately re-order the attributes of the workflow schema to match the relation's attributes and (d) we populate the target table.

Full-Partial Materialization

Whenever a workflow is executed for a certain peer and the produced results are successfully stored at the extent of the target virtual relation, we say that we have materialized these results. The fact that the results of a certain workflow for peer u_i have been materialized at the relation R of peer u is denoted as $(wf^{u_i.R}(u_i))$. Full materialization for a relation R of a peer u is the state of a query when all workflows for all the peers that have been selected to populate R have been successfully executed. We denote full materialization by $M(u.R)$. Assuming V^{all} as the set of these identified peers, we can formally define full materialization as $M(u.R)= U\ (wf^{u_i.R}(u_i))$, with $u_i \in V^{all}$.

Partial Materialization

For a relation R of a peer u is the state of a query when the workflows for a clean subset of the peers that have been selected to populate R have been successfully executed. We denote partial materialization by $M^p(u.R)$. Assuming V^{all} be the set of the peers that have been selected to participate in the population of R, and V^i as the set of the peers whose results have been successfully materialized, we can formally define partial materialization as $M(u.R)= U\ (wf^{u_i.R}(u_i))$, with $u_i \in V^i$, $V^i \subset V^{all}$.

SQL^P: An Extension of SQL for Ad Hoc P2P Networks

In this section, we discuss the extension of SQL that we introduce. The proposed language SQL^P (SQL for Peers) implements all the aforementioned requirements. *Figure 4* presents the general structure of an SQL^P query. We use [...] to refer to optional parts of the language and the expression **AND/OR** to signify that different clauses can be connected through one of these logical connectors.

Figure 4. The generic syntax of a query in SQL^P

```
SELECT <projected columns>
FROM <relations involved in the query>
[WHERE <conditions to be satisfied>]
[GROUP BY <grouper attributes>]
[HAVING <post-grouping condition>]
[ORDER BY <sorter attributes>]
[WITH
   [HORIZON <peer selection condition: LOCAL, or COMMUNITY <community
      name> or HOPS θ value θ ∈ {< , > , = , ≤, ≥}, or PEERS=[peer1,peer2,..., peern]>
      *AND / OR*]
   [AVAILABILITY <minimum availability of peers to be contacted>
   *AND / OR*]
   [RESPONSE_TIME <maximum tolerable response time>
   *AND / OR*]
   [CLASS <class name> *AND or OR*]]
   [AGE <maximum allowed caching time for previously collected tuples > *AND /
      OR* ]
   [TIMING <query timing: AD-HOC or CONTINUOUS > < tuning clause>
     If AD-HOC tuning clause refers to successful termination:
       PEERS_PERCENTAGE, or AMOUNT_TUPLES, or TIMEOUT
     If CONTINUOUS tuning clause refers to update mode:
       PULL_BASED WITH PERIOD, or PUSH_BASED>]
]
```

Querying the Graph of Peers

Assume a query Q submitted at node u at the time point T. Let $\{R_1, R_2, \ldots, R_n\}$ be the relations that participate in the FROM clause of the query. Then, we can write the query as: $Q(R_1, R_2, \ldots, R_n)$. Without loss of generality, we can assume that the first k relations $R_1, R_2, \ldots, R_k$, $k \leq n$, are virtual or hybrid. In order to be able to define the semantics of the query properly we need to materialize these relations and then, execute the query over their collected extent as usually. Nevertheless, before specifying this semantics, we need to define the following concepts.

Peers of Interest

The query Q, posed over peer u is divided in three parts. The first part is composed of the traditional SQL clauses, the second part comprises the clauses of our extension that occur after the keyword WITH that have the purpose of determining which peers are to be contacted, and the third part concerns the timing of the query.

The second part of the query depends on criteria like the horizon of the query of the graph of the viewpoint of peer u (HORIZON), QoS characteristics (AVAILABILITY, RESPONSE_TIME), the class of the peers (CLASS) and the age of the stored tuples in the virtual relations (i.e., if a peer has been recently contacted, as specified by the AGE clause, it is not necessary to contact it again). Remember that, due to the nature of the interaction among peers, it is not feasible to simply broadcast a request for tuples; on the contrary, specific Web service operations must be invoked on the specific port types of the peers.

In terms of semantics, we divide the second part into atomic conditions, logically connected through the connectors AND and OR. Assuming that these atomic conditions are $C_1, C_2, \ldots, C_r$, the non-traditional part of the query can be rewritten in a disjunctive normal form, that is, a disjunction of conjunctive conditions.

The interesting aspect of this part is that a preparatory query must be performed over the system catalog to determine specifically which peers must be contacted in order to materialize the virtual relations. Contacting a peer means that for each virtual/hybrid relation in the FROM clause of the query the execution of the appropriate workflow must be initiated. In terms of semantics, each atomic condition specifies a set of peers of the viewpoint of u that qualify to be contacted. Given an atomic condition C, we define the set of peers of interest $V^u(C)$ to be the set of peers that belong to the catalog of peer u that fulfill C. Specifically, given a time point T for a query Q containing C,

$$V^u(C) = \{ v \mid v \in viewpoint(u,T): C(v) = true \}.$$

We do not involve timepoint T to avoid overloading the notation. Having defined the peers of interest for an atomic condition, it is straightforward to obtain the set of peers of a composite condition in disjunctive normal form: the intersection of the peers of interest of the atomic conditions produces the peer sets of each conjunct; these sets are subsequently ORed to produce the final set of peers of interest of the query, which are to be contacted.

Now, we are ready to define the semantics of each individual clause concerning the determination of the peers of interest.

- **HORIZON**: The condition of the HORIZON clause determines the peers of interest on the basis of the position in the graph, or their semantical characteristics. The clause allows several possibilities to the users. Assuming that the condition of the HORIZON clause is C_1, and $VH^u(C_1)$ is the resulting set of peers of interest, we can specify $VH^u(C_1)$ for each of the following possibilities that SQL^P offers:

1. The only peer of interest is the local querying peer (C_l: LOCAL)
   $VH^u(C_l) = \{ u \}$
2. The peers of interest are the ones of a certain community of the peer (C_l: *COMMUNITY* <*C_NAME*>)
   $VH^u(C_l) = \{ v \mid v \in viewpoint(u,T): v \in community(C\_NAME,u) \}$
3. A radius of a certain number of hops dictates the peers of interest (C_l: *HOPS* θ *value*, with $\theta \in \{ =, <, \leq, >, \geq \}$)
   $VH^u(C_l) = \{ v \mid v \in viewpoint(u,T): distance(u,v)\ \theta\ value$, with $\theta \in \{ =, <, \leq, >, \geq \}\}$
4. A set of peer ids, that is, a set of specifically requested peers, determines the peers of interest (C_l: *PEERS*=$\{peer_1, peer_2, \ldots, peer_n\}$)
   $VH^u(C_l) = \{ v \mid v \in viewpoint(u,T): v \in \{peer_1, peer_2, \ldots, peer_n\}\}$

All the necessary information for the evaluation of any of the aforementioned atomic conditions is found in the system catalog of *u*.

- **Quality of service:** The clauses concerning the AVAILABILITY and RESPONSE TIME of the peers of interest aim to guarantee a certain level of quality of service for the peer posing a query.
- **CLASS:** It is possible that we only need to query the peers of a certain class. Classes carry both structural typing information (as they statically define the interface of their instances), but also semantic information (as collections of semantically—therefore structurally—similar instances). In SQL^P, it is easy to specify an atomic condition that restricts the peers of interest to a certain class, by giving a condition of the form C_4: *CLASS=class_name*. Assuming $VC^u(C_4)$ the result set of peers of interest, and *class*(*v*) a function that returns the class of each peer from the system catalog of the querying peer, the resulting set of peers of interest is formally defined as:

  $VC^u(C_4) = \{ v \mid v \in viewpoint(u,T): class(v) = class\_name \}$

- **AGE:** Apart from the constraining of peers where their properties are taken as criteria for their inclusion in the resulting set of peers of interest, we can perform some form of caching in the extents of the collected tuples for virtual or hybrid relations. In other words, assuming that a peer is frequently queried, it is not obligatory to pay the price of invoking its Web service operations, executing the data transformation workflow and materializing the same results again and again, but rather, it is resource efficient to cache its previous results. The AGE clause of SQL^P provides the possibility of specifying a maximum caching age for incoming tuples in a virtual/hybrid relation.
- **Query Timing**: Having clarified the general mechanism for the determination of peers of interest, we move on to provide the specification for the timing of queries. Fundamentally, we have two modes of operation: ad hoc or continuous. Each mode has its own tuning parameters:
  - If the query is continuous, this means that the user is continuously notified on the status of the query result.
  - If the query is ad hoc, the query eventually has to terminate. Differently from traditional query processing (which operates on finite sets of always available, locally stored tuples), we need to tune the conditions that signify termination of a query that has been late to complete its operation, either due to peer failures, or the size of the peer's graph. To capture these exceptions, we can terminate a query upon (a) the completion of a timeout period

of execution, (b) the materialization of a certain amount of tuples that the user judges as satisfactory for his information, or (c) the collection of responses from a certain percentage of peers that were initially contacted. In all these cases, the execution of the workflows whose results have not been materialized is interrupted, the rest of the query is executed as usual and the user is presented with a partial—still, non-empty—answer.

Query Execution

At this point we can describe the exact set of steps for executing a query. Suppose that at random time T, a query Q is performed by node u. Let $\{R_1, R_2, ..., R_n\}$ be the relations involved in query Q. Then the query can be written in the form: $Q(R_1, R_2, ..., R_n)$. We can assume that the relations $R_1, R_2, ..., R_k$, with $k \leq n$ are virtual or hybrid, without any impact on the generality. All tables $R_1, R_2, ..., R_k$ must be filled with tuples. The procedure is the same for all tables; therefore we will present it only for table R_1.

The first step is to determine the set of target peers for node u that performs the query ($V^u(C)$), by evaluating C over the set of peers belonging the viewpoint of u (*viewpoint*(u)). C comprises of the conditions located at the clauses AGE, HORIZON, AVAILABILITY, RESPONSE_TIME and CLASS.

Let $V^u(C) = \{ u_1, u_2, ..., u_m \}$. For each node of $V^u(C)$ the appropriate Web services are invoked in order to require the appropriate tuples. Let also $wf^{u.R1}(u_1)$, $wf^{u.R1}(u_2)$, …., $wf^{u.R1}(u_m)$, be the appropriate workflows of the peers belonging to $V^u(C)$.

The schema of each workflow is matched to the schema of relation R_1, which is the target relation. In the following, the clause TIMING is evaluated to determine the execution mode of the query (continuous or ad hoc) and the completion condition of the query. The next step is to attempt the execution of $wf^{u.R1}(u_1)$ and then perform a full or partial materialization of R_1, which is located in u, according to the query completion condition, which was mentioned before. Table R_1 is populated with the appropriate tuples and is ready to be queried. The same procedure is performed for all other virtual or hybrid tables. Therefore all tables of u are ready to be queried. At this point the query of u is performed over tables $R_1, R_2, ..., R_n$ based on traditional database methodology.

Examples

In the rest of this section, we will present examples of SQL^P. Assume a peer network of the topology of *Figure 5*, consisting of 5 peers, each representing a car in the highway. Queries are posed to peer p_1, that classifies the rest of the peers in two communities, (a) the community of dark-shaded close peers (*Distance_Under_5km*) and (b) the community of light-shaded, distant peers (*Distance_Over_5km*). Peer p_1 is informed on the existence and connectivity of the rest of the peers through the underlying routing protocol that operates as a black box in our setting.

Figure 5. Graph configuration for query posing

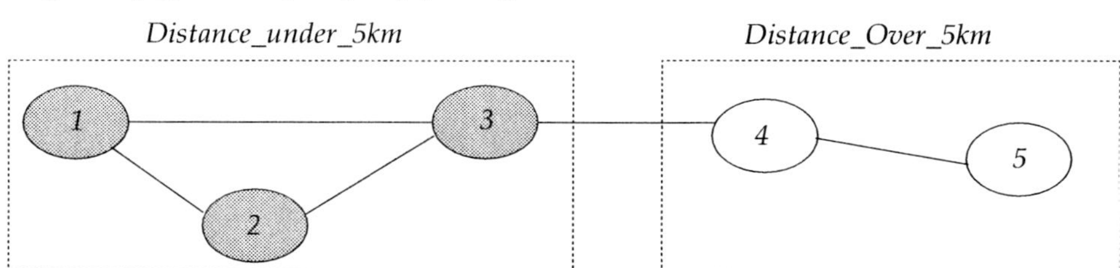

Peer p_l carries a database consisting of two relations with the following schemata:

CARS(ID, PLATE, BRAND, VEL)
BRANDS(BRAND, COUNTRY, METRIC_SYSTEM)

The first relation describes the information collected from the peers contacted (and mainly serves queries about the velocity of the cars in the context of the querying peer). This relation CARS is virtual: each time a query is posed, tuples must be collected from the context of peer p_l to populate it. The attribute BRAND is a foreign key to the relation BRANDS that is static and locally stored. Primary keys are underlined and the semantics of the attributes are the obvious ones. In the sequel, we give examples of SQL^P queries over the abovementioned environment.

Example 1

By this example we illustrate different situations where we can determine the peer nodes to which the query is addressed. Different strategies may be used for choosing the peers to query. In any case the decision is based on characteristics of the peers such as availability, response time, class of Web services implemented, and so forth. Peer p_l wishes to know the license number, velocity and manufacturing country of all cars belonging to its community. Furthermore, the peer that poses the query wishes to limit it to those peers that: (a) are located no more than 5 Km away (**Distance_Under_5km**), (b) their availability is more than 60%, (c) their response time is less than 4 seconds and finally, (d) implement the European class of Web services. The syntax of the examined query is depicted in *Figure 6.*

Example 2

Peer p_l wishes to know the license number, velocity and manufacturing country of all cars. The peer also wishes to complete the query when more than 70% percent of the target peers have replies successfully (*Figure 7*). To determine the target peers, the requesting peer selects the peers based on its catalog and according to their response time. The execution of the query stops when the requested percentage of 70% in our case is satisfied.

Example 3

Peer p_l wishes to know the license number, velocity and manufacturing country of all cars. The peer also wishes to complete the query when more than 5 tuples have been collected for the relation CARS (*Figure 8*). The requesting peer contacts each peer that appears in its catalog. This procedure

Figure 6. Query 1

```
SELECT          CARS.PLATE,CARS.VEL,BRANDS.COUNTRY
FROM            CARS, BRANDS
WHERE           CARS.BRAND=BRANDS.BRAND
WITH
HORIZON         COMMUNITY Distance_Under_5km AND
AVAILABILITY    > 60% AND
RESPONSE_TIME   < 4.0 AND
CLASS           = 'european'
```

Figure 7. Query 2

```
SELECT   CARS.PLATE,CARS.VEL,BRANDS.COUNTRY
FROM     CARS, BRANDS
WHERE    CARS.BRAND=BRANDS.BRAND
WITH
TIMING   AD-HOC PEERS_PERCENTAGE > 70%
```

Figure 8. Query 3

```
SELECT   CARS.PLATE,CARS.VEL,BRANDS.COUNTRY
FROM     CARS, BRANDS
WHERE    CARS.BRAND=BRANDS.BRAND
WITH
TIMING   AD-HOC AMOUNT_TUPLES > 5
```

Figure 9. Query 4

```
SELECT   CARS.PLATE,CARS.VEL,BRANDS.COUNTRY
FROM     CARS, BRANDS
WHERE    CARS.BRAND=BRANDS.BRAND
WITH
TIMING   AD-HOC TIMEOUT > 7
```

ends when the count of currently collected tuples becomes greater or equal to the posed limit.

Example 4

Peer p_l wishes to know the license number, velocity and manufacturing country of all cars. The peer also wishes to complete the query within a timeout of 7 seconds (*Figure 9*). The requesting peer contacts each peer that appears in its catalog. This procedure ends when the timeout is reached.

QUERY PROCESSING FOR SQL^P QUERIES

In this section, we deal with the problem of mapping the declarative SQL^P queries to executable query plans. As already mentioned, the execution of traditional SQL queries relies on their mapping to left-deep trees whose leaves are database relations, internal nodes are operators of the relational algebra and edges signify pipeline of the results of a node to another. Clearly, since we raise fundamental assumptions of traditional database querying, such as the finiteness and locality of tuples as well as the conditions under which a query terminates, we need to extend both the set of operators that take part in a query and the way the query tree is constructed. In this section, we start by introducing the novel operators for query processing. Next, we discuss how we algorithmically determine the set of peers of interest and, finally, we discuss the execution of a query.

Novel Operators

In this subsection, we start with the operators that participate in SQL^P query plans. We directly adopt the *Project*, *Select*, *Group*, *Order*, *Union*, *Intersection*, *Difference* and *Join* operators from traditional relational algebra and move on to define new operators. First, we discuss operators that are used to construct the set of peers of interest. Then, we present the operators that actually take part in a query plan.

- **Operators applicable to the catalog of a peer:**
  - **Check_Tables:** Operator *Check_Tables* determines whether the tables belonging to the FROM clause of a query are virtual, hybrid or local. The input to the operator is the FROM clause of the query and the output is the same list of tables, each annotated with the category to which it belongs.
  - **Check_Peers:** This is a composite operator that applies the procedure mentioned in the second section for the determination of a set of peers out of a condition in disjunctive normal form. All clauses of the form HORIZON, AVAILABILITY, RESPONSE_TIME and CLASS are evaluated over the catalog through a *Check_Peers* operator and the set of peers of interest is determined by combining the results of these operators through the appropriate *Unions* and *Intersections*.
  - **Check_Age:** The *Check_Age* operator is also an operator used to determine the set of peers of interest. For each relation that hosts transaction time and producing peer attributes, an invocation of the *Check_Age* operator scans the extent of the relation, and identifies the appropriate tuples and their peers. The output is passed to the appropriate Difference operator that subtracts the identified peers from the previously determined set of peers of interest.
- **Operators that participate in query plans:**
  - **Call_WS:** This operator is responsible for dynamically determining which Web service operation, over which port type, of a specific peer must be invoked. Each Web service of a peer to be invoked is practically wrapped by this operator. The result is collected and forwarded to the operator managing the execution of a workflow of Web services.
  - **Wrapper_Pop:** This operator is used in order to support the monitoring and execution of the workflow of Web services that populate a virtual or hybrid table. For each peer contacted in order to populate a certain virtual/hybrid relation, a *Wrapper_Pop* operator is introduced. Once the final XML result has been computed, its tuples are transformed to the schema of the target relation.
  - **Fill:** A *Fill* operator is introduced for each virtual relation. The operator takes as input all the results of the underlying *Wrapper_Pop* operators (one for each peer of interest) and coordinates their materialization. Also, *Fill* checks the necessary conditions concerning the timing and termination of the query and, whenever termination is required, it signals its populating operators appropriately.
  - **ExAg (Execute Again):** This operator is useful only in continuous queries and practically restarts query execution whenever the query period is completed.

Construction of the Query Tree

In this paragraph we discuss a simple algorithm to generate the tree of the query plan. Assume that a query is posed to peer p_1 and its viewpoint comprises n peers, specifically p_1, p_2, ..., p_n. The algorithm for the construction of the query tree is a bottom up algorithm that builds the tree from the leaves to the top and is described as follows:

1. We discover the virtual or hybrid relations that participate in the query. A specific sub-tree will be constructed for each of them.
2. We determine the set of peers of interest. For each peer that participates in the population of a certain relation, the leaves of the respective sub-tree are nodes representing the peer to be contacted. To keep the tree-like form of the plan, each peer can be replicated in each sub-tree to which it participates; nevertheless, each peer can also be modeled by a single node without any significant impact to the execution of the query.
3. We introduce a *Wrapper_Pop* for each peer that coordinates all the *Call_WS* operators that pertain to the operations of the peer. Between the peer node and the *Wrapper_Pop*, we introduce the appropriate *Call_WS* operators.
4. For each virtual or hybrid relations we introduce a *Fill* operator that combines the output of all the respective *Wrapper_Pop* operators; therefore it is their immediate ancestor.
5. Having introduced the *Fill* operators, the virtual or hybrid relations can be materialized and act as local ones. Therefore, the rest of the query tree is built as in traditional query processing.
6. If the query is continuous, we add an appropriate *ExAg* operator at the top.

Execution of a Query through the Query Tree

The execution of the query follows a simple strategy. First, we materialize the virtual/hybrid relations. Then, we execute the query as usual. Clearly, although this is not the best possible strategy for all cases (especially when only non-blocking operators are involved), we find that performing further optimizations is an orthogonal problem, already dealt in the context of blocking operators for streaming data (Babcock et al., 2002). Therefore, in this article we consider only this baseline strategy since all relevant results can directly be introduced in the optimizer module of a peer. Specifically, the set of steps to follow for the execution of the query are:

1. All the *Call_WS* operators are activated and the appropriate services are invoked.
2. The *Wrapper_Pop* operators collect the incoming XML results and queue them towards the appropriate *Fill* operators that further push them towards the extents of the relations in the hard disk. This is performed in a pipelined fashion.
3. Once all virtual/hybrid relations have been materialized, the rest of the query plan is a traditional left-deep tree that executes as usual.

Example

In the following, we discuss the construction of the query plan for the query of *Figure 10*.

1. Step 1: The query involves two tables, CARS and BRANDS. The application of the operator CHECK_TABLES over the two relations results in the determination that the first is a hybrid one and the second a locally stored one.
2. Step 2: The operator CHECK_PEERS is applied to the catalog of peer p_1, in order to

Figure 10. Query for which the plan is to be constructed

SELECT	CARS.PLATE,CARS.VEL,BRANDS.COUNTRY
FROM	CARS, BRANDS
WHERE	CARS.BRAND=BRANDS.BRAND
WITH	
AGE	< 5 AND
HORIZON	COMMUNITY DISTANCE_UNDER_5KM AND
TIMING	CONTINUOUS PULL_BASED_PERIOD = 7 AND
AVAILABILITY	> 60% AND
RESPONSE_TIME	< 3.0 AND
CLASS	= 'european'

determine the peers of interest of the query. Taking into consideration the age of tuples found in relation CARS and the system catalog, the peer p_1 decides that the peers of interest are peers *2* and *8*.

3. Step 3: The operator CALL_WS is applied over each peer of interest.
4. Step 4: For each peer over which a CALL_WS is applied, we apply the operator WRAPPER_POP to coordinate the execution of its operations.
5. Step 5: The operator FILL is applied for the result of each WRAPPER_POP.
6. Step 6: The rest of the query plan is constructed as usual, with the only difference that the sub-tree of relation CARS is the one constructed in the previous steps.

IMPLEMENTATION

Figure 12 shows the full-blown architecture required to support our approach for context-aware query processing in ad hoc environments of peers. The elements shown in the figure are divided with respect to the client and the server roles played by peers. To play the client role, a peer comprises a traditional query processing architecture, involving a parser, an optimizer and a query processor. A local database and the system catalog complement the ingredients of the client part of a peer. Playing the server role amounts in publishing a set of Web services, hosted by an application server, which is responsible for their proper execution. As usual, whenever a query is posed, the parser is the first module that is fired. The optimizer produces alternative plans out of which the best, with respect to a given cost model, is chosen. The query execution engine executes the query over the local database and returns the results.

Our first prototype implementation does not currently support the query optimizer subsystem. Instead, standard query plans are produced after parsing the user queries. The query execution subsystem includes a mechanism that allows visualizing the aforementioned plans. *Figure 11* gives a visualized execution plan through the Yed tool that graphically presents graphs.

Populating and updating the contents of the system catalog is done either statically, or dynamically. In the former case, the peer is responsible for updating the catalog through a catalog-specific API. The static update of the catalog takes advantage of the possible availability of peer-specific dynamic service discovery mechanisms. Such mechanisms may be exploited by the peer itself which takes further charge of updating the catalog accordingly.

The dynamic catalog update is realized by the catalog update subsystem, which relies on

Figure 11. Query plan for the aforementioned query of Figure 10

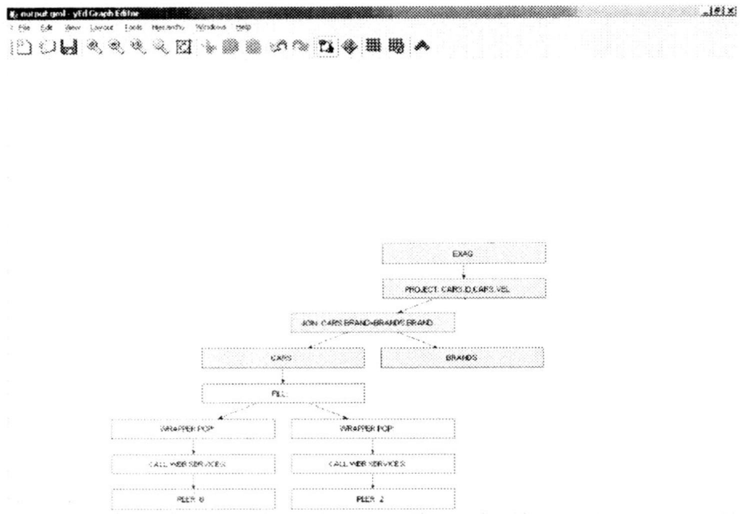

Figure 12. System architecture

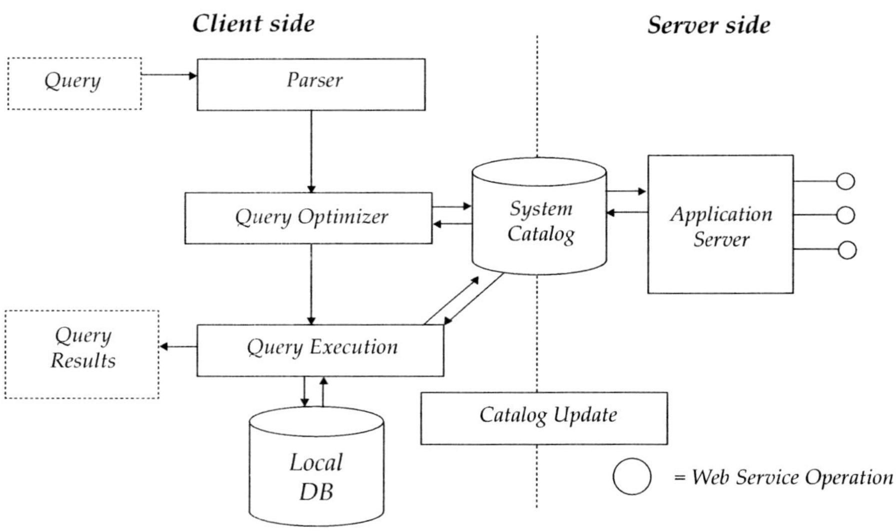

WSAMI, a middleware platform for mobile Web services (Issarny et al., 2005). WSAMI provides the naming & directory service that allows the dynamic discovery of Web services provided in mobile computing environments. Specifically, WSAMI is based on an SLP server – that is, an implementation of the standard SLP (http://www.openslp.com) protocol-- for the discovery of networked entities in mobile computing environments.

RELATED WORK

The work that is closely related with the proposed approach for context-aware query processing over ad hoc environments of peers can be categorized into work concerning the fundamentals of heterogeneous database systems, context-aware computing and approaches that specifically focus on context-aware service-oriented computing. The prominent approaches that fall in the aforementioned categories are briefly summarized in the remainder of this section.

Heterogeneous Database Systems

Our approach for querying of ad hoc environments of peers bears some similarity with the traditional wrapper-mediator architectures used in heterogeneous database systems (Roth & Schwarz, 1997; Haas et al., 1997). Such systems consist of a number of heterogeneous data sources. The user of the system has the illusion of a homogeneous data schema, which is actually realized by the wrapper-mediator architecture. In particular, each data source is associated with a wrapper. The wrapper encapsulates the data source under a well-defined interface that allows executing queries. Each user query is translated by the mediator into data source specific queries, executed by corresponding wrappers. As opposed to traditional heterogeneous database systems, in the environments we examine the roles of users and data sources are not discrete. Each peer is a heterogeneous data source offering information to other peers that play the role of the user. Therefore, each peer may eventually serve as a data source and a user issuing queries. Analogous to the wrapper elements in our case is the Web services that give access to peers playing the role of data sources. Analogous to the mediator element is the hybrid relation mapping procedure that executes workflows on Web services. In simple words, a traditional heterogeneous database system is a "1 mediator to N wrappers architecture." An ad hoc environment of peers in our case is an "N mediator to N wrappers architecture."

Another fundamental difference between the environments we examine and traditional heterogeneous database systems is that in our case the cardinality and the contents of the set of data sources may constantly change.

Context-Aware Computing and Infrastructures

In Dey (2001), context is defined as any information that can be used to characterize the interaction between a user and an application, including the user and the application. Several middleware infrastructures follow this definition toward enabling context reasoning and management (Fahy & Clarke, 2004; Chen, Finin, & Joshi, 2003; Chan & Chuang, 2003; Capra, Emmerich, & Mascolo, 2003; Gu, Pung, & Zhang, 2005; Roman et al., 2002). Amongst these approaches, CASS (Fahy & Clarke, 2004) bears some similarity with our approach, since context is modeled in terms of a relational data model. However, in our approach we do not assume centralized information management and virtual relations are dynamically compiled.

Context-Aware Service-Oriented Computing

In general, the integration of context-awareness and service-orientation just began to gain the attention of the corresponding research communities. In Keidl & Kemper (2004), for instance, the authors introduce ways for associating context to Web service invocations. In Maamar, Mostefaoui, & Mahmoud (2005) the authors go one step further by examining the problem of customizing Web service compositions with respect to contextual information. Web service execution is customized according to different types of context. Similarly, in Zahreddine & Mahmoud (2005) the authors propose a framework for dynamic context-aware service discovery and composition. Specifically, contextual information regarding the technical characteristics of user devices is used towards discovering services that match these characteristics.

CONCLUSION AND FUTURE WORK

In this article, we have dealt with context-aware query processing in ad-hoc peer-to-peer networks. Each peer in such an environment has a database over which users want to execute queries. This database involves (a) relations which are locally stored and (b) relations which are virtual or hybrid. In the case of virtual relations, all the tuples of the relation are collected from peers that are present in the network at the time when the query is posed. Hybrid relations involve both locally stored tuples and tuples collected from the network. The collaboration among peers is performed through Web services. The integration of the external data, before they are locally collected to a peer's database, is performed though a workflow of operations. We do not perform query processing in the traditional way, but rather, we involve context-aware query processing techniques that exploit the neighborhood of each peer and the Web service infrastructure that deals with the heterogeneity of peers. In this setting, we have formally defined the system model for SQL^P, an extension of traditional SQL on the basis of contextual environment requirements that concern the termination of queries, the failure of individual peers and the semantic characteristics of the peers of the network. We have precisely defined the semantics of the language SQL^P. We have also discussed issues of data integration, performed through workflows of Web services. Moreover, we have presented an initial query execution algorithm as well as the typical definition of all the operators that can take place in a query execution plan. A prototype implementation that is implemented is also discussed.

ACKNOWLEDGMENT

This research is co-funded by the European Union - European Social Fund (ESF) & National Sources, in the framework of the program "Pythagoras II" of the "Operational Program for Education and Initial Vocational Training" of the 3rd Community Support Framework of the Hellenic Ministry of Education.

REFERENCES

Abolhasan, M., Wysocki, T., & Dutkiewicz, E. (2004). A review of routing protocols for mobile ad hoc networks. *Ad Hoc Networks, 2*, 1-22.

Androutsellis-Theotokis, S., & Spinellis, D. (2004). A survey of peer-to-peer content distribution technologies. *ACM Computing Surveys, 36(4)*, 335-371.

Babcock, B., Babu, S., Datar, M., Motwani, R., & Widom, J. (2002, June). Models and issues in data stream systems. In *Proceedings of the 21st ACM SIGACT-SIGMOD-SIGART Symposium on Principles of Database Systems* (PODS'02) (pp. 1-16).

Capra, L., Emmerich, W., & Mascolo, C. (2003). CARISMA: Context - Aware Reflective Middleware System for Mobile Applications. *IEEE Transactions on Software Engineering, 29*(10), 929-945.

Chan, A.T., & Chuang, S.-N. (2003). MobiPADS: A reflective middleware for context-aware mobile computing. *IEEE Transactions on Software Engineering, 29*(10), 1072-1085.

Chen, H., Finin, T., & Joshi, A. (2003). An ontology for context-aware pervasive computing systems. *Knowledge Engineering Review, 18(3)*, 197-207.

Chlamtac, I., Conti, M., & Liu, J. J.-N. (2003). Mobile ad hoc networking: Imperatives and challenges. *Ad Hoc Networks, 1*(1), 13-64.

Dey, A.K. (2001). Understanding and using context. *Personal and Ubiquitous Computing, 5*(1), 4-7.

Fahy, P., & Clarke, S. (2004, June). CASS - Middleware for mobile context-aware applications. In *Proceedings of the 2nd ACM SIGMOBILE International Conference on Mobile Systems, Applications and Services* (MobiSys'04).

Gu, T., Pung, H.-K., & Zhang, D.-Q. (2005). A service-oriented middleware for building context-aware services. *Journal of Network and Computer Applications, 28*, 1-18.

Haas, L.M., Kossmann, D., Wimmers, E.L., & Yang, J. (1997, August). Optimizing queries across diverse data sources. In *Proceedings of 23rd International Conference on Very Large Data Bases* (VLDB'97) (pp. 276-285).

Issarny, V., Sacchetti, D., Tartanoglou, F., Sailhan, F., Chibout, R., Levy, N., & Talamona, A. (2005). Developing ambient intelligence systems: A solution based on Web services. *Journal of Automated Software Engineering, 12*(1), 101-137.

Keidl, M., & Kemper, A. (2004, March). A framework for context-aware adaptable Web services. In *Proceedings of 9th International Conference on Extending Database Technology* (EDBT '04) (pp. 826-829).

Maamar, Z., Mostefaoui, S., & Mahmoud, Q. (2005, January). Context for personalized Web services. In *Proceedings of 38th IEEE Hawaii International Conference on System Sciences* (HICSS'05) (pp. 166.2).

Madhavan, J., Bernstein, P.A., Doan, A., & Halevy, A.Y. (2005, April). Corpus-based schema matching. In *Proceedings of the 21st International Conference on Data Engineering* (ICDE 2005) (pp. 57-68).

Ozsu, T., & Valduriez, P. (1991). *Principles of distributed database systems*. Prentice-Hall.

Roman, M., Hess, C.K., Cerqueira, R., Ranganathan, A., Campbell, R.H., & Nahrstedt, K. (2002). Gaia: A middleware infrastructure to enable active spaces. *IEEE Pervasive Computing, 1*(4), 74-83.

Roth, M.T., & Schwarz, P.M. (1997, August). Don't scrap it, wrap it! A wrapper architecture for legacy data sources. In *Proceedings of 23rd International Conference on Very Large Data Bases* (VLDB'97) (pp. 266-275).

Zahreddine, W., & Mahmoud, Q.H. (2005, March). An agent-based approach to composite mobile Web services. In *Proceedings of 19th International Conference on Advanced Information Networking and Applications* (AINA 2005) (pp. 189-192).

This work was previously published in Journal of Electronic Commerce in Organizations, Vol. 6, Issue 1, edited by M. Khosrow-Pour, pp. 88-62, copyright 2008 by IGI Publishing (an imprint of IGI Global).

Chapter 5
The Role of Digital Service Encounters on Customers' Perceptions of Companies

Kristina Heinonen
Hanken Swedish School of Economics and Business Administration, Finland

ABSTRACT

The service encounter occurs whenever a customer interacts with a company personally or through technology through, for example, the Internet, e-mail, or telephone. Nowadays, customers frequently initiate the encounter as, for example, inquiries, information searches, and complaints are conveniently performed online. This article explores the role of digital service encounters on customer perceptions of companies. Digital service encounters in this article denote remote customer-company interactions via the Internet or e-mail. The focus is on active customers initiating interactions and on customer perceptions of company responses to these interactions. A conceptual framework that captures customer perceived service encounter value on two dimensions (responsiveness and personalization) is proposed. An empirical study exploring the value of company responses to digital contacts indicated that many contacts are responded to promptly and satisfyingly. However, there are also significant differences in the value of the service encounter. Some service encounters are perceived as unpersonalized, and some are even left without response.

INTRODUCTION

The service encounter or the "moments of truth" (Normann, 1991) occurs every time a customer interacts with a company. Customers can interact with companies in many ways, personally or through technology. Customer interaction and participation in the delivery of services has increased with the use of self-service technologies, such as online banking, automated filing of legal claims, or self-scanning at retail stores. The service encounter represents a building block from which the customers form their perceptions of the product or service and the company as a whole.

There are different types of service encounters, and traditionally the service encounter denoted the personal interaction between the service employee and the customer. In the personal service encounter, the service delivery process was highly controllable by the service employee. Nowadays, when customers contact the company through different technology interfaces it is frequently the customers that initiate and control the interaction. For example, inquiries, information searches, and complaints are conveniently performed on the Internet without the involvement of the service employee. These technology-based service encounters are challenging for companies, because technology extends the distance between the parties in the interaction and increases customer input in the service process. As a result, the interaction process becomes potentially more complex. Consequently, it becomes more difficult for the company to influence customers' perceptions of the company through the service encounter.

The purpose of this article is to explore the role of digital service encounters on customer perceptions of companies. Digital service encounters in this article denote remote customer-company interactions via the Internet or e-mail. The focus is on the active role of customers in initiating interactions and how well the customer perceives that the company responds to these interactions.

The article has the following structure. First, we discuss previous research on service encounters and illustrate how service encounters influence customer perceptions of companies. The focus is on the service management and e-service research. Then we propose a conceptual framework of service encounter value from a customer perspective. An empirical study is conducted to illustrate the practical use of the proposed framework and to give empirical findings on how successful companies are to respond to customer contacts through digital channels. We also discuss theoretical implications concerning the proposed framework as well as practical implications and challenges regarding how to evaluate and successfully manage digital service encounters. We conclude the article with future trends, suggestions for future research and concluding remarks that summarize the contribution of the article.

SERVICE ENCOUNTERS

It has been argued that perceptions of service encounters differ depending on the type of service encounter (Shostack, 1985). Service encounters have been categorised differently (Bitner, 2006; Lovelock, 1983, Shostack, 1985), but all classifications denote the level of customer input in the service process and the type of interaction between the customer and the service employee. At one end is the personal encounter between the customer and service employee and where the customer's input in the service process is low. At the other end is the remote service encounter where the interaction occurs through a technology interface and where the customer's input in the service process is extensive. For example, Shostack introduced three general types of service encounters: remote encounter, phone encounters, and face-to-face encounters. Remote interactions take place through ATM, the Internet, or a mobile phone. Similarly, Lovelock and Wirtz (2005) discuss differences between high-contact services and low-contact services, where the former emphasises encounters with service personnel and the latter involves encounters with equipment. Whereas the role of service employees is large for the people-dominant service encounters (Lewis & Entwistle, 1990), the remote service encounters enable the customer to more independently create the service encounter. The remote service encounter, that is, the service encounter that occurs through technology, is in focus in this article.

There is research on customer perceptions of service encounters although most research has focused on the *overall* quality or value of the service. For example, Walker (1995) studied service encounter satisfaction and developed a model of service encounter satisfaction based on three stages of disconfirmation. In the model, customers evaluate the perceived performance of both core service and peripheral service against their expectations. In other words, Walker separated the overall service encounter evaluation into different stages, but he did not conceptualise the sources of either satisfaction or dissatisfaction in each stage. Meuter, Ostrom, Roundtree, and Bitner (2000) studied customer satisfaction of technology-based service encounters, and they found four different sources of satisfaction and dissatisfaction of interactions with Internet-based services, automated phone services, kiosk services, and services delivered via CD or video technology. Chandon, Leo and Philippe (1997) studied different components of face-to-face service encounters that are relevant for assessing service quality. The face-to-face service encounter included interactivity, rituality, effectiveness, accessibility, and agent satisfaction, and they were explored with a dyadic perspective. Similarly, Jayawardhena (2007) developed a model of service encounter quality, encompassing professionalism, civility, friendliness and competence dimensions. The focus was on face-to-face service encounters within a business-to-business setting.

These examples of previous studies indicate the relevance of focusing on the service encounter instead of the overall service. The literature review suggests that research on the customer perceived *value* of service encounters is still needed, where most research on service encounters is focused on either satisfaction or quality.

SERVICE ENCOUNTER VALUE

We argue that service encounters contains value for the customers and that this value consists of both of the encounter outcome as well as the encounter process. Customers would thus evaluate the service encounter on these two dimensions: how quickly the company responds compared to expectations (responsiveness) and how personal the response is (personalization). These two dimensions are suggested to reflect the role of the interaction on customers' perceptions of the company and service.

In the figure below these two dimensions are combined in a matrix that represents the proposed conceptual framework of customer perceived service encounter value. The focus of the service encounter is on the digital or technology-mediated interaction between customers and service employees. The dimensions are categorized into three levels: low/negative, neutral, and high/positive, resulting in nine different potential outcomes. In the upper right-hand corner we have the optimal or best case as the service encounter is perceived as highly personalized and based on a prompt response. The opposite case is depicted by the lower left-hand corner where both responsiveness and personalization is low. The scale of personalization covers either presence or absence of personalization, with a neutral or standard approach in between. The quadrant in the middle represents a reasonable service encounter as perceived by the customers. The most challenging service encounters are those that are represented by the upper left-hand and the lower right-hand corners. The reason for this is that customers perceive that one of the dimensions is functioning well, whereas the other is working poorly.

Responsiveness

Accessibility of companies and responsiveness of services employees have been argued to be part of services and influence customers' perceptions of services. For example, response time was included as one part of service employees' contribution to accessibility in Grönroos' augmented service offering (Grönroos, 1987). Responsiveness was also included in the SERVQUAL model of service quality (Parasuraman, Zeithaml, & Berry,

1985), and it was defined as the willingness of service employees to help customers and respond to their requests, and to give prompt service. It includes timeliness of service, immediate mailing, responding to customer calls quickly, and giving prompt service.

Personalization

Personalization is a growing topic in the e-service research and it is used in many different contexts such as the public sector (Pieterson, 2007), e-services (Ansari & Mela, 2003), and mobile services (Grant & O'Donohoe, 2007; Searby, 2003). Personalization involves targeting content at users based on their interests or buying patterns (Searby, 2003). It has been argued that consumers value personalized content that facilitates the maintenance and development of personal relationships (Grant & O'Donohoe, 2007). Recent research has also attempted to conceptualise personalization (Vesanen & Raulas, 2006; Wind & Rangaswamy, 2001). For example, Miceli, Ricotta, and Costabile (2007) linked the personalization to the interaction between the service and the customer. They proposed a conceptual framework of e-customer profiling for interactive personalization based on content (expected customer benefits) and process (expected degree of interaction). The model was based on previous research and it included four significant drivers of online customer heterogeneity: value, knowledge, orientation, and relationship quality. The value of personalization was based on both the interface and interaction tools and content features. However, although there is much research on personalization, it has not previously been linked to the discussion on the *value* of service encounters.

EMPIRICAL STUDY

An empirical study was conducted in January-February 2006. The data were collected by 71 marketing students in a mystery shopper-type of approach. Each student chose three service organizations that he/she has used within the last 6 months and contacted the company electronically by e-mail or using an online evaluation/feedback form. The respondents independently chose different service companies to study, resulting in a convenience sample. The respondents did not communicate in the contact that the contact was part of a study. The contacts to the companies were of three types: an information request (e.g., concerning the company, contact information), a question concerning a specific service, and a complaint concerning a bad service experience. The sample included 217 contacts. The students were instructed to evaluate how well the company responded to the contact in terms of quality of response and response time. The respondents were from different nationalities, mainly Finnish, and the companies were mostly Finnish service companies. However, some international companies were also included. The table indicates the age spread as well as the gender distribution.

A service encounter value grid based on the proposed service encounter value dimensions was used (see Figure 1). The two dimensions were operationalized with three-point scales. The personalization dimension question was formulated as "How personal was the response?" and the alternatives were (1) The service encounter was personalised for me, (0) The service encounter was okay but it represented a standard response, and (-1) The service encounter did not meet my needs. The responsiveness question was formulated as "How responsive was the company?", and the alternatives were: (1) The company responded quickly, (0) The response time was reasonable, and (-1) The response time was long. The scale alternatives represented positive, mediocre, or negative perceptions.

Additionally, the respondents were to compare and contrast the different service contacts by answering the following questions: How correct/useful was the response? How easy was it

Figure 1. Conceptual model of service encounter value

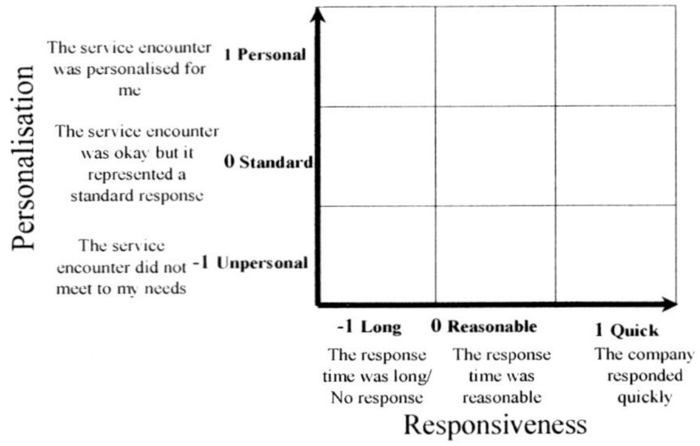

Table 1. Descriptive summary of the study

Descriptive details of the study		
Industries represented	Tourism and travel	14%
	Mobile operator	10%
	Bank	9%
	Electronics	7%
	IT	5%
	University service	5%
	Airline	4%
	Catering, restaurant	4%
	Transport	4%
	Retail	4%
	Consumer product	3%
	E-commerce	3%
	Hotels	3%
	Sports facilities	3%
	Beauty & health	2%
	Gasoline companies	2%
	Insurance	2%
	Postal services	2%
	Culture and Entertainment	2%
	Bookstore	1%
	Car rental	1%
	Cleaning services	1%
	Government department	1%
	University	1%
	Car sales	1%
	Communication	1%
	Library	1%
	Newspaper	1%
Respondents	Male	52%
	Female	48%
Age	Mean	23 yrs
	Min	20 yrs
	Max	31 yrs

to contact the company? How does the service encounter influence your perceptions of the service and company? What should the company have done differently?

FINDINGS

Table 2 provides descriptive details of the service encounters. The responses were mostly quick, as 65% of the responses came within one day. In fact, 84% of all responses came within two days. However, there were also companies that never replied to a contact from a customer, and complaints and requests accounted for the largest number of the nonresponses. Differences were dependent on the method of interaction. Online forms are also more often left without a response compared to an e-mail contact. Response time and personalization are often perceived as satisfying as is depicted in Table 2. However, there are also contacts that are answered slowly or they are perceived as mass-produced, as indicated by the high percentages of long response time and low level of personalization.

Figure 2 depicts service encounter value of all contacts. The service encounter value for the interactions was mostly high as 35% of all responses were perceived as being personalized and quick. However, 21% of all responses were perceived as being unpersonalized and slow. The spread of the contacts throughout the matrix indicates the relevance of both of dimensions in illustrating service encounter value. For example, 6% of the contacts are responded to quickly but they lack personalization and, vice versa, 4% of

Table 2. Descriptive summary of the service encounters

Description of the service encounters		n=217
Type of interaction	Questions	36%
	Complaints	32%
	Requests	32%
Type of contact method	E-mail	51%
	Online form	46%
	Other	3%
Response time	In one day	65%
	Between 1 and 2 days	19%
	Within one week	12%
	Within two weeks	3,5%
No response	Of total 217 contacts	19%
	Of all no-responses:	
	Interaction types	
	Complaints	44%
	Questions	42%
	Requests	15%
	Contact methods	
	Online forms	63%
	E-mails	34%
Perceptions of responsiveness	Long	29%
	Reasonable	16%
	Quick	56%
Perceptions of personalisation	Low	28%
	Neutral	26%
	High	46%

Figure 2. Service encounter value of all contacts

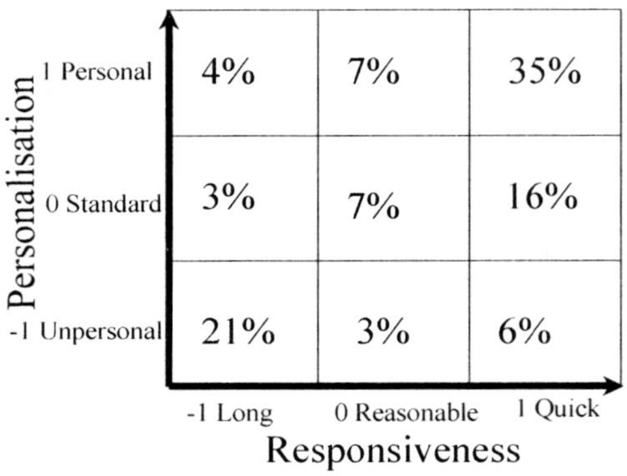

the contacts are highly personalized but they are delivered slowly.

There were differences in the service encounter value depending on the type of contact (Figure 3). The perceived service encounter value was high for questions, and they were perceived as mostly personalized and prompt, that is, located mostly in the upper right-hand corner. The responses to complaints were the most diverse of the three types and the responses are more evenly spread over the matrix compared to the other two contact types. This indicates the varying quality of responses to complaints. Requests were more frequently perceived as positive or negative, that is, personalised and quick or alternatively long response time/no response and unpersonalized, as indicated by the high percentages in the upper right-hand corner and lower left-hand corner.

Differences in the service encounter value also depended on the contact method used as illustrated in Figure 4. Contacts based on online forms were of lower value as a large number of the responses were

Figure 3. Service encounter value of questions, complaints and requests

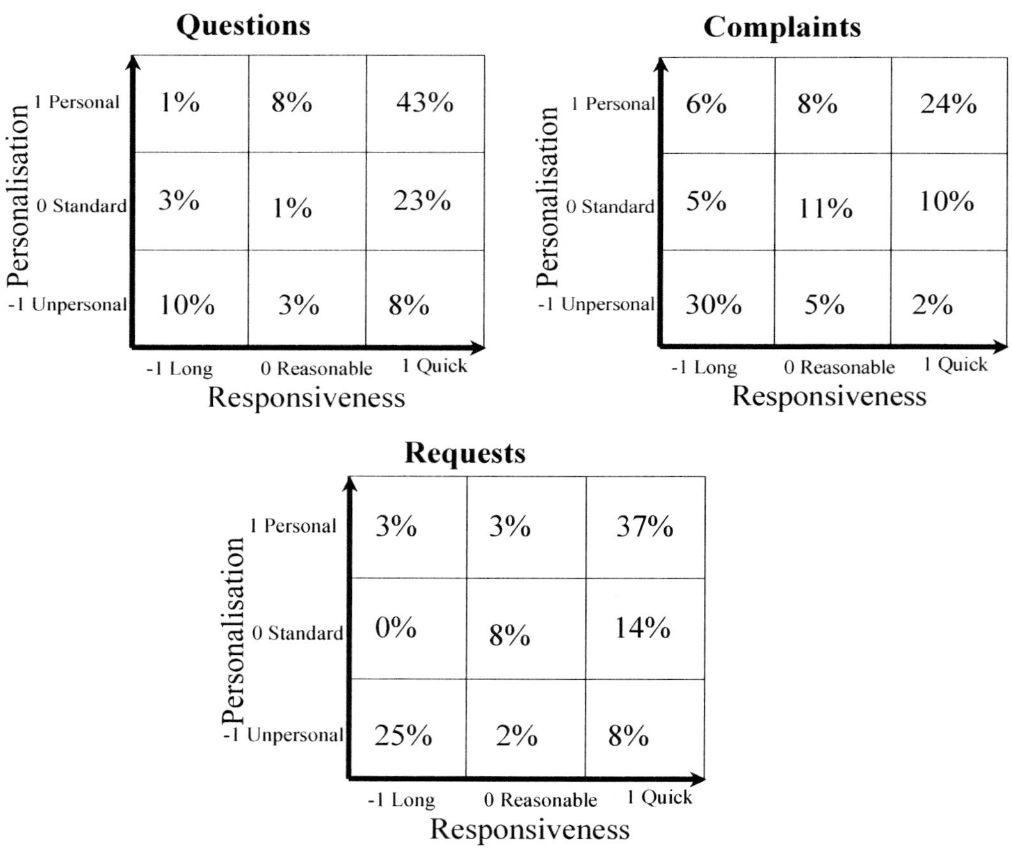

Figure 4. Service encounter value of online forms and e-mail

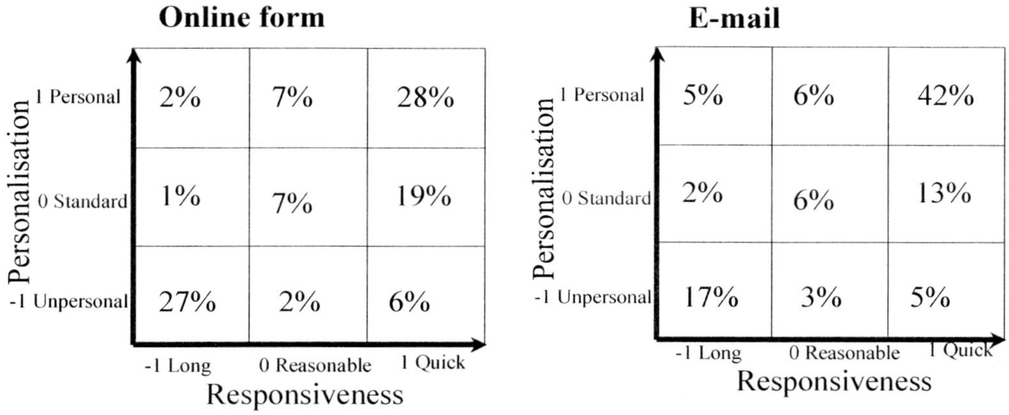

quick but standardized. E-mail interactions represented higher value and the responses were mostly quick and personal.

The qualitative findings showed issues that increased the level of perceived personalization, such as greeting, name, contact details, content, and style. The tone of voice and level of empathy were important. Also the importance of trying to understand where the consumer is coming from and anticipating customer needs were mentioned.

CONCLUSION AND IMPLICATIONS

The article describes how well companies respond to customers' contacts by technology interface such as email or online form. It introduces the term service encounter value and proposes a conceptual framework for monitoring service encounter value. The starting point is the active customer that takes the initiative to a service encounter and how well the service encounter performs according to the customer, that is, service encounter value. By drawing upon empirical data from contacts with different service companies, the article shows the value of remote service encounters. Different types of interaction as well as interaction methods were compared.

The findings indicate that many contacts are responded to promptly and satisfyingly. However, there are also significant differences in the value of the service encounter based on the type of interaction, that is, questions, complaints, and requests, and the interaction method, that is, online form and e-mail.

Theoretically, the article contributes to the research on service encounters or moments of truths, by proposing a conceptual model for capturing customers' perceptions of service encounter value based on the dimensions of responsiveness and personalization. It also contributes to the e-service research by exploring how customers perceive digital channels as a tool for contacting companies.

The practical contribution involves the empirical comparison of customers' perceptions of digital service encounters. More specifically, by empirically illustrating how well digital channels are used within companies, the article indicates some of the challenges involved in using digital channels in the interaction with customers. The study suggests that digital service encounters influence customers' images of the company. The proposed conceptual framework suggests that it is not only a question of responding quickly; rather, the value of the interaction is also based on the quality of the response. The empirical findings on how well companies respond to customers' contacts and initiations to interactions through digital channels suggest that companies need to improve their digital service encounters. There are differences depending on the contact channels used and the reason for contacting the company.

The practical implications of the study relate to the differences in the perceptions of service encounters. A managerial challenge is how to influence the value that is created in the service encounter. The findings suggest that companies are good at responding to simple questions by customers, however, the more complex the reason for contacting, for example, requests or complaints, the poorer the companies are at responding. Drawing on the findings, it is argued that companies need to improve their quality of responses especially for complaints and more complex requests. Complaints and requests usually require more personalization than a simple question. Also, a challenge is to improve the performance of the service encounter of less typical situations, that is, where either personalization or responsiveness is satisfactory but the other dimension works poorly. These situations can easily be neglected if the service encounter is not evaluated jointly with both personalization and responsiveness.

Digital service encounters are especially challenging when considering that remote encounters do not necessarily offer the opportunity for the service personnel to ask further information. The findings indicated that online forms are not as efficient a contact method as e-mail. This can be explained by the fact that further contacts are often needed after the initial contact initiative made by the customer, and replying to an e-mail is easier than responding to online forms.

The limitations of the study are related to the empirical implementation. A student sample is clearly a limitation; however, in this exploratory study where the main purpose is to illustrate the potential of the conceptual framework, a large and generalizable sample is not a prerequisite for valuable insight. Further studies need to address the model structure and test the dimension scales, and create a reliable scale for service encounter value. Moreover, the study can be extended in other respects. In this study only few digital service encounters were used. However, as customer can contact a company through many different interfaces, a more structured comparison of the methods could be done in further studies. This would also involve mobile telephone services as well as traditional human interaction. Furthermore, a more structured and descriptive comparison of differences between industries is needed. The focus in this article was on how service encounter value in general influences companies. In contrast to company/industry differences, the focus has been on showing the differences between contact types and interaction methods. In order to keep this focus, we have not specifically included differences based on firm size, industry, technology usage and so forth. These topics evidently represent an agenda for further research.

REFERENCES

Ansari, A., & Mela, C.F. (2003). E-customization. *Journal of Marketing Research*, *40*, 131-145.

Chandon, J.-L., Leo, P.-Y., & Philippe, J. (1997). Service encounter dimensions-a dyadic perspective: Measuring the dimensions of service encounters as perceived by customers and personnel. *International Journal of Service Industry Management*, *8*(1), 65-86.

Grant, I., & O'Donohoe, S. (2007). Why young consumers are not open to mobile marketing communication. *International Journal of Advertising*, *26*(2), 223-246.

Grönroos, C. (1987). Developing the service offering–a source of competitive advantage. In C.F. Surprenant (Ed.), *Add value to your service*. Chicago, IL: American Marketing Association.

Jayawardhena, C. (2007). Outcomes of service encounter quality in a business-to-business context. *Industrial Marketing Management, 36*(5), 575-588.

Lewis, B.R., & Entwistle, T.W. (1990). Managing the service encounter: A focus on the employee. *International Journal of Service Industry Management, 1*(3), 41-52.

Lovelock, C. (1983, Summer). Classifying services to gain strategic marketing insights. *Journal of Marketing, 47*, 9-20.

Lovelock, C., & Wirtz, J. (2007). *Services marketing: People, technology, strategy* (6th ed.). Upper Saddle River, NJ: Pearson Prentice Hall.

Meuter, M.L., Ostrom, A.L., Roundtree, R., & Bitner, M. (2000). Self-service technologies: Understanding customer satisfaction with technology-based service encounters. *Journal of Marketing, 64*(3), 50-64.

Miceli, G., Ricotta, F., & Costabile, M. (2007). Customizing customization: A conceptual framework for interactive personalization. *Journal of Interactive Marketing, 21*(2), 6-25.

Normann, R. (1991). *Service management strategy and leadership in service businesses* (2nd ed.). Chichester: John Wiley and Sons.

Parasuraman, A., Zeithaml, V., & Berry, L. (1985, Fall). A conceptual model of service quality and its implications for future research, *Journal of Marketing, 49*, 41-50.

Pieterson, W., Ebbersa, W., & van Dijk, J. (2006). Personalization in the public sector: An inventory of organizational and user obstacles towards personalization of electronic services in the public sector. *Government Information Quarterly, 24*(1), 148-164.

Searby, S. (2003). Personalisation—An overview of its use and potential. *BT Technology Journal, 21*(1), 13-19.

Shostack, G.L. (1985). Planning the service encounter. In J.A. Czepiel, M.R. Solomon, & C.F. Surprenant (Eds.), *The service encounter* (pp. 243-254). Lexington: Lexington Books.

Vesanen, J., & Raulas, M. (2006). Building bridges for personalization: A process model for marketing. *Journal of Interactive Marketing, 20*(1), 5-20.

Walker, J.L. (1995). Service encounter satisfaction: Conceptualized. *Journal of Services Marketing, 9*(1), 5-14.

Wind, J., & Rangaswamy, A. (2001). Customerization: The next revolution in mass customization. *Journal of Interactive Marketing, 15*(1), 13-32.

Zeithaml, V., Bitner, M.J., & Gremler, D. (2006). *Services marketing: Integrating customer focus across the firm* (4th ed.). Singapore: McGraw-Hill.

This work was previously published in Journal of Electronic Commerce in Organizations, Vol. 6, Issue 2, edited by M. Khosrow-Pour, pp. 1-10, copyright 2008 by IGI Publishing (an imprint of IGI Global).

Chapter 6
Contract Negotiation in E-Marketplaces: A Model Based on Dependency Relations

Larbi Esmahi
Athabasca University, Canada

ABSTRACT

E-transactions via shopping agents constitute a promising opportunity in the e-markets. In this article we will discuss the problem of contract negotiation in e-marketplaces. We succinctly present an overview of protocols commonly used to implement negotiation in e-markets. An analysis of the interaction process within e-markets according to different situation of individual and joint profit/cost is presented. We also present a case study of a marketplace for e-services using dependency relations within the negotiation process. The experimental results of this negotiation model show that a combination of utility functions and dependency relations increase the number of contacts and reduce the differences between agents' individual profit. Finally, we conclude the article with the introduction of some potential research problems related to e-markets, which will be explored within future extensions of this work.

INTRODUCTION

The advancement in distributed and intelligent computing has facilitated the use of software agents for implementing e-services. Most electronic market places offer to their customer's virtual agents that can do their bidding (i.e., eBay, onSale). E-transactions via shopping agents constitute a promising opportunity in the e-markets (Jennings, Faratin, Lomuscio, Parsons, Wooldridge, & Sierra, 2001; Chen, Vahidov, & Kersten, 2004). In fact, the software agents can be used in different stages of the e-business process: (i) attracting the customer, (ii) knowing how they buy, (iii) making transactions, (iv) perfecting orders, (v) giving effective customer service, (vi) offering customers recourse for problems such as breakage or returns, and (vii) providing a rapid conclusion such as electronic payment. In the distributed e-market paradigm, these functions are abstracted via agents representing both contractual parts. It becomes then critical to study and determine what kind of information and what kinds of bargain policies need to be used both by agents and by the market place. Many researchers in intelligent agents' domain have focused on the design of market architectures for electronic commerce (Fikes, Engelmore, Farquhar, & Pratt, 1995; Zwass, 1999; Schoop & Quix, 2001; Guan, 2002; Koning, 2002; Chiu, Cheung, & Till, 2003; Lamparter, Ankolekar, Studer, & Oberle, 2006; Ng, Chiu, & Hung, 2007), and on protocols governing the interaction of rational agents engaged in such transactions (Hogg & Jennings, 1997; Cheung, 2003; Kersten, 2005; Chiu, Cheung, Hung, Chiu, & Chung, 2005; Cheng, Hung, & Chiu, 2007; Neumann, Schnizler, Weber, & Weinhardt, 2007). New Web technologies such as ontologies and semantic Web are also becoming increasingly important components of e-marketplaces. In fact, ontologies provide machine-understandable semantics of data, which facilitate negotiation and automatic configuration of products and services. Many recent negotiation frameworks (Chiu, Poon, Lam, Tse, Sui, & Poon, 2005; Chiu, Cheung, Hung, & Leung, 2005; Chiu et al., 2005; Lee, Chiu, & Hung, 2006; Cheng et al., 2007) use ontology to facilitate matchmaking, recommendation, and negotiation.

While providing support for direct agent interaction, most of existing architectures for multi-agents virtual markets usually lack explicit facilities for handling negotiation protocols, since they do not provide such protocols as an integrated part of the framework.

In this article we will discuss the problem of contract negotiation in e-marketplaces. In the next section, we will present related models commonly used to implement negotiation in e-markets, game theory models, auction models, and contract-net protocols. The third section continues with interaction analysis by presenting the e-market bargaining process and identifying relevant types of dependency relationships between agents. It evaluates goal satisfaction under weak or strong dependency as well as the associated costs. The fourth section presents a case study of a marketplace for e-services where agents collaborate for the elaboration and provisioning of integrated services. We define the bargaining process that uses a negotiation protocol based on dependencies and a negotiation strategy based on risk evaluation. The fifth section presents some experimental analysis of the negotiation model implemented. The conclusion presented in the sixth section summarizes the article and paves the way for further research concerning the truth in the negotiation strategy and the use of temporal aspects on commitments and executions of contracts.

CONTRACT NEGOTIATION PROTOCOLS FOR E-MARKET PLACES

The interaction between agents inside the marketplace is managed by a negotiation protocol. In fact, the negotiation protocol defines a set of public rules that allow agents to set up transaction contracts or co-operation agreements. Previous work and significant achievements are reported on various related fields of research and concrete solutions. Most of the Internet based market-places use auction protocols, especially the English auction.

Hereafter, we present and evaluate some negotiation models developed either in some research works or implemented in some practical systems: game theory, auction models, contract-net protocols, and ontology-based negotiation.

Game Theory Models

Game theory models address many aspects of the agents' interaction: contract elaboration, profit repartition, and conflict resolution (Roth, 2000). Many negotiation models have been proposed in this topic (Rosenschein, & Genesereth, 1985; Genesereth, & Ginsberg, 1986; Zlotkin & Rosenshein, 1991; Ephrati & Rosenshein, 1992; Khedro & Genesereth, 1994; Kraus, Wilkenfeld, & Zlotkin, 1995). These models have some desirable properties, such as insuring the negotiation convergence, the Nash-equilibrium, and the pareto-optimality. The main representative works in this domain are those presented by Zlotkin and Rosenschein (Rosenschein & Genesereth, 1985; Zlotkin & Rosenshein, 1991; Rosenshein& Zlotkin, 1994). The authors propose a formal model that allows agents to select the pareto-optimal solution that maximize their utilities. The agents communicate their offers explicitly by exchanging messages and may accept concessions that allow them to elaborate contracts that satisfy their goals. A contract may concern task repartition (task oriented domains), utility value repartition (worth oriented domains), or decision making on the next state of the environment (state oriented domains). Different types of contracts have been studied: pure contracts, where the agent's role in the joint plan is fixed and mixed contracts, where the agent's role depends on a probability.

If we consider a negotiation between two agents, A1 and A2, the authors propose a protocol that can be summarized as follow:

1. At each step $t \geq 0$, both agents propose their deals $\delta 1(t)$ and $\delta 2(t)$ such that those deals satisfy two conditions:
   - The deals must be individually rational to their respective agents (• i.e., for every (Ai, δi), the utility $Ui(\delta i) \geq 0$); and
   - For each Ai member of {A1, A2}, • at each step $t > 0$ we should have $Ui(\delta i(t)) \leq Ui(\delta i(t-1))$.
2. The negotiation finishes at a step t when one of the two situations happens:
   - The agents agree on a deal. There exist $i \neq j$ members of {A1, A2}, such that $Uj(\delta i(t)) \geq Uj(\delta j(t))$.
   - The agents run on a conflict situation. For every Ai member of {A1, A2}, $Ui(\delta i(t)) = Ui(\delta i(t-1))$ (i.e., no more concession is possible for both agents).

The advantage of the proposed protocols in game theory consists of their suitability for rational cooperating agents that work for maximizing their profits. However, the main drawbacks of those models consist of: first, their inability to take into consideration the history of the negotiation process, and second, the fact that each step is processed as a stand-alone step. Furthermore, the agents are supposed to have complete information on their partners, especially by knowing all their matrix of profits. The agents are also supposed of being self-sufficient, while the complementarity and dependency between agents is ignored.

Auction Models

Auction theory analyzes the protocols and strategies used by agents during an auction sale. Many protocols have been proposed in auction theory (Klemperer, 1999, 2002; Rasmusen, 2001; Neumann et al., 2007):

- *English auction*: In the English auction, the bidding process is public, so each bidder has complete information about the auction. At any time, each agent is free to raise his bid. When no bidder is willing to raise anymore, the auction ends, and the highest bidder wins the item at the price of his bid. The agent's strategy consists of a series of bids, where the bidding value is function of his private value, his prior estimates of other bidder's valuations, and the past bids of others. An agent's dominant strategy is to always bid a small amount greater than the current highest bid, and stop when his maximum value is reached.
- *Sealed bid auction*: In the sealed-bid auction, each bidder submits one bid without know-

ing the others' bids. The highest bidder wins the item and pays the amount of his bid. The agent's strategy consists of determining his bid as a function of his private value and prior beliefs of others' valuations. In general, there is no dominant strategy for acting in this auction.

- *Dutch auction*: In the Dutch auction, the seller or the auction manager continuously lowers the price until one of the bidders takes the item at the current price. The Dutch auction is strategically equivalent to the sealed bid auction, because in both games, an agent's bid matters only if it is the highest and no relevant information is revealed during the auction process.
- *Vickrey auction*: The Vickrey auction is similar to the sealed bid auction with some details in exception. In fact, each bidder submits one bid without knowing the others' bids and the highest bidder wins, but he pays only the price of the second highest bid. The agent's strategy consists of determining his bid as a function of his private value and prior beliefs of others' valuations. The dominant strategy in Vickrey auctions is to bid one's true valuation. If an agent bids more than that and the increment hit the difference between winning or not, he will end up with a loss if he wins. If he bids less, there is a smaller chance of winning, but the winning price is unaffected. The dominant strategy result of Vickrey auctions means that an agent is better off bidding truthfully no matter what the other bidders are like: what are their capabilities, operating environments, bidding plans, and so forth. This has two desirable sides, namely (i) the agents reveal their preferences truthfully, which allows globally efficient decisions to be made and (ii) the agents need not waste effort in counter-speculating other agents, because they do not matter in making the bidding decision.

The auction models have two main drawbacks: they can be applied only for marketing domain and they are vulnerable to collusion. In the context of their vulnerability to collusion, the English auction presents the worst example. In fact, in the English auction the bidding process is public and the agents can verify if collusion is respected or not and in case of the collusion broke the agent can use all his purchase power without loosing anything. In the other protocols (sealed bid auction, Dutch auction, and Vickrey auction), the agents do not receive any information on the auction process and collusion formation is not so evident.

Contract-Net Protocol

In the contract-net protocol (CNP) (Smith & Davis, 1981; Smith, 1988), a contract is an explicit agreement between an agent who announces its tasks (the manager) and an agent who proposes to achieve those tasks (the contractor). The manager is responsible for tasks management and results processing, while the contractor is responsible for tasks processing and results transmission. The protocol proposes a set of performatives that the agents can use either for contract negotiation or contract execution. The following diagram summarizes the process for elaborating contracts.

In this first version of the CNP as presented, there are no mechanisms for evaluating offers. This means that the protocol can be used only in contexts with cooperative agents and modular tasks.

Many extensions have been proposed for the CNP in the last years. One of the most important works done in this topic is the extension proposed by Sandholm (1993, 1996, 2000), which is based on marginal costs. In fact, the agents process locally their marginal costs for achieving tasks and use those costs either for announcing a task, bidding, or awarding a contract. Partners' selection is based on profit evaluation using the marginal costs. The integration of utility has then extended

Figure 1. Contract-net protocol

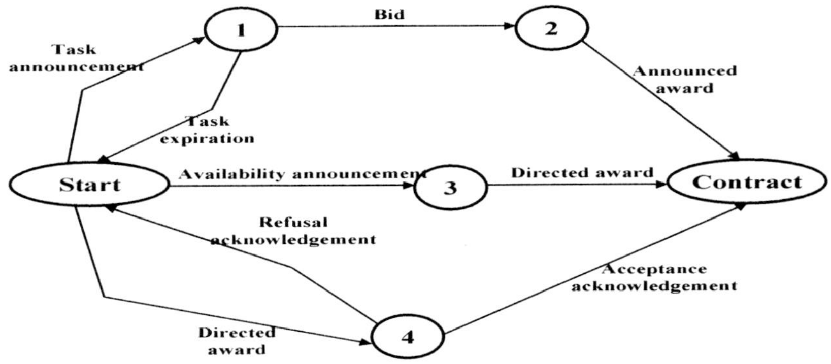

the protocol's domain of application to competitive and self-interest agents.

The CNP is one of the most complete models for cooperating agents. The notion of profit and utility recently added to it resolves the problem of motivation for competitive agents. However, the protocol still has some drawbacks, such as considering the agents as passive actors because no persuasion mechanisms can be used to convince partners for accepting an agreement. Also, dependency relations between agents are not considered.

Ontology-Based Negotiation

Semantic Web and ontology are two promising technologies for the development of e-market places. In fact, the semantic Web helps in making e-commerce interactions more flexible and automated, while ontology helps in addressing the problems of matchmaking, recommendation, and negotiation. Chiu et al. (Chiu et al., 2003, 2005, Cheung et al., 2006; Cheng et al., 2007) propose an ontology based model for e-negotiation using a three phase process: pre-negotiation phase, negotiation phase, and the post-settlement. During the pre-negotiation phase, issues are mapped into ontology concepts that are used to generate the graph of precedence dependencies. Based on the precedence graph, an efficient negotiation plan is determined. During the negotiation phase, the negotiating parties use a process similar to the contact net based on offers and counter-offers. The parties make or evaluate offers and counter-offers based on the plan elaborated during the pre-negotiation phase. An agreement is successfully created when all issues have been resolved. The post-settlement phase is for refining the policy or formalizing the agreement contract. In fact, the proposed negotiation model use an enhanced contact-net protocol but take profit of ontology to facilitate the matchmaking, recommendation, and generation of agreements using dependencies.

Ontology is also used by Neumann and Weinhardt (Neumann et al., 2002; Neumann & Weinhardt, 2002; Lamparter et al., 2006) for their e-negotiation framework. The proposed framework uses a knowledge-based system architecture based on three layers: domain layer, task layer, and inference layer. Since the proposed framework is meant to be domain independent, the domain layer represents all the static domain knowledge necessary to solve the problem. The domain knowledge is represented using ontology as a set of concepts and their relations. The task layer defines the ordering of the basic inference actions that control the flow of the inference layer. Two mean methods are used at this level, "heuristic classification" and "propose and revise." The inference layer defines the reasoning behaviour, which are the inference actions that constitute the problem solving strategy. An example of stock e-market place is presented to illustrate the use of this framework (Neumann et al., 2002).

INTERACTION PROCESS ANALYSIS IN E-MARKETS

The interaction in e-markets may be analysed according to different views. Since our concern is about contract negotiation, the main interaction to analyse in this context is the bargaining process inside the market place. The negotiation models presented in the previous section, deal with the interaction problem from the same point of view, which is the individual cost/profit for the agent. In our research we are studying other factors that may influence the interaction between agents, such as interdependency situations, agents' mental state[1], and trust factors. In this section, we will analyse the agents' interaction with respect to different dependency relations that may exist between agents.

Let us consider an environment that models the interaction between two agents, A1 and A2. Let Ci be the cost of agent Ai for the realization of his goal Gi (buy or sell a service/product), and Pi be its maximal acceptable cost for the realization of the same goal. If we assume that it exist a joint plan for A1 and A2 that allow both agents to satisfy both goals G1 and G2. Let us denote by T the global cost for the joint plan and Ti the cost assumed by agent Ai in the global cost (T =T1 + T2). Also, let Mri represents the cost for the minimal role that agent Ai can have in a joint plan satisfying both goals. The following dependency relations are thus defined:

- *Independency relation*: Two agents are said to be independent if each agent has an individual plan that satisfy his goal (i.e., Ci < Pi, i=1, 2) or there exists no joint plan that would allow the agents to improve their profits (i.e., Mri > Pi, i=1, 2). Figure 2 and Figure 3 illustrate these two situations.
- *Weak mutual dependency*: In such a situation, each agent has an individual plan that allows the satisfaction of his goal (i.e., Ci < Pi, i=1,2). However, there exist a joint plan that allows the two agents to enhance their profits (i.e., Ti < Ci, i=1,2) so that the two agents prefer to use the joint plan rather than using their individual plan (Figure 4).
- *Strong mutual dependency*: In such a situation, none of the two agents can satisfy his goal individually (i.e., Ci > Pi, i=1,2). However, there exists a joint plan that allows the two agents to satisfy their goals. Two situations may be distinguished: co-operative situation and compromise situation.

In the cooperative situation, the joint plan is acceptable for the two agents and gives them the best profit (i.e., Tci < Pi, i=1,2) (Figure 5).

In the compromise situation, the joint plan allows the agents to reduce the costs of satisfying their goals but the costs continue to be greater than the maximum accepted (Figure 6):

- *Unilateral dependency*: In this situation, one of the agents (A1) can satisfy his goal individually (i.e., C1 < P1), but the other agent (A2) does not have an individual plan that satisfies his goal (i.e., C1 > P1). However,

Figure 2. Independency relation (T non existent)

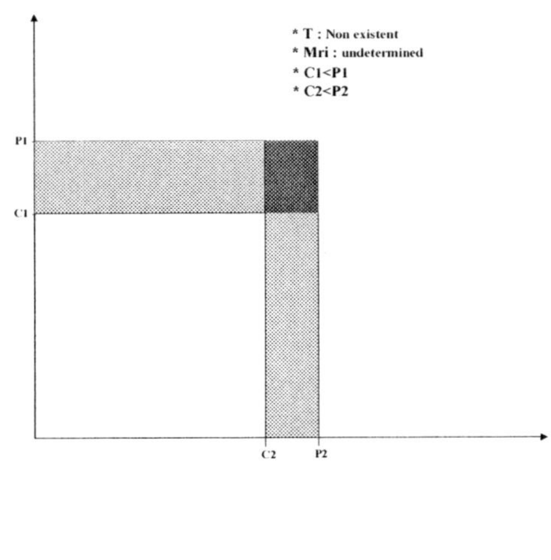

Figure 3. Independency relation (T not acceptable)

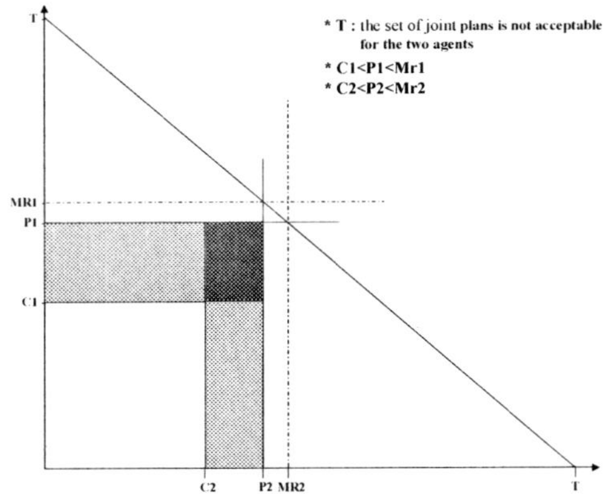

Figure 4. Weak mutual dependency

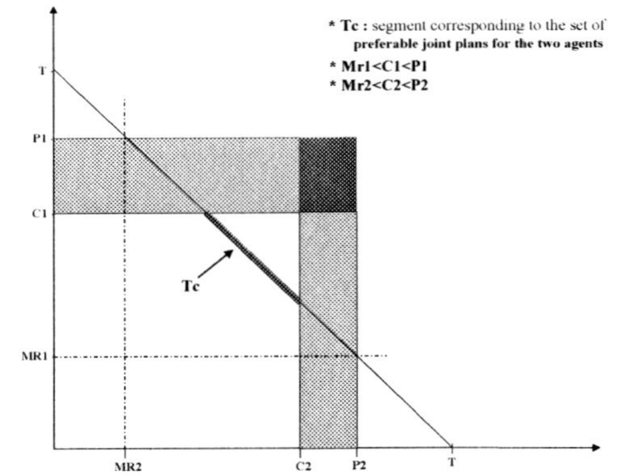

Figure 5. Strong mutual dependency (cooperative situation)

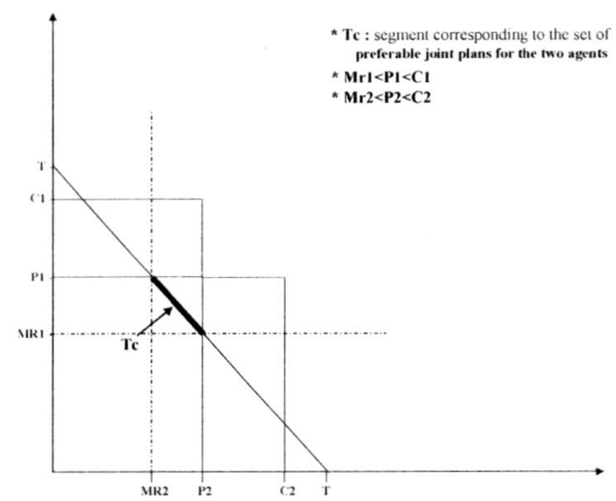

Figure 6. Strong mutual dependency (compromise situation)

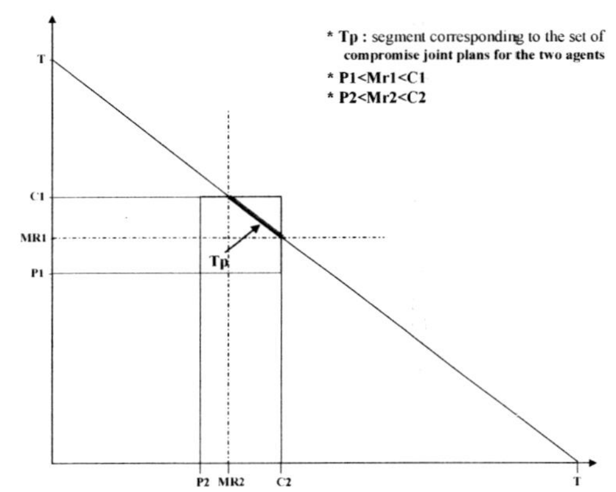

Figure 7. Unilateral dependency (cooperative situation)

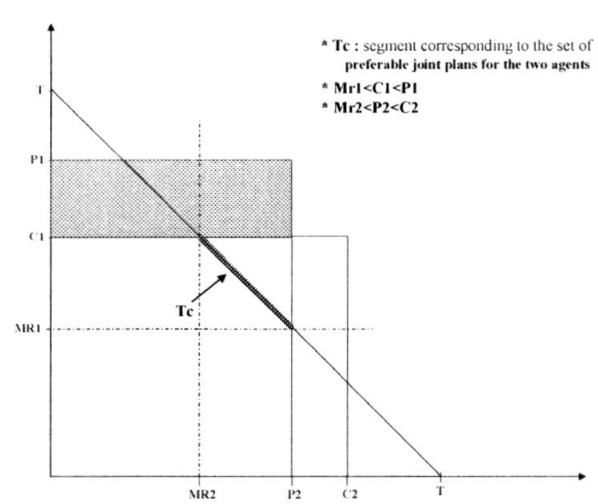

Figure 8. Unilateral dependency (compromise situation)

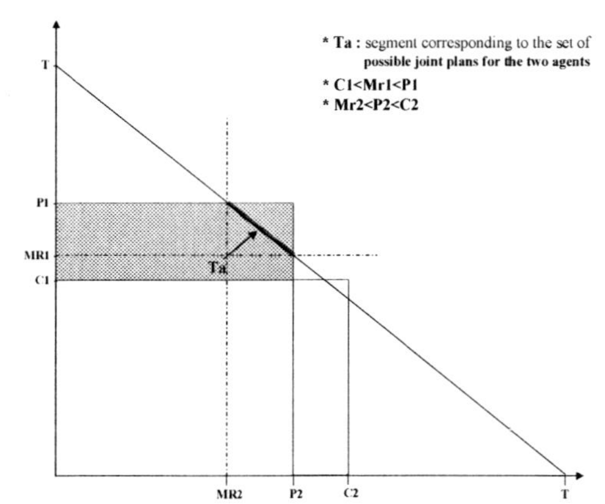

there is a joint plan that allows the realization of the two goals. So, agent A2 depends on the willingness of A1 to cooperate in a joint plan. Here, we can also distinguish two situations: co-operative situation and compromise situation.

In cooperative situations, the joint plan not only satisfies the two goals (i.e., $Ti < Pi$, i=1,2) but it also enhances the profit of A1 (i.e., $T1 < C1$). In this case, the two agents prefer to cooperate in a joint plan (Figure 7).

In compromise situations (Figure 8), the joint plan is acceptable for the two agents and satisfies both goals (i.e., $Ti < Pi$, i=1,2). However, agent A1 prefers its individual plan to the joint plan (i.e., $P1 > T1 > C1$). In this case, the joint plan cannot be used unless agent A1 accepts a reduction on this individual profit (i.e., $Mr1 > C1$).

TOWARD A NEGOTIATION PROCESS BASED ON DEPENDENCY RELATIONS

Two main components constitute the core of the negotiation process: the negotiation protocol and the negotiation strategy. In fact, the negotiation protocol defines the set of public rules that allow agents to set up transaction contracts or co-operation agreements, while the negotiation strategy defines the agent's behaviour during the negotiation process. In other terms, the negotiation protocol defines the interaction rules but the agent's offers and counter-offers are the result of the negotiation strategy.

Before starting the description of the proposed negotiation process in the market place, let us first describe the context of this work. Our scenario consists of building a market place infrastructure using mobile agents for providing electronic services. The marketplace plays a central role by providing an infrastructure for exchanging offers and requests. Potential business partners represented by their software agents can exhibit their services and available resources, search for offers from service providers, or send their own offers. Four type of agents act in the e-marketplace: the market place manager, the service providers, the virtual enterprise builders, and the service consumers.

Except for the market place manager whose role is specific (manage the marketplace services: registry, security, conflict mediation), the visiting agents may play at the same time many roles in the market place (i.e., an agent may be a service provider and a virtual enterprise builder at the same time). To clarify the idea about the virtual enterprise lets give this definition (Esmahi, Dini, & Bernard, 2000): A virtual enterprise (VE) may be defined as a temporary federation of autonomous service providers which acts as one unified entity in order to provide a multi-parties service (i.e., a complete travel service including transports, accommodations, car rental, entertainment, etc.). An agent who sees the opportunity of offering a multi-parties service creates a virtual enterprise. This agent contacts different service providers, negotiates a cooperation agreement, and becomes the VE builder. The VE builder monitors the process inside the virtual enterprise, especially negotiating contracts with clients and ensures the execution of those contracts by all the VE members. Since the VE is a federation of autonomous partners, the VE builder has to support distributed management.

The service providers are the owners of basic services. They use the market place to exhibit their services and announce their available resources and wait for a prospective buyer agent to contact them requesting for an offer. The service consumers are the end users of different services (basic services or multi-parties services). Those agents search the market place to find appropriate service providers and use the market facilities to manage the contract negotiation and execution.

Based on the VE model, our concern is about building a complex system for e-services and business process management, where agents need to negotiate in order to achieve their goals[2]. The agents' objectives do not consist only of minimizing cost or maximizing profit, but also the number of cooperative agreements built is of great importance since all contracts on multi-parties services are interdependent (i.e., a two parties contract cannot be executed if the VE builder does not reach an agreement with all participating agents to the final integrated service).

Consequently in this context, our system is neither of pure distributed problem solvers[3] type (Ephrati & Rosenshein, 1992; Khedro & Genesereth, 1994), nor of the pure autonomous agents[4] type (Rosenshein & Zlotkin, 1994; Kraus et al., 1995). Rather we are concerned with a hybrid case in which we wish to exploit the conceptual power of autonomous agents (as in the multi-agents system model), but also wish to ensure the overall system performs in a coherent manner (as in distributed problem solver model). Given these contradictory demands, we believe the best manner of building socially coherent multi-agents systems is to endow the individual autonomous agents with greater awareness about their interdependencies. This awareness enables the agents to explicitly propose offers and design contracts that exploit the dependency relations for satisfying a large number of goals. For that we are proposing a negotiation model that uses dependency relations in the negotiation process to improve the balance between the need of the individual agents and the needs of the overall system.

A Negotiation Protocol Based on Dependency Relations

Before describing the contract generation process based on dependencies, we present some useful definitions.

- *Strong dependency (S-Dep (A1, A2, I)):* A strong dependency occurs between a buyer, agent A1 and a seller, agent A2 on an item I, if agent A1 has requested to buy item I, and agent A2 is the unique agent in the market-place that has advertised himself to sell item I.

An example of that would be a company selling cell phone parts/services for a specific brand used only by the company commercializing that brand.

- *Weak dependency (W-Dep (A1, A2, I)):* A weak dependency occurs between a buyer, agent A1 and a seller, agent A2 on an item I, if agent A1 has requested to buy item I, and the agent A2 is one of the group of agents in the market-place that have advertised themselves to sell item I.

An example of that would be an agent buying a telephone service with many companies offering the service.

- *Multi-dependency (X-Dep (A1, A2, I1, I2)):* A buyer, agent A1 is said to be in multi-dependency towards a seller, agent A2 on the items I1and I2, If one of the following situations occurs:
    - S-Dep (A1, A2, I1) and S-Dep (A1, A2, I2)
    - S-Dep (A1, A2, I1) and W-Dep (A1, A2, I2)
    - W-Dep (A1, A2, I1) and W-Dep (A1, A2, I2)

An example of that would be a company commercializing a cell phone service with Internet connection, which needs to buy both the bandwidth and the internet service from one of the local telecom companies

- *Mutual dependency (M-Dep (A1, A2, I1, I2)):* Two agents are said to be in mutual dependency. If one of the following situations occurs:

- S-Dep (A1, A2, I1) and S-Dep (A2, A1, I2)
- S-Dep (A1, A2, I1) and W-Dep (A2, A1, I2)
- W-Dep (A1, A2, I1) and S-Dep (A2, A1, I2)
- W-Dep (A1, A2, I1) and W-Dep (A2, A1, I2)

An example of that would be two telecom companies having a network that covers complementary parts of the country. The two companies can serve the whole country only by buying services from each other.

For our negotiation model, we propose to use inter-dependencies between agents for enhancing the performance of the system. So based on the nature of the dependency relationships, different type of contacts can be elaborated:

- *Exchange contract*: Sometimes it is interesting to exchange some items between agents to enhance the global value of the contract, especially when the two agents have mutual dependency. In an exchange contract, the first agent subcontracts some items to the second agent, while the second agent, in turn, subcontracts some other items in compensation to the first agent. The value of exchange must be the same when measured in terms of loss or gain.

The exchange contract becomes more interesting with agents that have mutual dependencies. A general form of the exchange contract is when more than two items are exchanged between agents.

- *Grouping contract*: This type of contract is built by grouping a set of items in the same contract, in such a manner that the global cost of the contract becomes profitable for the two agents, even if the individual cost of some items is greater than the accepted price. The grouping contract becomes more interesting with agents that have multi-dependency relation. The interest for this type of contract grows if we remember that in real trading, the price of a group of items is always lower than the sum of their individual prices.
- *Multi-agent contract*: In this type of contract, a group of agents take a commitment to buy/sell a set of items. The contract is built in such a manner that is profitable for all participating agents. A multi-agent contract may include exchanging and grouping of items.

The marketplace manager acts as a facilitator that helps agents elaborating contracts. For its mediation process, the manager uses the classes of dependencies and contracts defined in the previous section. The mediation process uses first mutual dependencies between agents, in order to build a solution using an exchange contract. If no mutual dependencies are found, the manager then searches for multi-dependencies in order to build a grouping contract. In the case where neither mutual dependencies nor multi-dependencies can be found, the manager uses the graph of dependencies to construct multi-agent contract.

The general algorithm for elaborating contracts based on dependencies can be summarized as follows:

1. Construct and propose a exchange contract using mutual dependencies.
   - 1.2 search the graph for all mutual dependencies of the buyer towards the seller
   - 1.3 sort the dependencies based on their strength
   - 1.4 construct an exchange contract using the mutual dependency
   - 1.5 propose the exchange contract
   - 1.6 if the contract is accepted, then success; stop

   1.7 else select the next mutual dependency in the list and go to step 1.4
   1.8 if the set of mutual dependencies is exhausted without success then go to step 2
2. Construct a grouping contract using multi-dependencies.
   2.1 search the graph for all multi-dependencies of the seller towards the buyer
   2.2 sort the dependencies based on their strength
   2.3 select the first dependency in the list
   2.4 construct a grouping contract based on the multi-dependency relation
   2.5 propose the grouping contract
   2.6 if the contract is accepted, then success; stop
   2.7 else select the next multi-dependency in the list and go to step 2.4
   2.8 if the set of multi-dependencies is exhausted without success, then go to step 3
3. Construct a multi-agent contract using other agent's dependencies.
   3.1 search the graph for a balanced cycle of dependencies that closes the path[5] between the buyer and the seller
   3.2 construct a multi-agent contract using the cycle of dependencies
   3.3 propose the multi-agent contract to the group of agents
   3.4 if all agents agree on the contract, then success; stop
   3.5 else go to step 3.1
   3.6 if there is no cycle of dependencies between the buyer and the seller then go to step 4
4. Use other creative process to generate a solution (e.g., case-based reasoning, etc.).

A Negotiation Strategy Based on Risk Evaluation

Our negotiation protocol can be divided in two steps. First the potential partners exchange offers and counter-offers to negotiate a contract and if this first interaction is not successful then the marketplace manager tries to facilitate the contract elaboration using dependency relations as described. Hence, agents need a mechanism that helps them generate offers and counter-offers. The negotiation strategy defines such a mechanism.

Given that our marketplace adheres to the offer and demand laws, and the agents are supposed to be individually rational[6], it is clear that if none of them gives his partner an acceptable offer the negotiation runs into conflict. The conflictual situation can happen even when the space of possible deals includes profitable deals for both agents.

One way of thinking about which agent should concede at each step of the negotiation is to consider how much each has to lose by running into conflict at that point. In fact, if after step t the agent decides not to make a concession, he takes a risk that his partner will also not make a concession, and then they will run into a conflict. So, for our agents we have implemented a negotiation strategy based on risk evaluation. Precisely, we have adapted the Zeuthen strategy (Rosenshein & Zlotkin, 1994), so that we can deal with the lack of knowledge each agent has about his partner's utility[7]. In this adaptation, each agent uses the current price of the item in the market place as the minimum offer accepted by his partner. So, let us consider a buyer agent, Ai and his partner, Aj (seller) negotiating on item I, and consider the following variables:

γ_i : maximal cost acceptable for A_i
γ_j : minimal price acceptable for A_j
γ_m : current price in the market place, approximated by the manager using a level-headed average of all contracts' prices that have been realised until current time. In fact, if $\gamma_m(t)$

is the market price of item I until step t, and at $t+1$ there is sale of item I at price P then $\gamma_m(t+1)=(\gamma_m(t)+P)/2$.

For each step t, let $\delta_i(t)$, $\delta_j(t)$ be the deal offers made respectively by A_i and A_j.

In the buyer's point of view, the utility and risk is defined by:

The utility that agent A_i believes it will have by offering $\delta_i(t)$ is:

$U_i^i(\delta_i(t)) = max\ (\gamma_i - \delta_i(t),\ 0)$.

The utility that agent A_i believes it will have by accepting A_j's offer $\delta_j(t)$ is:

$U_i^i(\delta_j(t)) = max\ (\gamma_i - \delta_j(t),\ 0)$.

The utility that agent A_i believes that agent A_j will have by offering $\delta_j(t)$ is approximated by: $U_i^j(\delta_j(t)) = max\ (\delta_j(t) - \gamma_m,\ 0)$.
The utility that agent A_i believes agent A_j will have by accepting A_i's offer $\delta_i(t)$ is approximated by: $U_i^j(\delta_i(t)) = max\ (\delta_i(t) - \gamma_m,\ 0)$.
Hence, we can now define the degree of willingness to risk a conflict as follows:

The agent A_i's belief of willingness to risk a conflict is:

$$Risk_i^j(t) = \begin{cases} 1 & if \quad U_i^j(\delta_j(t)) = 0 \\ U_i^j(\delta_j(t)) - U_i^j(\delta_i(t)) / U_i^j(\delta_j(t)) & \end{cases}$$

The agent A_i's belief about the A_j's willingness to risk a conflict is:

$$Risk_i^i(t) = \begin{cases} 1 & if \quad U_i^i(\delta_i(t)) = 0 \\ U_i^i(\delta_i(t)) - U_i^i(\delta_j(t)) / U_i^i(\delta_i(t)) & \end{cases}$$

In the seller's point of view, the utility and risk is defined by:

The utility that agent A_j believes it will have by offering $\delta_j(t)$ is:

$U_j^j(\delta_j(t)) = max\ (\delta_j(t) - \gamma_j,\ 0)$.

The utility that agent A_j believes it will have by accepting A_i's offer $\delta_i(t)$ is:

$U_j^j(\delta_i(t)) = max\ (\delta_i(t) - \gamma_j,\ 0)$.

The utility that agent A_j believes that agent A_i will have by offering $\delta_i(t)$ is approximated by: $U_j^i(\delta_i(t)) = max\ (\gamma_m - \delta_i(t),\ 0)$.
The utility that agent A_j believes agent A_i will have by accepting A_j's offer $\delta_j(t)$ is approximated by: $U_j^i(\delta_j(t)) = max\ (\gamma_m - \delta_j(t),\ 0)$.
Hence, we can now define the degree of willingness to risk a conflict as follows:

The agent A_j's belief of willingness to risk a conflict is:

$$Risk_j^j(t) = \begin{cases} 1 & if \quad U_j^j(\delta_j(t)) = 0 \\ U_j^j(\delta_j(t)) - U_j^j(\delta_i(t)) / U_j^j(\delta_j(t)) & \end{cases}$$

The agent A_j's belief about the A_i's willingness to risk a conflict is:

$$Risk_j^i(t) = \begin{cases} 1 & if \quad U_j^i(\delta_i(t)) = 0 \\ U_j^i(\delta_i(t)) - U_j^i(\delta_j(t)) / U_j^i(\delta_i(t)) & \end{cases}$$

If t is not the last step in the negotiation, then the risk is always between 0 and 1, for both agents. $Risk_i^i(t)$ is an indication of how much A_i is willing to risk a conflict by sticking to his last offer. As $Risk_i^i(t)$ grows, agent A_i has less to lose from a conflict, and will be more willing to not concede and risk reaching a conflict. Intuitively, we propose the strategy where the agent with a smaller risk will make the next concession. Using this strategy the negotiation process can be summarized as follows: First, the buyer agent starts the negotiation by offering the seller the deal that is best for him among all possible deals. Next, at every subsequent step t, each agent calculates his risk (i.e., $Risk_i^i(t)$) and estimate his partner's risk (i.e., $Risk_i^j(t)$). If his risk is smaller or equal to that of its partner, then he must make an offer that involves the minimal[8] sufficient[9] concession from his point of view. Otherwise, he can offer the same deal that it offered previously. Of course, at every point the agents are only making offers within their acceptable set.

EXPERIMENTAL RESULTS

In order to evaluate the efficiency of the proposed model we have implemented three types of agents:

- *Basic agents*: that do not use any negotiation strategy, so they only accept offers that are greater than their strike-prices and reject the others.
- *Intermediate agents*: that use the negotiation strategy based on risk evaluation (cf. 4.2.) for their offers and counter-offers generation, but no dependency relations are taken into consideration.
- *Advanced agents*: that make use of all the models we have proposed, so they use the negotiation strategy for their offers generation and take into consideration the dependency relations for generating different types of contracts that help them to resolve conflicts.

As metrics we have considered two types of parameters: (1) the quality of the solution (contracts), measured by the individual profit and (2) the efficiency of the model, measured by the speed of goal satisfaction and the communication load.

- *Individual profit*[10] *(IP)*: The individual profit is defined for the seller agent as the difference between the contract price and his minimal acceptable price, and for the buyer as the differences between his maximal acceptable price and the contract price.
- *Speed of goal satisfaction*: The speed of goal satisfaction is calculated by the distribution of the cumulative number of goals satisfied on the simulation time (measured by the number of goal satisfied per second).
- *Communication Load*: The communication load is measured by the cumulated number of messages sent or received by the agent using communicative performatives. So that only performatives like "call for proposal, make-offer, counter-offer, mediate, ask, reply …" are considered, but performatives like "register, advertise, subscribe…" are not considered.

There are a number of important questions which need to be answered before it can be claimed that our negotiation strategy and mediation model implemented by what we call advanced agents, provide a good solution to the negotiation facilities needed within the presented context.

- How are the advanced agents better in terms of their individual profit, goal satisfaction, and communication load?
- How do advanced agents compare on an overall system perspective to other types of agents?
- What are the beneficial or undesirable effects on the overall system engendered by the use of the negotiation strategy and the dependency relations?

To answer the questions here, a number of experiments were undertaken. In these experiments two main dimensions were changed over time: the type of the monitored agent (basic, intermediate, and advanced) and the type of the partners' community.

To enable the comparison to take place, all the three types of agents were situated in identical simulation contexts. For this reason, the data for simulation are generated randomly using a uniform distribution. For each agent is generated a set of 200 goals, each goal consists of an item/service to sell or to buy with respectively the minimal accepted price for selling or the maximal cost for purchasing. So that each agent has two corresponding files: the wish list file containing items to buy and the inventory file containing items to sell. Each experiment involved precisely two members, the agent whose performance is being monitored and the rest of the society (an arbitrarily collection of partners of the same type). All agents had the same basic architecture (Esmahi, Dini &.Bernard, 1999), which is implemented using JAVA.

During the course of the simulation, the agents determine their candidate goals by taking it from their corresponding wish list file or by receiving a call for proposal relative to an item that is in their inventory file. The following sections describe the results of the simulations.

Analysis of the Agents' Performances

The analysis of the average results in all situations (by varying the partners' type) shows that when using the negotiation strategy, the profit has been increased by 33% but the goal satisfaction is reduced by 3% (Figure 9). However, when combining the dependency relations to the negotiation strategy, the profit is increased only by 15% but the goal satisfaction is increased by 25%. Hence, the use of dependency relations in the mediation process allow us to carry out a balance between satisfying more goals and making more individual profit, in other terms, balancing between serving the individual view of the agents by maximizing their individual profit or serving the global view of the system by satisfying more goals.

The principle of this balancing lead us to a comparison with the principle of social rationality[11] defined by Hogg (Hogg & Jennings, 1997), where agents strike to maximize the social benefit of the system by accepting all contracts that have a joint profit[12] greater than the joint loss, even if these contacts are not individually rational[13] for the contracting agent.

In this work we consider the principle of social commitment that can be defined as follows:

Figure 9. Agents' results on profit and satisfied goals

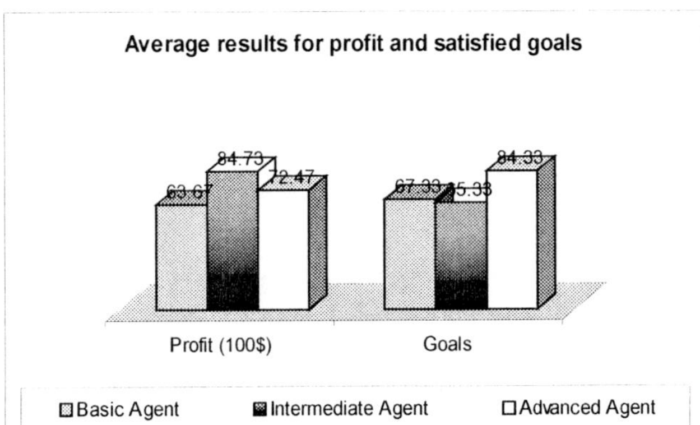

Principle of social commitment: If an agent has to choose between different individually rational contracts, then it must select the one which maximizes the joint profit and minimizes the difference between the agents' profits.

We have not set any explicit rules in the agent's knowledge base that guarantee the principle of social commitment, but we have improved it in an indirect way, that is, by the introduction of dependency relations in the mediation process and by allowing several types of contracts (exchange contract, multi-agent contract, and grouping contract). Figure 10 shows that the joint profit in negotiations that involves advanced agents is more than twice the joint profit that involves only basic or intermediate agents. Also, in Figure 11 we can see that in negotiations that involve advanced agents, the difference between agents' profit is at its minimal value (in average, less than the half of the profit difference made in case of basic or intermediate agents).

Using the negotiation strategy based on risk evaluation and the protocol where agents can exchange several propose and counter-propose results in an average increasing of 20% in the communication load. However, in order to analyze this increase in the communication load, we tried to relativize it to the number of goal satisfied. The following table shows the average rate of messages per goal.

Figure 10. Joint profit distribution

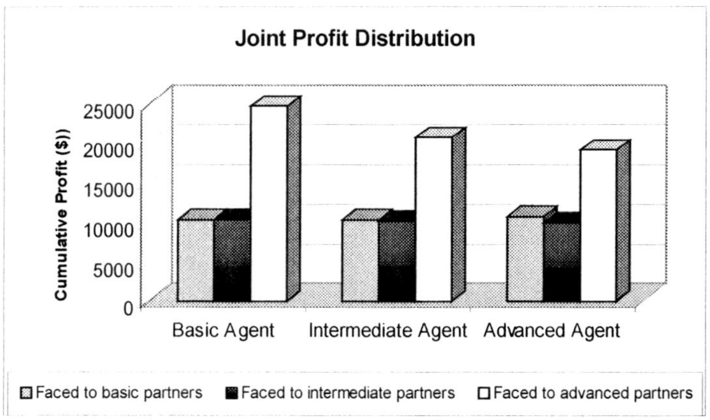

Figure 11. Deviation between agents' profit

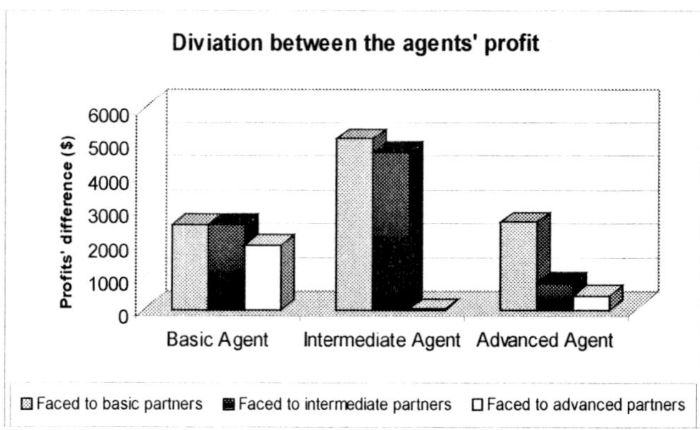

Table 1. Average rate of messages per goal

	Basic	Intermediate	Advanced
Partners are of basic type	13.12	21.72	19.13
Partners are of intermediate type	14.80	26.10	23.56
Partners are of advanced type	13.78	19.46	20.31
Global average	13.90	22.43	21.00
% of increasing compared to basic agent		61%	51%

Here we can see that the use of the negotiation strategy results in an increase of the number of messages per goal by 61% for the intermediate agent and by 51% for the advanced agent. So the use of dependency relations gives us an average decrease of 10%.

CONCLUSION AND OPEN PROBLEMS

The problem of negotiation process in e-markets is not so simple! The existing protocols and strategies do not accommodate the real world. Many other factors that influence the agents' negotiation must be taken into consideration: the agent's mental state, the capability of persuasion, and the degree of truth that an agent grants to his partners.

The main idea we are pursuing in this research consists of integrating to the agents' mental state a model of their world, especially a model of their partners, and to use this knowledge in the negotiation strategy.

Many research issues remain open for exploration in this project:

- Exploring the ways of integrating the truth component in the negotiation strategy. Two ways may be considered: representing the truth as knowledge in the mental state of the agent and using this knowledge for negotiation or representing truth as a value (may be fuzzy value) and using this value in the agent's utility function. The utility function is the base for evaluating offers, constructing counter-offers, and determining concession to do in each negotiation step.
- Extending negotiation model based on utility function by integrating the notion of time, temporal utility functions, and temporal penalty functions in order to accommodate delays in executing contacts and commitments and also in order to accommodate the real behaviour in the market place for prices and values.
- Exploring the use of social norms as arguments for convincing a partner to accept an offer or to change his position in a conflict. Social norms may be defined as some standard community rules or better try to get these rules emerging from the community interaction. In the last case, a hierarchical structure will be established for the emerging norms and a norm can move from one level to another according to the score granted by the community to this norm. The power of the norm as argument depends also on the level to which this norm belongs. Agents must not only respect norms in their interaction or propositions but also make sure that norms are respected inside the community. Those are socially committed agents!

REFERENCES

Chen, E., Vahidov, R., & Kersten, G. E. (2004). Agent-supported negotiations in the e-marketplace. *International Journal of Electronic Business, 3*(1), 28-49.

Cheng, V. S. Y., Hung, P. C. K., & Chiu, D. K. W. (2007). Enabling Web services policy negotiation with privacy preserved using XACML. In *Proceedings of the 40th Annual Hawaii International Conference on System Science (HICSS 2007)* (p. 33a).

Chiu, D. K. W., Cheung, S. C., Hung, P. C, K., Chiu, S. Y. Y., & Chung, A. K. K. (2005). Developing e-negotiation support with a meta-modeling approach in a Web services environment. *Decision Support Systems, 40*(1), 51-69.

Chiu, D. K. W., Cheung, S. C., Hung, P. C. K., & Leung, H. F. (2005). Facilitating e-negotiation processes with semantic Web technologies. In *Proceedings of the 38th Annual Hawaii International Conference on System Science (HICSS 2005)* (p. 36a).

Chiu, D. K. W., Cheung, S. C., & Till, S. (2003). A three-layer architecture for e-contract enforcement in an e-service environment. In *Proceedings of the 36th Annual Hawaii International Conference on System Science (HICSS 2003)* (p. 74a).

Cheung, S. C., Hung, P. C. K., & Chiu, D. K. W. (2003). On the e-negotiation of unmatched logrolling views. In *Proceedings of the 36th Annual Hawaii International Conference on System Science (HICSS 2003)* (p. 29).

Chiu, D. K. W., Poon, J. K. M., Lam, W. C., Tse, C. Y., Sui, W. H. T., & Poon, W. S. (2005). How ontologies can help in an e-marketplace. In *Proceedings of the 13th European Conference on Information Systems, Information Systems in a Rapidly Changing Economy (ECIS 2005)* (p. 57).

Ephrati, E., & Rosenschein, J. S. (1992). Constrained intelligent action: Planning under the influence of a master agent. In *Proceedings of the Tenth National Conference on Artificial Intelligence*, San Jose, California, USA, (pp. 263-268).

Esmahi, L., Dini, P., & Bernard, J. C. (1999, June). Towards an open intelligent market place for mobile agents. In *Proceedings of the Eighth IEEE International Workshops on Enabling Technologies: Infrastructure for Collaborative Enterprises (WET-ICE'99)*, Stanford University, California, USA, (pp. 279-286).

Esmahi, L., Dini, P., & Bernard, J. C. (2000). MIAMAP: A virtual market place for intelligent agents. In *Proceedings of the 33rd Hawaii International Conference on System Sciences (HICSS-33)* (Vol. 8, p. 8018).

Fikes, R., Engelmore, R., Farquhar, A., & Pratt, W. (1995). *Network-based information broker.* Stanford University: Knowledge System Laboratory. Retrieved December 15, 2006, from: http://www-ksl.stanford.edu/KSL_Abstracts/KSL-95-13.html

Genesereth, M. R., Ginsberg, M. L., & Rosenschein, J. S. (1986, August). Cooperation without communication. In *Proceedings of the Fifth National Conference on Artificial Intelligence*, Philadelphia, PA, (pp. 51-57).

Guan S. U., & Zhu, F. (2002). Agent fabrication and its implementation for agent-based electronic commerce. *International Journal of Information Technology and Decision Making, 1*(3), 473-490.

Hogg, L. M., & Jennings, N. R. (1997, November 8-10). Socially rational agents. In *Proceedings of the AAAI Fall symposium on Socially Intelligent Agents*, Boston, Mass., (pp. 61-63).

Jennings, N. R., Faratin, P., Lomuscio, A. R., Parsons, S., Wooldridge, M. J., & Sierra, C. (2001). Automated negotiations: Prospects, methods and challenges. *Group Decision and Negotiation, 10*(2), 199–215.

Kersten, G. E., & Lai, H. (2005). *Satisfiability and completeness of protocols for electronic negotiations* (Research paper INR01/05). Retrieved January 4, 2007, from http://interneg.concordia.ca/interneg/research/papers/2005/01.pdf

Khedro, T., & Genesereth, M. R. (1994). Progressive negotiation for resolving conflicts among distributed heterogeneous cooperating agents. In *Proceedings of the 12th National Conference on Artificial Intelligence*, Seattle, Washington, USA, (pp. 381-386).

Klemperer, P. (1999). Auction theory: A guide to the literature. *Journal of Economic Surveys, 13*, 227-286.

Klemperer, P. (2002). What really matters in auction design. *Journal of Economic Perspectives, 16*(1), 169-189.

Koning, J. L. (2002). A review on the interaction issues in agent-based marketplaces. *International Journal of Information Technology and Decision Making, 1*(3), 457-472.

Kraus, S., Wilkenfeld, J., & Zlotkin, G. (1995). Multiagent negotiation under time constraints. *Artificial Intelligence, 75*(2), 297-345.

Lamparter, S., Ankolekar, A., Studer, R., Oberle, D., & Weinhardt, C. (2006). A policy framework for trading configurable goods and services in open electronic markets. In *Proceedings of the 8th International Conference on Electronic Commerce: The New E-commerce: Innovations for Conquering Current Barriers, Obstacles and Limitations to Conducting Successful Business on the Internet* (pp. 162-173).

Lee, K. K. K., Chiu, D. K. W., & Hung, P. C. K. (2006). Web-service based information integration for e-financial planning system matchmaking decision support. In *Proceedings of the IEEE/WIC/ACM International Conference on Web Intelligence and International Agent Technology Workshops, (WI-IAT 2006)* (pp. 259-262).

Neumann, D., Schnizler, B., Weber, I., & Weinhardt, C. (2007). Second-best combinatorial auctions—the case of the pricing-per-column mechanism. In *Proceedings of the 40th Annual Hawaii International Conference on System Science (HICSS 2007)* (p. 33b).

Neumann, D., Gimpel, H., Holtmann, C., & Weinhardt, C. (2002). Agent-based bidding in electronic markets. In *Proceedings of the Second IFIP Conference on E-commerce, E-business, E-government (I3E 2002)* (pp. 553-567).

Neumann, D., & Weinhardt, C. (2002). Domain-independent e-negotiation design: Prospects, methods, and challenges. In *Proceedings of the 13th International Workshop on Database and Expert Systems Applications (DEXA 2002)* (pp. 680-686).

Ng, L. L. N., Chiu, D. K. W., & Hung, P. C. K. (2007). Tendering process model (TPM) implementation for B2B integration in a Web services environment. In *Proceedings of the 40th Annual Hawaii International Conference on System Science (HICSS 2007)* (p. 143b).

Rasmusen, E. (2001). *Games and information: An introduction to game theory* (3rd ed.). Blackwell Publishers.

Rosenschein, J. S., & Genesereth, M. R. (1985, August). Deals among rational agents. In *Proceedings of the Ninth International Joint Conference on Artificial Intelligence*, Los Angeles, California, (pp. 91-99).

Rosenschein, J. S., & Zlotkin, G. (1994). *Rules of encounter*. MIT Press.

Roth, E. (2000). Game theory as a tool for market design. In F. Patrone, I. Garcia-Jurado, & S. Tijs (Eds.), *Game practice: Contributions from applied game theory* (pp. 7-18).

Sandholm, T. W. (1993). An implementation of the contract net protocol based on marginal cost calculations. In *Proceedings 11th National Conference on Artificial Intelligence* (pp. 256-262).

Sandholm, T. W. (1996). *Negotiation among self-interested computationally limited agents.* Unpublished PhD dissertation, Department of Computer Science, University of Massachusetts. Retrieved February 15, 2006, from: http://www.cs.cmu.edu/~sandholm/dissertation.ps

Sandholm, T. W. (2000). Agents in electronic commerce: Component technologies for automated negotiation and coalition formation. *Autonomous Agents and Multi-Agent Systems, 3*(1), 73-96.

Schoop, M., & Quix, C. (2001). DOC.COM: A framework for effective negotiation support in electronic marketplaces. *Computer Networks, 37*(2), 153-170.

Smith, R. G. (1988). The contract net protocol: high-level communication and control in a distributed problem solver'. In A. H. Bond & L. Gasser (Eds.), *Readings in distributed artificial intelligence* (pp. 357-366). Morgan Kauffmann Publishers.

Smith, R. G., & Davis, R. (1981). Frameworks for cooperation in distributed problem solving. *IEEE Transactions on Systems, Man and Cybernetics, 11*(1), 61-70.

Zlotkin, G., & Rosenschein, J. S. (1991). Cooperation and conflict resolution via negotiation among autonomous agents in noncooperative domains'. *IEEE Transactions on Systems, Man and Cybernetics, 21*(6), 1317-1324.

Zwass, V. (1999). Structure and macro-level impacts of electronic commerce: From technological infrastructure to electronic marketplaces. In K. E. Kendall (Ed.), *Emerging information technology* (pp. 517-526). Sage Publications.

ENDNOTES

1 The agent's mental state is the agent's beliefs about his environment (always represented as facts), intentions, and objectives at each time point.

2 In the marketplace, a goal corresponds to a selling or purchase of an item/service according to some defined characteristics.

3 In pure distributed problem solvers, the only concern of the designer is with the overall performance of the system and the performance of the individual agent is not important.

4 In pure autonomous agents systems, the concern of the designer is with the performance of the individual agent and the system level performance is left to emerge from the agents interactions without taking into consideration the interdependency between the system's components.

5 A path, X to Y in the graph of dependency, constitutes a causal chain that provides an amount of transaction that must be balanced in the other direction from Y to X, in order to determine a cycle.

6 An agent is individually rational if it accepts only deals that give him non-negative utility.

7 In the Zeuthen strategy, it is assumed that the two agents have complete and correct knowledge about each other, a thing that we believe is unrealistic to suppose and implement in a market place.

8 A sufficient concession is one that changes the balance of risk between the agent and

his partner. For example, after a sufficient concession made by Ai, we will have Riskii(t) > Riskij(t).

9 The minimal sufficient concession is the sufficient concession that the agent believes will give his partner the least utility value.

10 In this context, the term « profit » covers positive, negative, and indifferent utility values.

11 The principle of social rationality as was stated by Hogg and Jennings as: "if a socially rational agent can perform an action whose joint benefit is greater than its joint loss, then it may select that action."

12 Joint profit (JP): the sum of the individual profit or individual loss for all agents participating in the contract. If N agents participate in a contract then $JP = \Sigma_{i=1}^{N} Pi$ where Pi is the agent's Ai profit (positive value) or agent's Ai loss (negative value).

13 A contract is individually rational for an agent if it gives him a non-negative profit.

This work was previously published in Journal of Electronic Commerce in Organizations, Vol. 6, Issue 3, edited by M. Khosrow-Pour, pp. 74-91, copyright 2008 by IGI Publishing (an imprint of IGI Global).

Chapter 7
Designing Reputation and Trust Management Systems

Roman Beck
Johann Wolfgang Goethe University, Germany

Jochen Franke
Johann Wolfgang Goethe University, Germany

ABSTRACT

This article analyzes the handling of customer complaints after shipping ordered goods by applying automated reputation and trust accounts as decision support. Customer complaints are cost intensive and difficult to standardize. A game theory based analysis of the process yields insights into unfavorable interactions between both business partners. Trust and reputation mechanisms have been found useful in addressing these types of interactions. A reputation and trust management system (RTMS) is proposed based on design theory guidelines as an IS artifact to prevent customers from issuing false complaints. A generic simulation setting for analysis of the mechanism is presented to evaluate the applicability of the RTMS. The findings suggest that the RTMS performs best in market environments where transaction frequency is high, individual complaint-handling costs are high compared to product revenues, and the market has a high fraction of potentially cheating customers.

INTRODUCTION

The continued demand for automated interorganizational business processes to reduce transaction costs in supply chains has provided a strong demand for extensive information systems (IS) support. While areas for the application of IS in supply chain management are growing rapidly, the management and automation of personal relationships in impersonal electronic business relations is still an area that has not been adequately served by existing IS research and development. In this article, we describe how a reputation and trust management system (RTMS) for an automated evaluation of business relationships in supply chains can be designed and implemented. As

RTMS research domain, we have chosen the management of customer complaints since it is also a largely unexplored, yet promising application area. While empirical research and data are limited in this area, two cases provide an indication of how much money can be saved by an improved complaint-handling process: Eastman Chemicals saved $2 million after improving its business processes associated with investigating and responding to complaints by cutting expenses for waste removal and rework caused by off-quality products or incorrect paperwork (Hallen & Latino, 2003). The second example provides a more accurate view on the de facto costs of handling customer complaints manually: According to Schilling and Sobotta (1999), a medium-sized enterprise with approximately €5 million annual revenue calculated the average processing costs as €837.47 for each complaint handling process in 1997.

The need for human interaction and decision (e.g., to check complaints or to prevent opportunistic customer behavior) historically has been a major impediment to increasing the degree of automation. Since handling of complaints is costly for both suppliers and customers, only 5% to 10% of all dissatisfied customers decide to complain at all (Tax & Brown 1998). Dissatisfied customers are likely to switch providers, which usually leads to future revenue losses higher than the costs caused by complaints in the first place (Fornell & Wernerfelt, 1987). Therefore, suppliers face two dilemmas: First, they cannot automate or standardize the complaint-handling process, since opportunistic customers may benefit from this lack of human diligence. Second, dissatisfied customers, having switched to another supplier, may never notify the errant supplier, since the manual complaint-handling process is too expensive in comparison to the value of the defective or missing delivery.

This article proposes an RTMS-based complaint-handling solution, not only to provide benefits from the efficiency of computer-based customer complaint management but also to prevent opportunistic behavior and customer losses in relevant market environments. We provide a mechanism that allows increasing the role of automated business processes while concurrently mitigating incentives for opportunistic behavior in business-to- business as well as business-to-consumer relationships. We believe that this approach is a contribution to IS literature, since reputation and trust management research from behavioral science has not yet been expatiated adequately in existing IS research.

After describing the problem relevance, the theoretical background of the article presents foundations of reputation and trust as well as transaction cost theory. Since we strive to contribute to knowledge by following a design science approach, the guidelines provided by Hevner, March, and Park (2004) and further IS design science contributions are related to this research in the theoretical section. Next, we detail the (predominantly) existing defective product handling or customer complaint process after receiving defective articles or failing to receive articles. A game-theoretical model of supplier and customer motivations is introduced providing the formal representation and logic for process redesign. Afterward, we modify the customer complaint-handling process by introducing RTMS to minimize the number of manual interactions. To evaluate our solution, results of a simulation model are provided for demonstrating the utility and efficacy of the proposed design artifact. The validity of the sociotechnical approach is discussed and scenarios are identified where this IT artifact may yield higher benefits for suppliers. The article closes with a short summary of our findings and a discussion of the design problems.

THEORETICAL BACKGROUND

The need for efficient relationship management arises whenever independent business partners have to coordinate interdependent activities (Malone & Crowston, 1994). When engineering a rigorous RTMS that meets design science requirements, we must consider reputation and trust as well as economic demands. Both will provide the theoretical foundation upon which this research rests. Before digging deeper into the theoretical foundations, basic guidelines for engineering artifacts according to design science requirements are given.

Design Science and Artifact Engineering

According to Walls, Widmeyer, and El Sawy (1992), design theory is different from grand theories (e.g., as propagated by Popper). Serving human purposes by improving process performance, building and evaluating constructs, models, methods, and instantiations are typical design science research activities (March & Smith, 1995). This differentiates design theory from, for example, grounded theory (Eisenhardt, 1989; Glaser & Strauss, 1967), which uses an empirical inductive approach and qualifies design theory to be part of middle-range theories (Merton, 1968). Nevertheless, design theory is suggested to utilize grand theories deductively as kernel theories. In this article, reputation and trust, as well as economic theories, serve as these so-called kernel theories. According to Merton (1968), emerging disciplines should develop special theories with limited conceptual ranges that function as stepping stones or middle-range theories on the way toward a total conceptualization or grand theory. In this epistemological context of middle-range theorizing, Walls et al. (1992) postulated that "the IS discipline needs to articulate and develop a class of 'design theories' and provide examples where goal-oriented theorizing has successfully led to executive information systems (EIS), management information systems (MIS), decision support systems (DSS) (Walls et al., 1992), or emergent knowledge process systems (EKPS) (Markus, Majchrzak, & Gasser, 2002)." Inspired by the idea of developing theories unique to the IS discipline, Hevner et al. (2004) articulated seven guidelines on how to evaluate and present rigorous design science research. We use these guidelines to create a purposeful RTMS artifact and, more specifically, a method (guideline 1) for the trust and reputation management in customer complaint handling, which, as outlined before, represents a relevant organizational problem (guideline 2). The RTMS was evaluated by applying a simulation approach (guideline 3) to reengineer and automate the customer complaint handling to contribute to a more effective and efficient customer complaint process (guideline 4). Regarding research rigor (guideline 5), the RTMS has been informed by kernel theories, such as theories on reputation and trust and transaction cost economics, and subsequently defined and formally represented as a game theoretical problem. Simulated artificial market scenarios are developed to find the limitations of the RTMS artifact (guideline 6). Finally, the solution is communicated in this article to allow for a thorough discussion in the scientific community (guideline 7). In the following sections, the kernel theories applied in this research to comply with Hevner's fifth guideline are introduced.

Reputation and Trust

In the business world, a supplier's reputation reflects an aggregate ratio incorporating multiple factors: quality of merchandise, reliability of financial transactions, and/or level of customer service. It is often observed that reputation and trust acquire fundamental importance in long-term business-to-business (B2B) relations. According to Mui, Mohtashemi, and Halberstadt (2002), reputation is a "perception that an agent creates through past actions about its intentions and norms" and trust is a "subjective expectation an agent has about another's future behavior based on the history of their encounters." It has been shown that reputation reduces the complexity of the decision process (Wigand, Picot,& Reichwald, 1997) by better estimating the likelihood of failed orders and through a reduction in the number of quality tests needed for a product (Marsh, 1992).

It is important to distinguish between the individual and social dimensions of reputation (Sabater & Sierra, 2002). This article focuses on the individual dimension of reputation relevant for direct interactions between two business partners. Experience of transactions with a partner is directly reflected in an assigned reputation value. The social dimension of reputation relies on intermediates to propagate common reputation assessments and must be aggregated through standardized processes. Due to the specific setting of bilateral supplier-customer relationships, the social aspect of reputation can be neglected because, typically, only two partners are involved in the complaint-handling process at hand.

Models of reputation and trust have been developed extensively in agent-based computational economics. A broad overview of approaches to the use of reputation in multiagent systems is provided by Mui, Halberstadt, and Mohtashemi (2002). Sabater and Sierra (2001) introduced a reputation model, taking the individual and social dimension of reputation into account for a multiagent society. Others propose a formalization of reputation for multiagent systems, applying the sociological concept of role fulfillment for establishing a positive reputation and for examining the link between reputation and trust (Carter, Bitting, & Ghorbani, 2002). The role of trust in supply relationships and the underlying implications were addressed by Lane and Bachmann (1996) in an empirical study of business relationships in Germany and U.K. (Lane & Bachmann, 1996). As they pointed out, trust relations are highly dependent on stable social, institutional, and legal structures. Moorman, Zaltman, and Deshpande (1992) investigated the specific relationship between providers and users of market research reports, providing a reasonable introduction to the role of trust in relationships (Moorman et al., 1992).

Das and Teng (1998) argued that trust and control are the two pivotal sources of confidence in the cooperative behavior of business partners in strategic alliances. Both sources of confidence are highly interdependent. A large amount of control reflects a low amount of trust and vice versa. Without any control, the trusting party assumes the risk of the trustee's opportunistic behavior. As described, trust and control are inherently different approaches to business relationships. The costs to control the behavior of business partners can be extremely high. If reputation or trust is not established and the threshold to behave in an undesirable manner is low, the defrauded partner's control costs can be higher than the value of the goods, and consequently one may accept—to a certain degree—some fading in deliveries. Business partners are anticipating that control is difficult (e.g., in the case of defective, low-value goods, where shipping them back to the vendor is more expensive than accepting to discard them by the customer). Such behavior is more likely in new business relations and more anonymous markets, such as electronic marketplaces, where no face-to-face contact is established.

Reputation Mechanisms and Transaction Costs

Increasing the level of control by establishing contracts or mechanisms to prevent opportunistic behavior can result in higher transaction costs so that, in the worst case, the handling of an order might be more costly than the expected benefit. In the context of reputation and trust, ex-post transaction costs are of particular importance (Williamson, 1975, 1985). Ex-post transaction costs refer to costs that emerge after the order has been shipped and before the transaction cycle is completed. Ex-post transaction costs will increase if the trust level decreases. In other words, the monitoring and enforcement costs to prevent ex-post bargaining will be higher if the incentive for opportunistic behavior increases (Dahlstrom & Nygaard, 1999). For suppliers, such costly uncertainties are based on unanticipated changes in the behavior of business partners (Noordewier, George, & Nevin, 1990). The greater the level of uncertainty, the more difficult it is to formulate, negotiate, and enforce a contract to reduce the risk of being a victim of opportunistic behavior. In long-term relations, expensive tracking and monitoring instruments may be replaced by mutual trust; however, trust and reputation must be effectively managed in an automated way when the number of business partners increases.

A REPUTATION AND TRUST MANAGEMENT SYSTEM FOR CUSTOMER COMPLAINT-HANDLING PROCESSES: DESIGNING AN ARTIFACT

In our RTMS, extensive control in the customer complaint-handling process is replaced by trust to reduce costs for suppliers and customers. A supplier utilizing RTMS assigns individual reputation values to its customers and tracks past actions in complaint issues to assess the probability of future opportunistic behaviors. The supplier can use this reputation measure to decide whether to trust the customer and accept the complaint without validating the claim, or to pursue a detailed investigation. In the following sections, we will elaborate on the proposed automated system in detail and introduce the artifact, referring to Hevner's first guideline for design science.

Customer Complaint Alternatives and Implications

Many business processes are not yet fully automated. In order to discuss the complaint process, both on the customer and supplier sides in more detail, the alternatives and relevant business cases are depicted in the following. Drawing from the *exit, voice, and loyalty* model provided by Hirschman (1970), and the customers *problem impact tree* framework of Rust, Subramanian, and Wells (1992), a problem tree of *voice a complaint or exit without making a complaint* is utilized. According to Hirschman, customers have two potential feedback options: (1) to voice complaints and thereby express the dissatisfaction directly to the supplier or (2) to stop buying and exit the relation. Both options have different but always unfavorable impacts on suppliers, who must respond with adequate defensive strategies to overcome those problems. To elaborate, all possible customer complaints scenarios are first described briefly: After submitting an order and receiving a delivery note from the supplier, the incoming orders are checked by the customer's receiving

Figure 1. Customer action alternatives

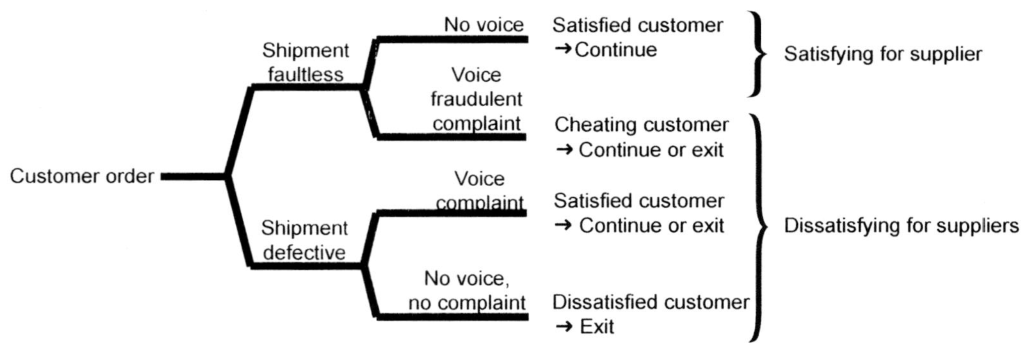

department. In the case of a faultless shipment, one expects that customers have no reason to complain (see the upper branch of Figure 1). This is true in nearly all cases: Customers receiving correct deliveries will be satisfied, continue with the supplier, and will not place any complaints. The situation is slightly different if complaints are not too costly and the supplier does not ask for the defective items to be sent back in order to validate the complaint. If customers do not perceive the recall of defective items as a credible threat, then they might be tempted to cheat and complain about faultless shipments. Avoiding such an incentive is a pivotal element when designing an automated customer complaint-handling solution.

In the case of defective or partially missing items in the shipment (see the lower branch of Figure 1), the supplier must be contacted and/or the broken parts sent back. Afterward, the supplier sends the defective parts again and the customer tracks the complaint until all replacement parts are received. If the supplier handles the complaint satisfactorily, the customer will buy again. If this is not the case and the customer is dissatisfied with the process management, then the exit strategy might be chosen. In the latter case, the supplier has no chance to contact the dissatisfied customer if a defective shipment is delivered and the customer decides not to complain. This can be the case if the complaint process is more costly then the value of the defective products. Dissatisfied with the delivered quality, it is likely that such a customer will discontinue the business relationship.

As Figure 1 reveals, dissatisfying scenarios can emerge for suppliers, even when the shipment was faultless. A solution to the dissatisfying results for customers and suppliers might be offered by an automated reputation-based system where customers do not have to prove that parts of a shipment are damaged or missing. Instead, the supplier simply believes the customer based on the reputation the customer has acquired in past transactions and trusts him or her in the case of complaints.

Designing an RTMS-Based Automated Customer Complaint-Handling Solution

In this section, a simplified customer complaint process is described to reduce the handling costs for suppliers and customers. It will be shown that from a game-theoretical point of view, the simplified customer complaint-handling process dominates the conventional process if customers are always truthful. If truthful customers cannot be assumed, a reputation mechanism is introduced to inhibit cheating. Before digging deeper into the conventional and the simplified complaint-handling process from a game-theoretic perspective, the assumptions our model is based on are delimited:

- Neither supplier nor customer knows the exact value of the defective ratio d.
- The exact quality of the products en route is not known (e.g., due to unknown conditions during the shipment).
- There is a long-term recurring business relationship between supplier and customer. Products are exchanged frequently between both of them.
- The value of a single order is relatively low, as can be observed for raw materials or office supplies.
- The customer complaint-handling costs of the new simplified process are ignored. In the simplified process, the customer only has to send an electronic notification to the supplier without shipping the defective items; the supplier does not have to perform a manual check of the incoming goods and thus is assumed to cause no relevant costs compared to the conventional scenario, where the customer has to process the defective shipment for physically returning it to the supplier.
- There are no limitations referring to legal issues.

Table 1. Conventional customer complaint process cost matrix

Shipment defect?	Customer complains? Yes	Customer complains? No
Yes (d)	Customer: complaint costs (c_C^C) Supplier: {0; fraction of defective shipment (v)+} complaint costs (c_S^C)	Customer: {0; fraction of defective shipment (v)} Supplier: 0
No (1-d)	Customer: complaint costs (c_C^C) Supplier: complaint costs (c_S^C)	Customer: 0 Supplier: 0

We use a game-theoretical design approach to analyze the trade situation for the conventional and the simplified complaint-handling process. In a conventional complaint-handling process, the customer checks the shipment, and if there are defects, the defective parts of the shipment are sent back freight forward to the supplier. The supplier checks whether the complaint is justified. Both partners have expenses due to the manual processing and shipment of products. Table 1 depicts the cost matrix in a game with a conventional customer complaint process.

If the shipment is indeed defective and the customer decides to reclaim (see the upper left cell in Table 1), both customer and supplier pay for manual handling of the customer complaints c_C^C and c_S^C, respectively. Additionally, the supplier will not be paid for its defective products, and the value v (ranging from 0 to the total value of the shipment if all parts are defective) of these parts is lost. When the customer decides not to reclaim the defective products (see the upper right cell in Table 1), his or her loss equals the value of the defective shipped products v. If the shipped products have only minor defects, the consumer may be able to use the products partially, thereby reducing his or her loss to a fraction of v, indicating the shipment's remaining utility. Nevertheless, compared to flawless products, the consumer encounters loss ranging from a cost of 0 for minor defects to the value of the shipment v for major defects. If the shipment is not defective and the customer decides to issue a complaint (see the lower left cell in Table 1), both partners will have to pay complaint costs c_C^C and c_S^C. After the order is sent back, the supplier checks the products and finds them nondefective and may reship them or sell them to another customer. Thus, there are no further costs, despite the complaint processing costs. In cases where the shipment is not defective and the customer does not decide to reclaim (see the lower right cell in Table 1), the transaction is completed as originally intended with no additional cost outside the regular transaction process.

Now an RTMS-supported, simplified customer complaint-handling process is implemented, reducing complaint costs for both partners. In cases when the customer decides to complain about a shipment, the supplier trusts the customer, assuming the products are indeed defective without the need for validation. The customer subtracts the invoice accordingly or a new shipment is

Table 2. Simplified customer complaint process cost matrix

		Customer complains?	
		Yes	No
Shipment defect?	Yes (d)	Customer: 0 Supplier: fraction of defective shipment (v)	Customer: {0; fraction of defective shipment (v)} Supplier: 0
	No (1-d)	Customer: - fraction of defective shipment (-v) Supplier: fraction of defective shipment (v)	Customer: 0 Supplier: 0

immediately scheduled and the supplier does not audit the complaint further. This new setting is described in Table 2.

If the shipment is not defective and the customer decides not to reclaim (see the lower right cell in Table 2), the situation is unchanged. In cases where the products are defective and the customer does not complain (see the upper right cell in Table 2), the situation is unchanged, despite the lack of complaint costs. The critical case is a cheating customer who lodges a complaint for a shipment that is not defective at all (see the lower left cell in Table 2). In this case, the customer does not pay for the faultless products. She/he immediately *earns* the value of the products ("negative loss costs (-*v*)"). On the other hand, the supplier loses the value of the products shipped.

Comparing both situations reveals that for defective product shipments, the second scenario with a simplified customer complaint process is advantageous. If supplier-side complaint costs are less than the value of the shipment, only the lower left quadrant of the cost-matrix is disadvantageous. This outcome, which implies a cheating customer, should be avoided.

As we have seen, the costs of shipping and handling complaints in a specific market are important for the viability of the simplified customer complaint process. In the case of low or negligible shipping and complaint-handling costs, it might be rational to always return defective shipments, depending on the relationship of total complaint costs to the individual value of a shipment. However, if total complaint costs are high in relation to the shipment's value, the simplified complaint process can realize substantial cost savings.

The Reputation and Trust Management System to Inhibit Fraudulent Behavior

In the case of accurate shipments, there is a significant difference between the conventional and simplified scenario. If the customer decides to complain for faultless shipment, then she/he will not have to pay for the faultless products and immediately gains the value *v*. Concurrently, the supplier loses the equivalent value because it trusts the customer and does not perform a quality check on the reclaimed products that would expose a cheating customer. If there is no additional monitoring or control structure, the customer will always reclaim the delivered shipments, regardless of the actual status (whether it is indeed defective or not) in the scenario with the new system. It is a weakly dominant strategy for the customer always to complain. Thus, the supplier always loses the equivalent value of the shipment if no mechanism is applied to counter cheating behavior.

In an idealized world, customers would always tell the truth to reduce transaction costs. Both partners could improve their respective position in all cases, because only the upper left and lower right sections in Table 2 would be relevant. Assuming a customer who is always telling the truth reveals that the conventional complaint-handling mechanism is dominated by the simplified automated complaint handling. Both parties benefit from the reduction of transaction costs when processing complaints. Nevertheless, the world is not ideal, and the customer might be tempted to complain about defective products even if it is not justified. The pivotal question here is how to assure that the customer has no interest in cheating. One solution is to apply an inexpensive incentive mechanism enforced by a RTMS.

Reputation in this context is based on business transactions with a certain customer in the past. The more orders successfully processed in the past, the higher the reputation account (and the higher the level of trust). Otherwise, the customer withdraws from hid or her reputation account on the supplier side if transactions failed in the past. In the simplest case, the supplier could estimate the defection rate d of its products r and adjust the customer's reputation account if his or her complaint rate significantly differs from the estimated quality (e.g., by applying a χ^2 test).

The supplier's credible threat is to switch back to the conventional customer complaint-handling mechanism, imposing complaint-processing costs on future transactions. This threat only works for infinitely repeated games, as are assumed in this model. This assumption seems appropriate for our setting, since B2B relationships are often characterized as long-term relationships with frequently recurring transactions. The supplier can implement several strategies to ensure that the customer is truthful. The following strategies can be applied, if the supplier knows the defection rate d with reasonably high accuracy:

- The supplier can randomly select reclaimed shipments and request the customer to return the products for an intensive test. If the products are faultless, the customer cannot be trusted and is removed from the simplified customer complaint-handling process. The process is immediately switched back to the traditional handling process. This *grim trigger* strategy is potentially suboptimal if the customer accidentally complains about products that are not defective.
- The supplier can switch back to the conventional complaint-handling process if the ratio of complained products significantly exceeds the defectiveness ratio d. This mechanism only works if the supplier knows the defectiveness ratio d with high accuracy.
- Each customer receives a reputation account for a given period, calculated as the product of the mean ordered value and the quality parameter d. If a customer reclaims a shipment, the shipment's value is subtracted from this account and if the account is exhausted, the customer has to justify his or her behavior. This mechanism also relies strongly on the accuracy of the parameter d.

The threshold for identifying cheating behavior on the part of a customer should be chosen according to the accuracy with which d is known. If d is not known and is subject to change, this threshold should be increased and vice versa.

If the supplier does not know the defectiveness ratio d, it can improve the reputation mechanism by taking into account the responses of all other customers for each product. Each customer has individual reputation values for each product. If a customer reclaims a shipment, the value of this shipment is subtracted from his or her reputation account for the product in question. Afterward, the reputation values of all customers receive a bonus. This bonus for product r and customer i is calculated as an adjusted ratio of the mean quantity ordered by the customer. This value can be regularly recalculated for all orders of a given period (e.g., monthly). The following equation calculates the reputation bonus for each customer i and product r.

$$bonus_r^i = \frac{q_r^i}{\sum_{j=1}^{n} q_r^j} * p_r q_r^d$$

p_r: price of product r

q_r^i: aggregated quantity of product r ordered by customer i in a given period

q_r^j: aggregated quantity of product r ordered by customer j in a given period

n: number of customers with reputation accounts

q_r^d: quantity of defective product r that is reclaimed

The RTMS works as follows: If all customers are acting truthfully, the individual reputation accounts for every product will be zero on average. A simple example should illustrate the mechanism: A defectiveness ratio d of 10%, a price of 1 for a given product r and three customers are assumed. The first customer regularly orders 1,000 units, customer 2 orders 50 units and customer 3 orders 200 units. Each customer reclaims truthfully 10% of the shipments. When the first customer reclaims 100 units, his or her reputation account is immediately reduced by 100, equivalent to the total value of the complaint. Afterwards, all customers' reputation accounts are given a bonus (including the customer initiating the claim), resulting in 80 bonus points for customer 1, 4 bonus points for customer 2, and 16 bonus points for customer 3. This process is also applied for the complaints of the other customers, leading to neutral reputation accounts at the end of the selected period.

If one of the customers decides to cheat and complains with a higher ratio (e.g., 15%), then his or her reputation account will be negative while the accounts of the other customers will be positive. If the first customer complains 15% of his or her shipments and the other customers complain 10%, their respective reputation accounts for the illustrative example will be -10, +2, and +7.2. Customers with a higher complaint ratio than other customers can be identified by their negative reputation accounts. The first cheating customer will put him- or herself into an inferior position compared to truthful customers. This system can only be cheated if all customers collude to produce a consistent and artificially inflated complaint ratio. Furthermore, the mechanism does not work with a small number of customers. If there were only one customer, then the reputation value would never deviate.

EXPERIMENTAL EVALUATION OF THE PROPOSED REPUTATION AND TRUST MANAGEMENT SYSTEM

To evaluate the developed solution as suggested in Hevner's third guideline regarding design science, we constructed a simulation to conduct sensitivity analyses for different transaction frequencies and fractions of potential cheaters in the market. For simplification and computational reasons, we assume that the structure of relationships remains unchanged within each simulation run—customers are always able to correctly assess the quality of the delivered products (faultless or defective), and that the production capacities of the suppliers' facilities are not limited. Further, we assume that there are no shortages and arbitrary amounts of products ordered may be delivered. The following section describes the dynamic behavior of the simulation and explains the core processes performed by the simulated agents. In the subsequent section, specific simulation settings are described and the results are discussed.

Model Evaluation and Simulation Setting

The simulation implements the proposed IS-based reputation model and assesses environmental conditions where suppliers using the proposed solution would outperform comparable suppliers without it. For the simulation, an idealized trading situation between suppliers and customers is assumed. An arbitrary number of suppliers and customers can be simulated, including truthful acting, as well as cheating customers. The transaction starts with the customer who generates an order. The receiving supplier executes and ships the ordered goods to the customer who is checking the incoming delivery. A random percentage of products in the suppliers' shipments is defective. The customers check the shipments and decide whether to complain or not. All suppliers receive identical orders in order to compare different parameterizations of the reputation mechanism.

Figure 2. Course of action of a simulated period

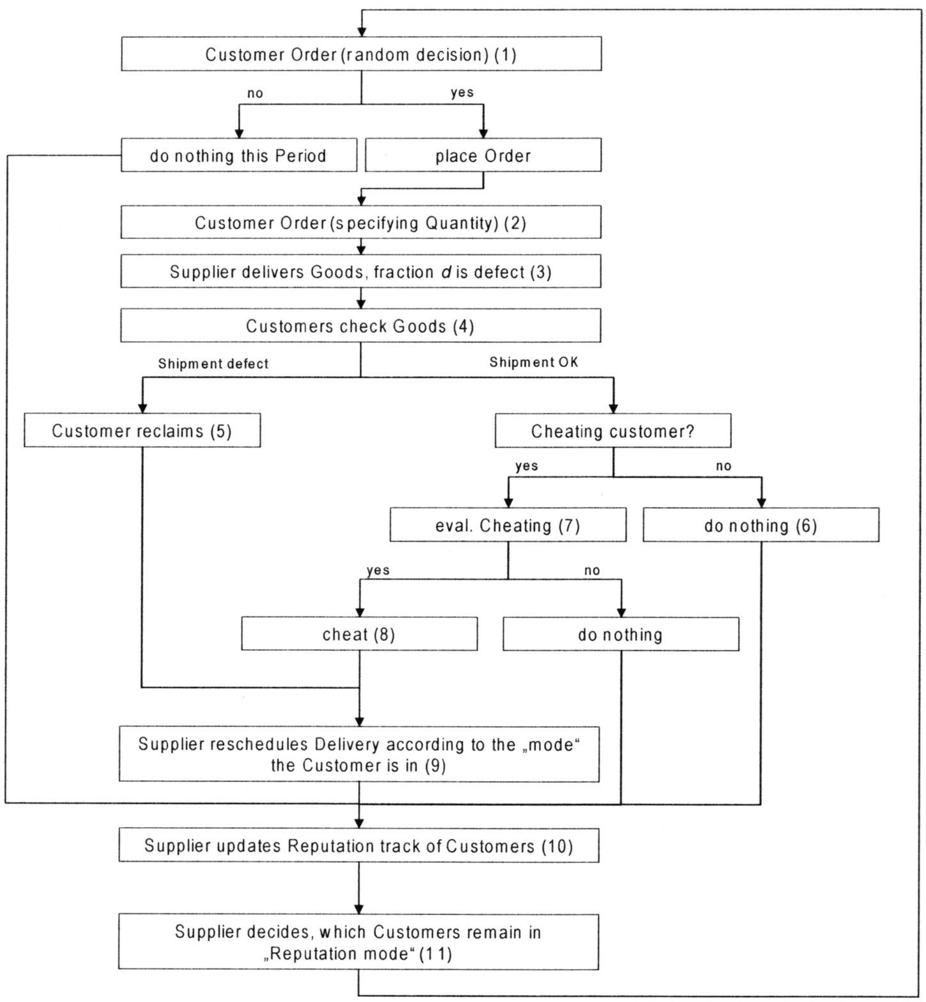

If the specified supplier implements the reputation system, a new shipment will be scheduled immediately after a customer complaint is lodged—if the customer's reputation value is high enough (in accordance with the reputation mechanism outlined in section 0). Furthermore, the system will also update reputation values of all customers. If a customer exceeds a prespecified reputation threshold on the lower bound, the supplier will switch back to conventional mode and check all complained products. Although we assume every supplier is deploying a quality management program to ensure high standards in production, a small but unavoidable ratio *d* of defective products leaves every company unnoticed. For our study, this defective ratio follows a normal distribution but can be freely configured in the model. The performance of each supplier is assessed by the operating profit resulting from the difference between revenues and costs. Revenues are calculated for faultless shipped and paid products that do not result in a customer complaint. Occurring costs are (1) variable costs for each product shipped (independent of faultless or defective) and (2) costs imposed by processing customer complaints if no RTMS is in place. Customers in our simulation approach randomly issue identical orders to all suppliers. They also check all shipments arriving from the suppliers. If they are truthful customers, they will only complain if the shipment is indeed defective. Cheating customers, in contrast, may also reject a fraction of shipments that are not defective. Simulation time is discrete and a fixed number of processes are executed for all agents in every simulated period (see Figure 2).

As an initial condition, all suppliers will designate all customers as *trusted.* If a customer exceeds his or her reputation threshold, she/he will be removed from *trusted* status, requiring him or her to resend the shipment, thus generating complaint handling costs.

At the beginning of a period, each customer randomly decides with a prespecified probability whether she/he issues an order in this period or not (1). By varying the order likelihood of a customer, the transaction frequency between supplier and customer can be adjusted. If the customer decides to order in this period, she/he calculates an order quantity drawn randomly from a normal distribution and issues identical orders to all suppliers (2). Mean and standard deviation are prespecified in the simulation setting. After receiving orders from all customers, suppliers process orders and ship goods according to the quantities requested. A randomly drawn fraction of products shipped is defective. The defective ratio is normally distributed; the mean defective ratio and the standard deviation are input parameters of the simulation (3). After all goods have been shipped, customers check the received shipments to determine whether they have received defective products (4). If defective products are included in the shipment, the customer immediately issues a complaint message to the supplier specifying the amount of defective products (5). If the shipment is faultless and the customer is configured to act truthfully, nothing happens (6). If the shipment is faultless but the customer is configured as a potential cheater, she/he decides whether to cheat or not (7). If she/he decides to cheat, a complaint message is issued (8). In the next step, suppliers process all complaints received and act according to the trust status of the customer. If the customer is in trusted mode, the shipment is rescheduled without further checks and the reputation value of the customer is adjusted. If the customer is not in trusted mode, the shipment must be sent back by the customer to allow the supplier to verify the claim. If there are indeed defective products in the shipment, a new shipment is scheduled containing faultless products; otherwise nothing happens (9). Finally, all suppliers update the reputation values of all customers who are in trusted mode (10) and decide which customers to keep in trusted mode for the next period (11). Step 11 marks the

Table 3. Parameterization of different market scenarios

Market scenarios	Low transaction frequency, low fraction of cheating consumers (LFLC)	Low transaction frequency, high fraction of cheating consumers (LFHC)	High transaction frequency, low fraction of cheating consumers (HFLC)	High transaction frequency, high fraction of cheating consumers (HFHC)
	Parameterization			
Transaction frequency (order probability)	5%	5%	30%	30%
Fraction of cheating consumers	5%	25%	5%	25%

conclusion of the simulation period after which a new period begins with customers deciding whether to order.

To compare the different simulation runs, some settings are kept constant throughout all simulation runs. Each market simulation consists of 1,000 consumers and four suppliers each, trading for 1,000 periods. Each simulation run is repeated 50 times. Furthermore, all suppliers produce with an equal ratio of defective products (mean 0.02, standard deviation 0.05) in all simulations. The four simulated suppliers differ in terms of (1) reputation account thresholds and (2) usage of the reputation account mechanism:

- Suppliers 1, 2, and 3 differ in their threshold for determining whether a consumer regularly cheats or not.
- Supplier 1 applies a very high threshold, which means that she/he will apply the reputation account all the time (all customers will always be in trusted mode).
- Supplier 2 applies a medium threshold.
- Supplier 3 applies a low threshold.
- Supplier 4 does not apply the reputation account at all and marks the "bottom line" of a supplier without the proposed mechanism.
- Therefore, Suppliers 1 and 4 will mark the two extremes of the scale, with 1 always trusting all consumers and 4 never using reputation accounts (and therefore literally distrusting all consumers).

The 1,000 consumers share consistent overall parameters, differing only in their attitude toward "cheating". A fraction of the 1,000 consumers will never cheat, while others will consider cheating, the proportion of whom will be varied in the simulation runs. When placing an order, all consumers share the same normal distribution of order quantity (mean = 100, *SD* = 75). They also will always complain if there is at least one defective item in a given shipment. If a consumer belongs to the group of cheating consumers, she/he will try to cheat with a likelihood of 15%. If she/he decides to cheat, she/he will always try to complain 20% of the original (faultless) shipment.

To assess in which market settings the RTMS will be advantageous, different idealized markets are simulated. In the following, the impact of transaction frequency and the impact of different fractions of cheating consumers on the reputation system will be investigated (cf. Table 3).

Sensitivity Analysis of the RTMS

To analyze the results, the average number of cheated products per 1,000 items shipped was calculated. Figure 3 provides the results for the four simulated market scenarios. The headers depict the type of market scenario (e.g., HFHC stands for *high frequency of transactions, and high fraction of cheating consumers*; see also Table 3).

As expected, Supplier 4, who always distrusts all customers and does not apply the reputation mechanism, does not experience loss through cheating customers since, even if there is complaint, it will always check whether the claim was valid. On the other hand, Supplier 1, who always trusts everyone and employs the new system, has a ratio of approximately 12 cheated items per 1,000 shipped in the scenarios with a high fraction of cheating consumers, and 3 cheated items per 1,000 shipped in the low cheating scenario. For the other suppliers, the fraction of cheated products not detected ranges between those extremes. Therefore, it can be stated that for the given settings,

Figure 3. Simulation results

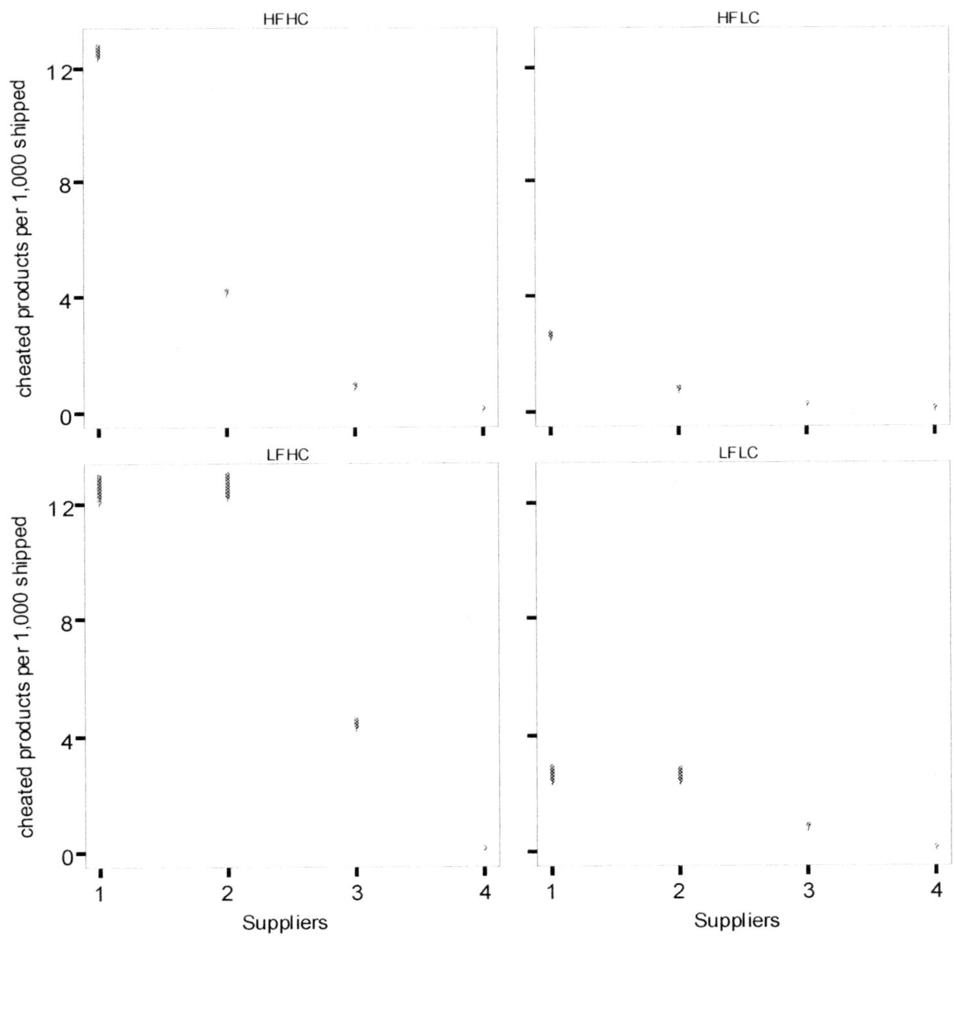

the reputation account system is able to identify cheating customers and to eliminate them from the trusted mode system (cp. Suppliers 2 and 3). In the case of markets with low transaction frequency, Supplier 2 is unable to achieve a better result than Supplier 1. In these cases, the reputation account system takes more time to identify the cheating customers. The system works best in markets with a high transaction frequency. In markets with low transaction frequency, the system will fail. In low-transaction-frequency scenarios, the threshold ratios must be set lower to ensure that cheating consumers are identified. In scenarios with a low transaction frequency and a low fraction of cheating customers (cp. LFLC), the effect of the reputation account system is small.

Profitability Analysis of the RTMS

We now look at the profits of suppliers depending on the customer complaint-handling costs. The absolute number of complaints is independent of the costs associated with the complaint. Based on the mean values of the simulation runs, it is feasible to calculate the financial flows in each market scenario. For the HFHC market scenario, the results for different complaint-handlings costs levels are depicted in Figure 4. The main tendency can also be found in the other scenarios, but it is most clearly visible in this scenario. If customer complaint-handling costs are high compared to variable production costs, the reputation account solution is always advantageous (cp. Supplier 4 without deploying a reputation account solution

Figure 4. Profits (complaint-handling costs: upper left = 100, upper right = 50, lower left = 2, lower right = 1)

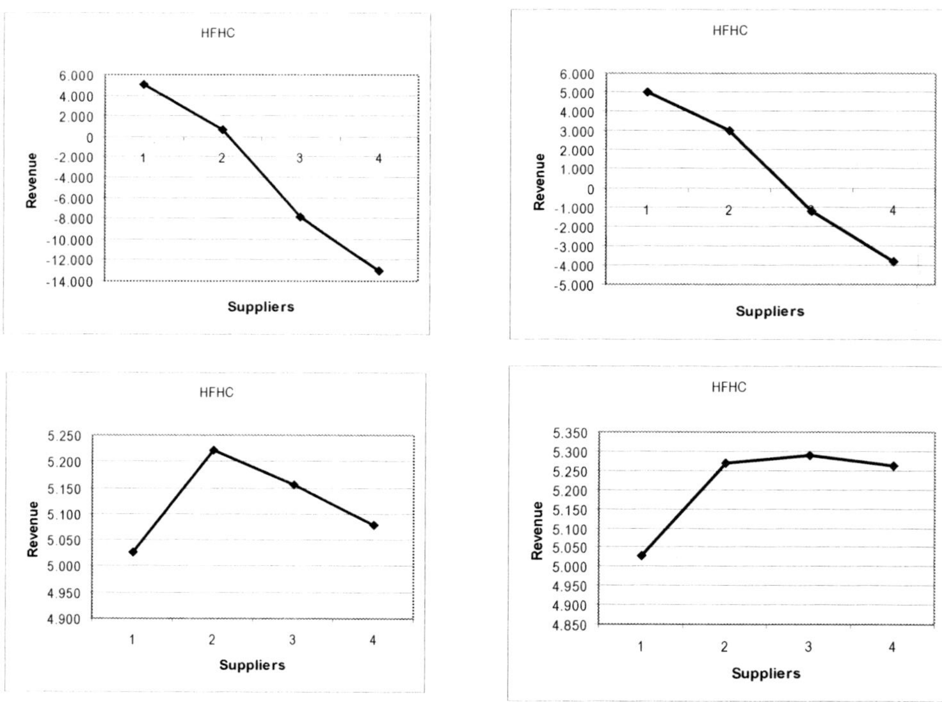

has the highest losses of all suppliers in the upper two diagrams of Figure 4). Not until customer complaint handling costs nearly equal product revenues (see Figure 4, lower left diagram) or are below product revenues (see Figure 4, lower right diagram), does the supplier without the reputation account mechanism become profitable. In these scenarios depicted in the lower two diagrams, the reputation account mechanism is not always the best solution, especially the "always trust" strategy of Supplier 1 should not be applied.

In summary, the proposed reputation account mechanism is especially advantageous in settings where (a) the transaction frequency is high, (b) the individual complaint-handling costs are high compared to product revenues, and (c) the market has a high fraction of potentially cheating consumers. In markets where complaint-handling costs are low compared to the individual production costs, the reputation account mechanism should not be deployed.

SUMMARY AND CONCLUSIONS

The combination of information systems and game-theory inspired reputation and trust accounts in a RTMS establishes new solutions to automate business transactions where human decisions were formerly necessary. Through the reduction of manual handling and shipping costs, quality of the complaint-handling process may be increased both for customers and suppliers, resulting in higher customer retention. A game-theoretic analysis of the order and customer complaint process has yielded insights into undesired outcomes of the interaction of suppliers and customers. While faulty deliveries will always remain a problem, costs associated with customer complaint-handling can be reduced significantly if substituting human decision competence with an automated information system. Thus, we believe that an economic interpretation of existing information systems may help to uncover as-yet unrealized potential for computer-mediated applications and offer the RTMS as an example for this claim. The RTMS allows firms to deploy a simplified customer complaint-handling process while preventing customers from acting opportunistically. The RTMS has been developed according to the guidelines put forward for design science approaches and has been tested in an agent-based artificial setting, indicating its strength in specific market environments. In more detail, the RTMS has been found applicable in market environments where (a) the transaction frequency is high, (b) the individual complaint-handling costs are high compared to product revenues, and (c) the market has a high fraction of potentially cheating consumers.

REFERENCES

Carter, J., Bitting, E., & Ghorbani, A. A. (2002). Reputation formalization within information sharing multiagent architectures. *Computational Intelligence, 18*(4), 515-534.

Dahlstrom, R., & Nygaard, A. (1999). An Empirical investigation of ex post transaction costs in franchised distribution channels. *Journal of Marketing Research, 36*(2), 160-171.

Das, T. K., & Teng, B.-S. (1998). Between trust and control: Developing confidence in partner cooperation in alliances. *Academy of Management Review, 23*(3), 491-512.

Eisenhardt, K. M. (1989). Building theories from case study research. *Academy of Management Review, 14*(4),532-550.

Fornell, C., & Wernerfelt, B. (1987). Defensive marketing strategy by customer complaint management: A theoretical analysis. *Journal of Marketing Research,* 24(4), 337-346.

Glaser, B. G., & Strauss, A. L. (1967). *The discovery of grounded theory: Strategies for qualitative research.* New York: Aldien de Gruyter,.

Hallen, G., & Latino, R. J. (2003, June). Eastman Chemical's success story. *Quality Progres*, 50-54.

Hevner, A. R., March, S. T., & Park, J. (2004). Design science in information systems research. *MIS Quarterly, 28*(1), 75-105.

Hirschman, A. O. (1970). *Exit voice and loyalty: Responses to decline in firms, organizations, and states.* Cambridge, MA: Harvard University Press.

Lane, C., & Bachmann, R. (1996). The social constitution of trust: Supplier relations in Britain and Germany. *Organization Studies, 17*(3), 365-395.

Malone, T. W., & Crowston, K. (1994). The interdisciplinary study of coordination. *ACM Computing Survey, 26*(1), 87-119.

March, S. T., & Smith, G. F. (1995). Design and natural science research on information technology. *Decision Support Systems, 15*(4), 251-266.

Markus, L. M., Majchrzak, A., & Gasser, L. (2002). A design theory for systems that support emergent knowledge processes. *MIS Quarterly, 26*(3), 179-212.

Marsh, S. (1992). *Trust and reliance in multiagent-systems: A preliminary report.* University of Stirling, Department of Computer Science and Mathematics,.

Merton, R. K. (1968). *Social theory and social structure.* New York: The Free Press.

Moorman, C., Zaltman, G., & Deshpande, R. (1992). Relationships between providers and users of market research: The dynamics of trust within and between organizations. *Journal of Marketing Research,* 29(3), 314-328.

Mui, L., Halberstadt, A., & Mohtashemi, M. (2002). Notions of reputation in multiagents systems: A review. In *Proceedings of the First International Joint Conference on Autonomous Agents and Multiagent Systems* (pp. 280-287)l New York: ACM Press.

Mui, L., Mohtashemi, M., & Halberstadt, A. (2002). A computational model of trust and reputation. In *Proceedings of the 35th Hawaii International Conference on System Science (HICSS)* (p. 188). Big Island, HI:

Noordewier, T. G., George, J., & Nevin, J. R. (1990). Performance outcomes of purchasing arrangements in industrial buy-vendor relationships. *Journal of Marketing Research, 54,* 80-93.

Rust, R. T., Subramanian, B., & Wells, M. (1992). Making complaints a management tool. *Marketing Management, 1*(3), 40-45.

Sabater, J., & Sierra, C. (2001). REGRET: A reputation model for gregarious societies. In *Proceedings of the Fourth Workshop on Deception Fraud and Trust in Agent Societies* (pp. 61-70). Montreal, Canada:

Sabater, J., & Sierra, C. (2002). Reputation and social network analysis in multiagent systems. In *Proceedings of the First International Joint Conference on Autonomous Agents and Multiagent Systems* (pp. 475-482). New York: ACM Press.

Schilling, V., & Sobotta, A. (1999). Prozesskostenrechnung der mittelständischen industrie. *Betriebswirtschaftliches Forschungszentrum für Fragen der mittelständischen Wirtschaft e.V.* Bayreuth, Germany.

Tax, S. S., & Brown, S. W. (1998). Recovering and learning from service failure. *Sloan Management Review, 40*(1), 75-88.

Walls, J. G., Widmeyer, G. R., & El Sawy, O. A. (1992). Building an information system design theory for Vigiliant EIS. *Information Systems Research, 3*(1), 36-59.

Wigand, R. T., Picot, A., & Reichwald, R. (1997). *Information, organization and management: Expanding markets and corporate boundaries.* Chichester, England: Wiley.

Williamson, O. E. (1975). *Markets and hierarchies: Analysis and antitrust implications. A Study in the economics of internal organization.* New York: The Free Press.

Williamson, O. E. (1985). *The economic institutions of capitalism: Firms, markets, relational contracting.* New York: The Free Press.

This work was previously published in Journal of Electronic Commerce in Organizations, Vol. 6, Issue 4, edited byM. Khosrow-Pour, pp. 8-29, copyright 2008 by IGI Publishing (an imprint of IGI Global).

Chapter 8
An Empirical Study on the Integrated Framework of e-CRM in Online Shopping:
Evaluating the Relationships Among Perceived Value, Satisfaction, and Trust Based on Customers' Perspectives

Changsu Kim
Yeongnam University, Korea

Weihong Zhao
Jiangxi Normal University, China

Kyung Hoon Yang
University of Wisconsin-La Crosse, USA

ABSTRACT

Based on customer cognitive, affective and conative experiences in Internet online shopping, this study, from customers' perspectives, develops a conceptual framework for e-CRM to explain the psychological process that customers maintain a long-term exchange relationship with specific online retailer. The conceptual framework proposes a series of causal linkages among the key variables affecting customer commitment to specific online retailer, such as perceived value (as cognitive belief), satisfaction (as affective experience) and trust (as conative relationship intention). Three key exogenous variables affecting Internet online shopping experiences, such as perceived service quality, perceived product quality, and perceived price fairness, are integrated into the framework. This study empirically tested and supported a large part of the proposed framework and the causal linkages within it. The empirical results highlight some managerial implications for successfully developing and implementing a strategy for e-CRM.

INTRODUCTION

Customer relationship management in electronic commerce (e-CRM) is one of the fastest growing management techniques adopted by online enterprises. It is premised on the belief that developing long-term relationships with customers is the best way to gain customer loyalty (Reichheld & Schefter, 2000). Much research has been done on such topics as e-CRM management (Romano & Fjermestad, 2003), e-CRM marketing techniques (Jackson & Wang 1995; Pan & Lee, 2003), the adoption of e-CRM in organizations (Wu & Wu, 2005), and e-CRM applications that facilitate Internet business (Wang & Head 2001; Adebanjo, 2003). Still, many online enterprises encounter difficulties implementing effective e-CRM because they tend to overlook customer perspectives on e-CRM issues (Woodcock & Starkey, 2001).

This study suggests that online retailers should have a substantial understanding of factors that can influence a customer to maintain a long-term relationship with them. Therefore, the objective of this study is to develop an integrated e-CRM framework by investigating the psychological process that occurs when a customer maintains a long-term relationship with an Internet online retailer. We developed a theoretical model based on Oliver's four-stage cognitive-affective-conative-action model (1980). Each cognitive, affective, and conative factor is explained by the theory of reasoned action (TRA), the expectation confirmation theory (ECT), and the commitment trust theory (CTT), respectively. By highlighting key factors, we empirically investigate a series of dynamic linkages among these factors affecting customer long-term relationship orientation. Finally, managerial implications and limitations are discussed in the conclusion.

THEORETICAL BACKGROUND

Overview of Web-Based Customer Relationship Management

E-CRM is a newly developing customer-oriented business philosophy that reorients online enterprise operations in order to improve customer satisfaction, loyalty, and retention (Adebanjo, 2003; Pan & Lee, 2003). E-CRM is not only about technology but also about aligning business processes with customer strategies (Rigby, Reichheld, & Schefter, 2002). Anton (1996) emphasizes the management of customer relations with "continuous improvement or re-engineering" for better service and competitive positioning. Behavioral and affective factors are also regarded as playing a major role in e-CRM (Vellido, Lisboa, & Meehan, 2000; Romano & Fjermestad, 2003; Wu & Wu, 2005; Raihan & Hamid, 2005).

Many facets discussed in the literature about e-CRM correspond to stages in Oliver's cognitive-affective-conative-action model for loyalty (1980). The first stage in this model is cognitive loyalty. In this stage, a consumer may continue to patronize a store for a compelling reason. However, loyalty at this stage is driven purely by functional characteristics, and the consumer's commitment towards the store may not be very strong. The next stage is affective loyalty. This is stronger than cognitive loyalty because the consumer's affective loyalty is a function of both prior attitudes about the store followed by satisfaction. The third stage is conative loyalty. At this phase the consumer has a commitment to buy. This stage is exemplified by repurchase intentions and actions such as recommending the store to others. The final stage is action loyalty, in which the customer purchases items from a store on an ongoing basis.

In the next three sections, we will explain e-CRM in terms of these cognitive, affective, and conative stages that lead to the final stage of "action loyalty."

Explanation of Cognitive Belief

Purchasing stages of Internet online shopping can be classified into five phases: information research, ordering, requesting post-purchase service, taking delivery, and paying online (Nour & Fadlalla, 2000). Through these experiences, customers have a cognitive response related to the perceived benefits and costs of a purchase through a specific retailer (Zeithaml, 1988). Cognition can also be described as awareness, knowledge, or beliefs that may or may not have been derived from previous shopping experiences (Fishbein, 1967).

According to the theory of reasoned action (TRA) (Fishbein & Ajzen, 1975; Ajzen & Fishbein, 1980), an individual's performance is determined by his or her behavioral intentions, which are jointly determined by cognitive factors such as attitudes and subjective norms. As an extended model of TRA, the theory of planned behavior (TPB) was derived by adding perceived behavioral control as a determinant of behavior (Ajzen, 1985). Davis (1986) proposed the technology acceptance model (TAM), based on TRA and TPB, to explain and predict user acceptance of information systems or information communication technology (ICT). In TAM, cognitive beliefs such as perceived usefulness and perceived ease of use are counted as key factors for technology acceptance. The three theories (TRA, TPB, and TAM) have been widely validated and are widely used for predicting or explaining cognitive behavior is social psychology. In addition, many information systems studies have been conducted based on TAM (Shih, 2004; Yu et al., 2005).

Explanation of Affective Experience

The expectation-confirmation theory (ECT) has suggested that satisfaction is the primary motivation for the continued purchase of a product or service (Oliver, 1980). Satisfaction depends on the extent to which customers perceive their prior expectations of a product or service to be confirmed during actual use (Zhang & Prybutok, 2005).

The majority of previous studies consider satisfaction to be an affective response to an expectancy confirmation involving a cognitive process (Pascoe, 1983; Melone, 1990; Taylor, 1994; Oliver, 1997). Anderson and Srinivasan (2003) suggest that customer satisfaction should be evaluated as a positive, indifferent, or negative feeling following the customer's initial experience with the service. This affective evaluation is identical to the notion of attitude in the IS-use literature (Melone, 1990), and the attitude-intention association validated in IS-use research provides additional support for the association between satisfaction and choice or continued usage (Davis, 1989; Mathieson, 1991; Taylor & Todd 1995).

Explanation of Conative Behavior

Geyskens, Steenkamp, Steenkamp, Scheer, and Kumar (1996) describe commitment as a customer's long-term orientation toward a business relationship. Morgan and Hunt (1994), Kalafatis and Miller (1997), and Wu and Cavusgil (2006) consider commitment as the crucial factor in determining long-term customer retention. Morgan and Hunt's (1994) empirically validated and widely accepted commitment-trust theory (CTT) claims that long-term relationships are built on the foundation of mutual "trust-commitment," which is similar to the process of creating long-term traditional buyer-seller relationships (Wu & Cavusgil, 2006; Pan, Pan, Newman, & Flynn, 2006). Because of the connection between customer commitment and the buyer-seller relationship, inducing customer commitment is a crucial issue for the development and implementation of an e-CRM strategy.

On the basis of CTT, some researchers have analyzed the importance of trust in online relationships as a conative response to cognitive beliefs and affective experiences (Lee & Turban, 2001; McKnight & Chervany, 2002). Lack of trust has been regarded as one of the greatest factors inhibiting online business (Marti & Garcia-Molina, 2006). On the other hand, Business Week (2001) reported that customers are willing to buy repeatedly from the most trusted sites. Thus, online retailers rely on strong trust to build committed, conative customer behavior.

INTEGRATED FRAMEWORK AND RESEARCH HYPOTHESES

Based on the theoretical background, this study suggests that e-CRM should satisfy customers' psychological needs and induce them to commit to long-term relationships.

Previous studies from manifold research areas such as technology and human factors have contributed to successful e-CRM. The majority of studies have focused on factors that affect customers' long-term relationship orientation (Cole, O'Keefe, & Siala, 1999; Sukpanich & Chen 1999; Liu & Arnett 2000; Phau & Poon 2000; Vellido et al., 2000). Less research has been centered on an integrated conceptual model to explain why customers commit to a long-term exchange relationship with a specific online retailer.

To address this gap in the research, this study concentrates on customers' cognitive beliefs, affective experiences, and conative behaviors in order to propose an integrated framework for e-CRM that identifies a series of linkages among the psychological variables of perceived value, satisfaction, and trust. In addition, three key exogenous variables affecting shopping experiences are considered for the integrated framework (Figure 1).

As shown in Figure 1, trust is regarded as the key factor affecting customer relationship commitment (Morgan & Hunt, 1994; McKnight & Chervany, 2002; Flavian, Guinaliu, & Gurrea, 2006). Perceived value (as a cognitive belief) and satisfaction (as an affective experience) directly and indirectly influence trust (as conative behaviors) and commitment. According to related literature, however, this cognitive belief or affective experience alone is insufficient; both are critical antecedents to behavioral intention, which can be considered a conative behavior (Grewal & Shrama, 1999; Cronin, Brady, & Hult, 2000; Parasuraman & Grewal, 2000; Sirdeshmukh, Singh, & Sabol, 2002).

In examinations of exogenous factors influencing customer commitment, most studies have focused on service quality (Jarvenpaa & Todd, 1997; Sharma & Patterson, 1999; Zeithaml, Parasuraman, & Malhotra, 2000; Venetis & Ghauri, 2004). Much attention has been paid to the intangibility of service quality (Bebko, 2000; Elliot & Fowell, 2000; Crosby & Johnson, 2004). They also insisted that service quality is a necessary but not sufficient factor leading to customer commitment to a long-term relationship. Perceived product quality is also regarded as an important exogenous factor related to customer commitment (Elliot & Fowell, 2000; Crosby & Johnson, 2004). Price, which has been investigated in previous studies (Voss, Parasuraman, & Grewal, 1998; Crosby & Johnson, 2004; Homburg, Hoyer, & Koschate, 2005) is another important factor influencing customer commitment.

In Figure 1, we use solid lines to represent relationships supported by previous research and dotted lines to represent our own hypotheses, which are based on the literature but not on specific research.

Commitment

Though commitment has been described in many ways, it is usually defined as the explicit or implicit pledge of relational continuity between exchange partners (Dwyer, Schurr, & Oh, 1987); and a commitment is described as an enduring desire to maintain a valued relationship (Moorman, Zaltman, & Deshpande, 1992). Moreover, Morgan and Hunt (1994) describe commitment as an ongoing relationship that is maintained through maximum efforts from both partners. Others define commitment as the perception of interdependence in which outcomes are expected to benefit both parties in the long run (Wu & Cavusgil, 2006). Still others define commitment as a "psychological attachment," meaning an affective disposition reflecting the desire and willingness of the parties to devote considerable effort to maintaining a relationship (Gruen, Summers, & Acito, 2000). Despite the diversity of definitions of commitment, all refer to the likelihood of continuing the relationship in the future.

This study's focus is on commitment as an attitude that guides or mediates an individual's overt response or behavioral intention regarding an object (Ajzen & Fishbein, 1980). Thus, commitment is defined as the customer's behavioral intention to continue a business relationship. It refers not only to a customer's future transactional intentions, but also to the deliberateness of these intentions, distinguishing a committed relationship from a mere transactional one. It reflects the affective and cognitive motivations that maintain a long-term relationship and the tendency to resist changing preferences (Pan et al., 2006; Wu & Cavusgil, 2006).

Trust

Based upon previous literature, this study defines trust as a set of specific relationship intentions dealing primarily with integrity, benevolence, competence, and predictability of an Internet online retailer (Gefen, Karahanna, & Straub, 2003). The majority of literature has emphasized the role of trust as being central to the success of building customer relationships (Morgan & Hunt, 1994; Lee & Turban, 2001; McKnight & Chervany, 2002; Siau & Shen, 2003). Online retailers have to make customers trust them; without trust, customers will avoid shopping online (Gefen, 2000). Some studies have suggested that online customers generally stay away from online retailers that they do not trust (Jarvenpaa & Tractinsky, 1999). The online shopping customer overcomes an exploratory stage before carrying out commercial transactions with specific Internet retailers, and the customer's experience with the retailer will determine his or her commitment and level of trust (Reichheld & Schefter, 2000). This study speculates that customer trust in specific Internet retailers leads to trust-related attitudinal commitment. Based on these observations, the following hypothesis is proposed:

H1: *Trust will positively affect commitment to a relationship.*

Perceived Value

In the literature, perceived value has been identified as having four components (Grewal, Monroe, & Krishnan, 1998; Parasuraman & Grewal, 2000): Acquisition value is the benefit buyers believe they are getting by acquiring a product; transaction value is the pleasure of getting a good deal; in-use value is the utility derived from using the product; and redemption value is the price of the product at the time of trade-in or end-of-life. The attributes of a product serve as the logical driver of perceived value within the framework of these four components inasmuch as customers place value on product attributes that are directly related to their wants or needs. Customer perceived value is also important because it can lead to brand loyalty (Chiou, 2004; Li, Xu, & Li, 2005).

Consistency between product attributes and customer values (i.e., positive perceived value) reduces uncertainty and helps the customer build trust in the form of reliable expectations of the provider in ongoing exchanges (Carver & Scheier, 1990). Previous research supports this association between perceived value and trust (Sirdeshmukh et al., 2002; Anderson & Srinivasan, 2003).

Consistent with the theory of reasoned action (Fishbein & Ajzen, 1975), this study speculates that a customer's positive perceived value of both products and service can lead to trusting beliefs and result in intentions to commit to a long-term relationship with an online retailer. Based on these observations, the following hypotheses are proposed:

H2a: *Perceived value will affect trust.*

H2b: *Perceived value will affect commitment to a relationship.*

Satisfaction

Oliver (1997) defines satisfaction as the consumer's fulfillment response, hence satisfaction involves a minimum of two stimuli—an outcome and a comparison referent. This study conceptualizes customer satisfaction as an evaluation of emotions reflecting the degree to which the customer believes the service provider evokes positive feelings (Cronin et al., 2000). In addition, customer satisfaction has traditionally been regarded as a fundamental determinant of long-term consumer behavior (Zhang & Prybutok, 2005).

In relational sales contexts, trust is a cumulative process that develops over the course of repeated satisfactory interactions with the salesperson. Satisfaction from past transactions provides the customer with confidence in the retailer (Siau & Shen, 2003). Empirical studies have shown that satisfaction is an antecedent of trust (Garbarino & Johnson, 1999; Siau & Shen, 2003). Likewise, many previous studies have hypothesized and empirically validated the link between satisfaction, customer retention and word-of-mouth (Rucci, Kirn, & Quinn, 1998; Bansal & Taylor, 1999; Cronin et al., 2000). Indeed, this kind of link is fundamental to the business concept that holds that satisfying customer needs and wants is the key to repeat purchases. Many empirical researchers have shown that customer satisfaction leads to buyer anticipation of future interactions with and commitment to the salesperson (Johnson, Barksdale, & Boles, 2001; Zhang & Prybutok, 2005). Therefore, this study speculates that a customer's satisfaction from experience with a specific online retailer leads to trust and commitment to a relationship. Based on these observations, the following hypotheses are proposed:

H3a: *Satisfaction will positively affect trust.*

H3b: *Satisfaction will positively affect commitment to a relationship.*

Perceived Service Quality

Service quality is considered to be one of the key determinants of online retailer success (Jarvenpaa & Todd, 1997; Zeithaml et al., 2000). Berry (1981) defined perceived service quality as the customer's judgment about a firm's overall excellence or superiority. This definition suggests that perceived quality is similar to an individual's general attitude toward the firm (Bitner, 1990). Zeithaml et al. (2000) defined Web site service quality as the extent to which a Web site facilitates efficient and effective shopping, purchase, and delivery of products and services.

Online shoppers perceive the benefits and convenience of obtaining information about products directly from the Internet rather than through an off-line store (Ariely, 2000; Zhang & Prybutok, 2005). The reduction in time needed to research product costs and product-related information has also been identified as a key benefit of online shopping (Ariely, 2000; Lynch & Arielty, 2000). Being able to search quickly for information about a product increases customer satisfaction and improves customer intentions to revisit and repurchase from a specific online retailer (Lynch & Ariely, 2000). Customer perceptions of convenience directly affect perceptions of service quality and satisfaction in online shopping (Zhang & Prybutok, 2005). Similarly, Wolfinbarger and Gilly (2001) found that fulfillment ratings were the strongest predictor of customers' satisfaction and intentions to repurchase items or service from an online retailer. In these studies and others, there is a consensus among researchers that favors superior customer service as having the greatest impact on customer satisfaction and loyalty (Rust & Lemon, 2001; Winer, 2001; Grewal, Gopalkrishnan, & Levy, 2004).

Therefore, this study speculates that perceived service quality of an online retailer is positively associated with perceived value and customer satisfaction. Based on the online business and service quality literature, this study defines perceived service quality in online shopping as the customer's overall evaluation and judgment of the excellence and quality of the online shopping experience. Thus, a perception of service quality delivered by an online retailer is hypothesized as an antecedent of perceived value and satisfaction.

Based on these observations, the following hypotheses are proposed:

H4a: *Perceived service quality will positively affect perceived value.*

H4b: *Perceived service quality will positively affect satisfaction.*

Perceived Product Quality

According to Garvin (1983), perceived product quality is the customer's judgment about the superiority or excellence of a product. For both online and off-line shoppers, perceived product quality could be a key factor, especially in maintaining long-term customer relationships (Snoj, Korda, & Mumel, 2004; Petre, Minocha, & Roberts, 2006). Perceived product quality plays a crucial role affecting purchasing choices (Brucks, Zeithaml, & Naylor, 2000). Similarly, many previous studies suggest that perceived product quality is positively associated with perceived value (Sebastianelli & Tamimi, 2002; Snoj et al., 2004; Story & Loroz, 2005; Petre et al., 2006). Perceived product quality, as a form of overall evaluation of a product, is a relatively global value judgment (Sebastianelli & Tamimi, 2002; Crosby & Johnson, 2004; Story & Loroz, 2005). As the perception of value comes from a tradeoff between a "give" component (perceived sacrifice) and a "take" component (in the form of products and services), a higher level of perceived product quality will lead to a higher level of perceived value.

In addition, product quality is the core concern in purchasing decisions. Product quality and service quality are of the same importance in affecting customer satisfaction (Parasuraman, Zeithaml, & Berry, 1994). However, because customers are unable to see the actual product in an online shopping context, their decisions are based on expectations about product quality rather than certain knowledge. Satisfaction occurs when the product quality is greater than initially expected. After consumption, each customer evaluates the quality of the purchased product and updates his or her expectations about the quality of future purchases (Athanassopoulos, 2000; Hellier, Geursen, Carr, & Rickard, 2003). Thus, product quality is a key factor affecting shopping satisfaction, especially in maintaining long-term customer relationships.

This study, therefore, speculates that perceived product quality is positively associated with perceived value and satisfaction in online shopping. Based on these observations, the following hypotheses are proposed:

H5a: *Perceived product quality will positively affect perceived value.*

H5b: *Perceived product quality will positively affect satisfaction.*

Perceived Price Fairness

For customers, the perceived price, which includes time, effort, and search costs, is more meaningful than the actual monetary price of an item or service. The customer usually judges price and service quality based on the concept of "equity" and generates his or her satisfaction or dissatisfaction level based on that concept (Oliver, 1997). In other words, in an evaluation of overall price fairness, the customer considers both the monetary and non-monetary costs of acquiring the product or service (Athanassopoulos, 2000; Cronin et al., 2000; Homburg et al., 2005).

Fornell, Johnson, Anderson, Cha, and Bryant (1996) also found that price perceptions affect customer satisfaction in seven industry sectors. Similarly, in an experimental setting involving a hotel check-in scenario, Voss et al. (1998) suggested that price perception affects satisfaction. When shopping on the Internet, customers cannot actually see or handle the product, so price perceptions play a critical role in post-purchase satisfaction and intentions to return to the online retailer for future purchases (Jarvenpaa & Todd, 1997; Liu & Arnett, 2000; Babin et al., 2003; Bolton, Warlop, & Alba, 2003). Furthermore, studies show that perceived price fairness is important in determining the duration of the provider-customer relationship (Bolton & Lemon, 1999; Bolton et al., 2000).

This study, therefore, speculates that perceived price fairness, in terms of monetary and non-monetary costs to the consumer, is positively associated with perceived value and satisfaction in the online shopping experience. Based on these observations, the following hypotheses are proposed:

H6a: *Perceived price fairness will positively affect perceived value.*

H6b: *Perceived price fairness will positively affect satisfaction.*

DATA ANALYSIS

Measures

Seven constructs were measured in this study: perceived service quality, perceived product quality, perceived price fairness, perceived value, satisfaction, trust, and commitment. All constructs were measured using multiple items fully anchored and seven-point Likert scales ranging from "strongly disagree" (1) to "strongly agree" (7). To ensure content validity of the scales, items were selected to represent the concept about which generalizations were made. Therefore, items selected for the constructs were adapted from prior studies to ensure content validity. These scales, with their related literature, are presented in Table 1.

Data Collection and Sample Characteristics

Two-hundred fifty survey questionnaires were distributed in three universities in Korea, and 232 were returned. Among them, seven were discarded because of errors and 225 survey questionnaires were used. The demographic characteristics are given in Table 2. The respondent group ranged in age from 20 to 35, consisting of 41.5% male and 58.5% female; 96% were university students; 50.2% had more than 12 months of online shopping experience; 75.1% had an online shopping frequency of 1~5 times per month; and 13.4% had an online shopping frequency of greater than 5 times per month in the last 2 years.

RESULTS

Measurement Assessment

Construct validity determines the extent to which a scale measures a variable of interest. In this study, to test construct validity in terms of convergent validity and discriminant validity, a principle component factor analysis with varimax rotation was conducted for distinctions of variables. As shown in Table 3, seven factors emerged with no cross-construct loadings above 0.5, indicating good discriminant validity. The instrument also demonstrated convergent validity with factor loadings exceeding 0.5 for each construct except for one item measuring perceived service quality, which was deleted because its loading value was under 0.5, and one item measuring perceived value, which was deleted because of cross-construct loading. Consequently, after deleting these two items, the results confirm that each of the seven constructs is unidimensional and factorially distinct and that all items used to operationalize a particular construct are loaded onto a single factor.

In addition, we performed a confirmatory factor analysis (CFA), shown in Table 4. As a result, the factor loadings of each scale item range from 0.544 (PSQ1) to 0.925 (PPF3) and are greater than the recommended level of 0.40, which is based on 200 samples and 0.05 significance level (Hair, Anderson, Tatham, & Black, 1998).

Finally, reliability was evaluated using Cronbach's alpha to assess the internal consistency of the items representing each construct. The reliability of each construct, as shown in Table 5, ranged from 0.861 to 0.926, exceeding the common threshold value (0.60) recommended by Nunnally (1978).

Analysis of Research Hypotheses

Structural equation modeling was performed to test the proposed hypothesized relationships in the integrated framework using AMOS 5.0. As presented in Figure 2, the results of testing the structural model indicate a favorable fit of the model. The chi-square statistic (χ^2 = 838.918, degrees of freedom = 450, p =0.000) is significant because of the sensitivity of the sample size. GFI is 0.82, AGFI is 79, RMR is 0.076, NFI is 0.86, and CFI is 0.93; and the RMSEA value of 0.062 (<0.08) indicates an acceptable fit of the data according to Arbuckle and Wothe (2000) cutoff criteria. Thus, overall, the data indicates a favorable fit for our proposed model.

Trust influences commitment remarkably (H1), and perceived value and satisfaction influence trust remarkably (H2a and H3a). Satisfaction weakly influences commitment (H3b), but perceived value was not found to have a significantly direct impact on commitment (H2b). Perceived service quality influences satisfaction (H4b), but it does not influence perceived value (H4a). Perceived product quality influences satisfaction strongly (H5b) and perceived value reasonably (H5a). Perceived price fairness was found to have a significantly direct impact on perceived value (H6a), but a weak impact on satisfaction (H6a). Results are summarized in Table 6.

DISCUSSION

The results indicate that trust plays a crucially intervening role in the relationship among perceived value, customer satisfaction, and commitment in online shopping.

The results demonstrate that perceived value and satisfaction stem from online retailer attributes that lead to trusting beliefs and result in intentions to engage in trust-related behaviors such as attitudinal commitment and long-term relationship orientation. Although Snoj et al. (2004) suggest that perceived value leads to store and brand loyalty, the relationship between the two constructs in this study's results is not straightforward.

Perceived value in the online shopping context shows direct impact on trust and an indirect impact on commitment (which is developed through trust). This is because perceived value, as a cognitive belief, is based on experience. This result is also consistent with the cognitive-affective-conative behavioral model in the reasoned action theory (Fishbein, 1967).

In addition, for the three exogenous factors, perceived service quality has an impact on satisfaction (path coefficient 0.44) but not on perceived value, perceived product quality has a stronger impact on satisfaction (path coefficient 0.50) than on perceived value (path coefficient 0.17), and perceived price fairness has a stronger impact on perceived value (path coefficient 0.55) than on satisfaction (path coefficient 0.20). The results suggest that perceived price fairness has a strong impact on perceived value, while perceived product quality and perceived service quality have a strong impact on satisfaction.

These results not only provide empirical validation of the proposed integrated framework for e-CRM in this study, but also provide managerial implications for online retailers to successfully develop and implement e-CRM.

CONCLUSION AND IMPLICATIONS

Based on customers' cognitive, affective and conative experiences in online shopping, this study, from customers' perspectives, proposes an integrated e-CRM framework identifying a series of linkages among the main psychological variables affecting customer commitment. Moreover, most of the linkages proposed in the integrated framework are found to be significant.

There are managerial implications for these findings. First, the integrated framework shows factors that explain why customers maintain long-term exchange relationships. The integrated e-CRM framework proposed in this study can provide interrelated explanations about what brings customers back. As presented in this framework, successful e-CRM needs to effectively enhance customer value management, customer satisfaction management, and customer trust management through ICT in order to induce customer commitment.

Second, the integrated framework suggests that perceived value and satisfaction, as cognitive and affective antecedents, influence online trust directly; however, they influence customer commitment only indirectly through trust. Trust is the core factor that directly affects relationship commitment. Thus, it is critically important for online retailers to build customer trust in the shopping context. Providing superior value and a satisfying shopping experience are two effective ways to build customer trust.

Third, the results clearly suggest that superior value and customer satisfaction originate from superior perceived service or product quality, and from perceived price fairness. Thus, for online retailers, a successful implementation of e-CRM should integrate business and ICT to work closely together to effectively offer superior customer value and satisfaction via customized services or products, and competitive pricing.

This study has some limitations that can be addressed in future research. First, the investigation of the e-CRM integrated framework based on customers' perspectives is relatively new to e-commerce researchers. In the future, additional research efforts are needed to evaluate the validity of the e-CRM integrated framework. Second, the constructs in the framework need to be more completely measured with multi-dimensional instruments. Finally, the research findings gathered from university students need greater generalization in B2C e-commerce. There may be limitations related to using university students in the survey (Lumayem & Hirt, 2003). However, online access and purchasing power among students is high.

As we stated previously in this article, the focus of the study is on the development of an integrated framework. Thus, additional research efforts are needed to evaluate the validity of the proposed models and our findings.

ACKNOWLEDGMENT

The authors thank the associate editor and four anonymous referees for their comments on earlier drafts of the article. Their comments and suggestions were helpful in improving the quality of the article. All errors are those of the authors.

REFERENCES

Adebanjo, D. (2003). Classifying and selecting e-CRM applications: an analysis-based proposal. *Management Science*, *41*(6), 570-577.

Ajzen, I., & Fishbein, M. (1980). The prediction of behavior from attitudinal and normative variables. *Journal of Experimental Social Psychology*, *6*(4), 466-487.

Anderson, R. E., & Srinivasan, S. S. (2003). E-satisfaction and e-loyalty: A contingency framework. *Psychology and Marketing*, *20*, 123-138.

Anton, J. (1996). *Customer relationship management: Making hard decisions with soft numbers*. Englewood Cliffs, NJ: Prentice-Hall.

Arbuckle, J. L., & Wothe, W. (2000). *Amos 4.0 user's guide*. Small Waters Corporation.

Ariely, D. (2000). Controlling the information flow: Effects on consumers' decision making and preferences. *Journal of Consumer Research*, *27*(2), 233-248.

Athanassopoulos, A. D. (2000). Customer satisfaction cues to support market segmentation and explain switching behavior. *Journal of Business Research*, *47*(3), 191-207.

Bansal, H. S., & Taylor, S. F. (1999). The service provider switching model (SPSM): A model of consumer switching behavior in the service industry. *Journal of Service Research*, *2*(2), 200-218.

Bebko, C. P. (2000). Service intangibility and its impact on consumer expectations of service quality. *Journal of Services Marketing*, *14*(1), 9-26.

Berry, R. W. (1981). The employee as customer. *Journal of Retail Banking*, *3*(March), 33-40.

Bitner, M. J. (1990). Evaluating service encounters: The effects of physical surroundings and employee responses. *Journal of Marketing*, *54*(2), 69-81

Bolton, R. N., Kannan, P. K., & Gramlett, M. D. (2000). Implications of loyalty program membership and service experiences for customer retention and value. *Journal of the Academy of Marketing Science*, *28*(1), 95-108.

Bolton, R. N., & Lemon, K. N. (1999). A dynamic model of customers' usage of services: Usage as an antecedent and consequence of satisfaction. *Journal of Marketing Research, 36*(2), 171-186.

Bolton, L., Warlop, L., & Alba, J. W. (2003). Consumer perceptions of price (un)fairness. *Journal of Consumer Research*, *29*(4), 474-491.

Brucks, M., Zeithaml, V. A., & Naylor, G. (2000). Price and brand name as indicators of quality dimensions for consumer durables. *Journal of Academy of Marketing Science*, *28*(3), 359-374.

Business Week. (2001, December 3). *It's all about trust*.

Carver, C. S., & Scheier, M. F. (1990). Origins and functions of positive and negative affect: A control-process view. *Psychological Review*, *97*(1), 19-35.

Chiou, J.-S. (2004). The antecedents of consumers' loyalty toward Internet service providers. *Information & Management*, *41*, 685-695.

Cole M., O'Keefe, R. M., & Siala, H. (1999). From the user interface to the consumer interface. *Information Systems Frontiers*, *1*(4), 349-361.

Cronin, J., Brady, M. K., & Hult, G. T. (2000). Assessing the effects of quality, value, and customer satisfaction on consumer behavioral intentions in service environments. *Journal of Retailing*, *76*(2), 193-218.

Crosby, L. A., Evans, K. R., & Cowles, D. (1990). Relationship quality in services selling: An interpersonal influence perspective. *Journal of Marketing*, *54*, 68–81.

Crosby, L. A., & Johnson, S. L. (2004). Making the intangible...tangible. *Marketing Management, 13*(3), 12-13.

Davis, F. D. (1989). Perceived usefulness, perceived ease of use, and user acceptance of information technology. *MIS Quarterly, 13*(3), 319-340.

Dwyer, F. R., Schurr, P. H., & Oh, S. (1987). Developing buyer-seller relationships. *Journal of Marketing, 51*, 11-27.

Elliot, S., & Fowell, S. (2000). Expectations versus reality: A snapshot of customer experiences with Internet retailing. *International Journal of Information Management, 20*, 323-336.

Faniel, I. M. (2005). Enhancing perceived value of cross-departmental knowledge on innovation via KM technology features. *Academy of Management Proceedings*, 1-16.

Fishbein, M. (1967). *Readings in attitude theory and measurement.* New York: Wiley.

Fishbein, M., & Ajzen, I. (1975). *Beliefs, attitude, intention and behavior:*

An introduction to theory and research. Reading, MA: Addison-Wesley.

Flavian, C., Guinaliu, M., & Gurrea, R. (2006). The role played by perceived usability, satisfaction and consumer trust on Website loyalty. *Information & Management, 43*, 1-14.

Fornell, C., Johnson, M. D., Anderson, E. W., Cha, J., & Bryant, B. E. (1996). The American customer satisfaction index: Nature, purpose, and findings. *Journal of Marketing, 60*(October), 7-18.

Garbarino, E., & Johnson, M. S. (1999). The different roles of satisfaction, trust, and commitment in customer relationships. *Journal of Marketing, 63*(April), 70–87.

Garvin, D. A. (1983). Quaity on the line. *Harvard Business Review, September-October*, 65-75.

Gefen, D. (2000). E-commerce: The role of familiarity and trust. *Omega, 28*(6), 725-737.

Gefen, D., Karahanna, E., & Straub, D. W. (2003). Trust and TAM in online shopping: An integrated model. *MIS Quarterly, 27*(1), 51-90.

Geyskens, I., Steenkamp, J., Steenkamp, E. M., Scheer, L. K., & Kumar, N. (1996). The effects of trust and interdependence on relationship commitment: a trans-atlantic study. *International Journal of Research in Marketing, 13*, 303-317.

Grewal, D., Gopalkrishnan, R., Iyer, & Levy, M. (2004). Internet retailing: Enablers, limiters, and market consequences. *Journal of Business Research, 57*(7), 703–713.

Grewal, D. Levy, M., & Lehmann, D. R. (2004). Retail branding and customer loyalty: an overview. *Journal of Retailing, 80*(4), 249-263.

Grewal, D., Monroe, K. B., & Krishnan, R. (1998). The effects of price comparison advertising on buyers' perceptions of acquisition value and transaction value. *Journal of Marketing, 62*, 46-59.

Grewal, D., & Sharma, A. (1999). Value-based marketing strategies. In *Proceeding of the Annual Conference of the Academy of Marketing Science*. Coral Gables, FL.

Gruen, T. W., Summers, J. O., & Acito, F. (2000). Relationship marketing activities,

commitment, and membership behaviors in professional associations. *Journal of Marketing, 64*(July), 34-49.

Hair, J. R., Anderson, R. E., Tatham, R. L., & Black, W. C. (1998). *Multivariate date analysis with readings* (5th ed). Englewood Cliffs, NJ: Prentice-Hall.

Hellier, P., Geursen, G., Carr, R., & Rickard, J. (2003). Customer repurchase intention: A general structural equation model. *European Journal of Marketing, 37*(11), 1762-1800.

Homburg, C., Hoyer, W. D., & Koschate, N. (2005). Customers' reactions to price increases: Do customer satisfaction and perceived motive fairness matter? *Journal of the Academy of Marketing Science, 33*(1), 36-49.

Jackson, R., & Wang, P. (1995). *The convergence of database marketing and interactive media networks.* In E. Forrest & R. Mizerski (Eds.), *Interactive marketing: The future present* (pp. 355-372). NTC Publishing Group.

Jarvenpaa, S. L., & Todd, P. A. (1997). Consumer reactions to electronic shopping on the world wide Web. *International Journal of Electronic Commerce, 1*(2), 59-88.

Jarvenpaa, S. L., & Tractinsky, N. (1999). Consumer trust in an Internet store: A cross-cultural validation. *Journal of Computer Mediated Communication, 5*(2), 1-35.

Johnson, J. T., Barksdale, H. C., Jr., & Boles, J. S. (2001). The strategic role of the salesperson in reducing customer defection in business relationships. *Journal of Personal Sell and Sales Management, 21*(Spring), 123–134.

Kalafatis, S. P., & Miller, H. (1997). A re-examination of the commitment-trust theory. In H. G. Gemünden, T. Ritter, & W. Achim (Eds.), *Relationships and networks in international markets.* (pp. 213-227). Oxford, UK: Pergamon.

Lassar, W., Mittal, B., & Sharma, A. (1995). Measuring customer-based brand equity. *Journal of Consumer Marketing, 12*(4), 11-19.

Lee, M., & Turban, E. (2001). Trust model for consumer Internet shopping. *International Journal of Electronic Commerce, 6*(1), 75-91.

Lewis, J. D., & Weigert, A. (1985). Trust as a social reality. *Social Forces, 63*, 967-985.

Li, C., . Xu, Y., & Li, H. (2005). An empirical study of dynamic customer relationship management. *Journal of Retailing and Consumer Services, 12*, 431-441.

Liu, C., & Arnett, K. P. (2000). Exploring the factors associated with Web site success in the context of electronic commerce. *Information and Management, 38*(October), 23-34.

Lumayem, M., & Hirt, S. G. (2003). Force of habit and information systems usage: Theory and initial validation. *Journal of the Association for Information Systems, 4*, 65-95.

Lynch, J. G., & Ariely, D. (2000). Wine online: Search costs affect competition on price, quality and distribution. *Marketing Science, 19*(1), 83-103.

Marti, S., & Garcia-Molina, H. (2006). Taxonomy of trust: Categorizing P2P reputation systems. *Computer Networks, 50*, 472-484.

Mathieson, K. (1991). Predicting user intentions: Comparing the technology acceptance model with the theory of planned behavior. *Information Systems Research, 2*(3), 173-191.

Mathwick, C., Malhotra, N. K., & Rigdon, E. (2002). The effect of dynamic retail experiences on experiential perceptions of value: An Internet and catalog comparison. *Journal of Retailing, 78*(1), 50-51.

McKnight, D., & Chervany, N. (2002). What trust means in e-commerce customer relationship: an interdisciplinary conceptual typology. *International Journal of Electronic Commerce, 6*(2), 35-59.

Melone, N. P. (1990). A theoretical assessment of the user-satisfaction construct in information systems research. *Management Science, 36*(1), 76-81.

Moorman, C., Zaltman, G., & Deshpande, R. (1992). Relationships between providers and users of market research: The dynamics of trust within and between organizations. *Journal of Marketing Research, 29*(3), 314-328.

Morgan, R. M., & Hunt, S. E. (1994). The commitment-trust theory of relationship marketing. *Journal of Marketing, 58*(July), 20-38.

Nour, M. A., & Fadlalla, A. (2000). A framework for Web marketing strategies. *Information Systems Management,* 41-50.

Oliver, R. L. (1997). *Satisfaction: A behavioral perspective on the consumer*: New York: McGraw-Hill.

Oliver, R. L. (1980). A cognitive model of the antecedents and consequences of satisfaction decisions. *Journal of Marketing Research, 17*(November), 460-469.

Pan, S. L., & Lee, J. (2003). Using e-CRM for a united view of the customer. *Communication of the ACM, 46*(4), 95-99.

Pan, G., Pan, S. L., Newman, M., & Flynn, D. (2006). Escalation and de-escalation of commitment: a commitment transformation analysis of an e-government project. *Information Systems Journal, 16*, 3-21.

Parasuraman, A., & Grewal, D. (2000). The impact of technology on the quality-value-loyalty chain: A research agenda. *Journal of the Academy of Marketing Science, 28*(1), 168-174.

Parasuraman, A., Zeithaml, V. A., & Berry, L. L. (1994). Reassessment of expectations as a comparison standard in measuring service quality: Implications for further research. *Journal of Marketing, 58*, 111-124.

Parthasarathy, M., & Bhallachorjee, A. (1998). Understanding post-adoption behavior in the context of online service. *Information Systems Research, 9*(4), 362-378.

Pascoe, G. C. (1983). Patient satisfaction in primary health care: A literature review and analysis. *Eval Program Plann, 6*, 185-197.

Petre, M., Minocha, S., & Roberts, D. (2006). Usability beyond the Website: An empirically-grounded e-commerce evaluation instrument for the total customer experience. *Behavior & Information Technology, 25*(2), 189-203.

Phau, I., & Poon, S. M. (2000). Factors influencing the types of products and services purchased over the Internet. *Internet Research, 10*(2), 102-113.

Rai, A., Lang, S. S., & Welker, R. B. (2002). Assessing the validity of IS success models: An empirical test and theoretical analysis. *Information Systems Research, 13*(1), 50-69.

Raihan, N., & Hamid, A. (2005). E-CRM: Are we there yet? *The Journal of American Academy of Business, March*, 51-57.

Reichheld, F. F., & Schefter, P. (2000). E-loyalty: Your secret weapon on the Web. *Harvard Business Review, July-August*, 105-13.

Rigby, D. K., Reichheld, F., & Schefter, P. (2002). Avoid the four perils of CRM. *Harvard Business Review, February*, 101-110.

Romano, N. C., & Fjermesad, J. (2003). Electronic commerce customer relationship management: A research agenda. *Information Technology and Management, 4*, 233-258.

Rucci, A. J., Kirn, S. P., & Quinn, R. T. (1998). The rmployee-customer profit chain at SEARS. *Harvard Business Review, January/February*, 82-97.

Rust, R. T., & Lemon, K. N. (2001). E-service and the consumer. *International Journal of Electronic Commerce, 5*, 85–101.

Sebastianell, R., & Tamimi, N. (2002). How product quality dimensions relate to defining quality. *International Journal of Quality & Reliability Management, 19*(4), 442-453.

Sharma, N., & Patterson, P. G. (1999). The impact of communication effectiveness and service quality on relationship commitment in consumer professional services. *The Journal of Services Marketing, 13*(2), 151-170.

Siau, K., & Shen, Z. (2003). Building customer trust in m-commerce. *Communications of the ACM, 46*(4), 91-94.

Sirdeshmukh, D., Singh, J., & Sabol, B. (2002). Consumer trust, value, and loyalty in relational exchanges. *Journal of Marketing, 66*, 15-37.

Snoj, B., Korda, A. P., & Mumel, A. (2004). The relationships perceived quality, perceived risk and perceived product value. *Journal of Product & Brand Management. 13*(3), 156-167.

Story, J., & Loroz, P. S. (2005). Technological congruence and perceived quality of brand extensions. *Journal of Product & Brand Management, 14*(7), 438-447.

Sukpanich & Chen (1999). Antecedents of desirable consumer behaviors in electronic commerce. In *Proceedings of 5th Americas Conference on Information Systems*. Milwaukee, WI: Omnipress.

Taylor, S.A. (1994). Distinguishing service quality from patient satisfaction in developing health care marketing strategies. *Hosp Health Serv Adm, 39*(2), 221-36.

Taylor, S. A., & Todd, P. A. (1995). Understanding information technology usage: A test of competing models. *Information Systems Research, 6*(2), 144-176.

Vellido, A., Lisboa, P. J. G., & Meehan, K. (2000). Quantitative characterization and prediction of online purchasing behavior: A latent variable approach. *International Journal of Electronic Commerce, 4*(4).

Venetis, K. A., & Pervez, N. G. (2004). Service quality and customer retention: Building long-term relationships. *European Journal of Marketing, 38*(11/12), 1577-1598.

Voss, G. B., Parasuraman, A., & Grewal, G. (1998). The roles of price, performance, and expectations in determining satisfaction in service exchanges. *Journal of Marketing, 62*(October), 46-61.

Wang, F, & Head, M. (2001). Designing Web-based information systems for e-commerce. In *Proceedings of the 224 World Congress on the Management of Electronic Commerce*. Hamilton, Canada.

Wolfinbarger, M., & Gilly, M. C. (2001). Shopping online for freedom, control and fun. *California Management Review, 43*(2), 34-55.

Woodcock, N., & Starkey, M. (2001). I wouldn't start from here' finding a way in CRM projects. *Journal of Database Marketing, 18*(1), 61-74.

Wu, F., & Cavusgil, S. T. (2006). Organizational learning, commitment, and joint value creation in interfirm relationships. *Journal of Business Research, 59*, 81-89.

Wu, I. L., & Wu, K. W. (2005). A hybrid technology acceptance approach for exploring e-CRM adoption in organizations. *Behavior & Information Technology, 24*(4), 303-316.

Zeithaml, V., (1988). Consumer perceptions of price, quality, and value: A means-end model and synthesis of evidence. *Journal of Marketing, 52*(3), 2-22.

Zeithaml, V., Parasuraman, A., & Malhotra, A. (2000). E-service quality: Definition, dimensions and conceptual model. Working Paper. Cambridge, MA: Marketing Science Institute.

Zhang, X., & Prybutok, V. (2005). A consumer perspective of e-service quality. *IEEE Transactions on Engineering Management, 52*(4), 461-477.

This work was previously published in Journal of Electronic Commerce in Organizations, Vol. 6, Issue 3, edited by M. Khosrow-Pour, pp. 1-19, copyright 2008 by IGI Publishing (an imprint of IGI Global).

Chapter 9
Improving M-Commerce Services Effectiveness with the Use of User-Centric Content Delivery

Panagiotis Germanakos
National & Kapodistrian University of Athens, Greece

Nikos Tsianos
National & Kapodistrian University of Athens, Greece

Zacharias Lekkas
National & Kapodistrian University of Athens, Greece

Constantinos Mourlas
National & Kapodistrian University of Athens, Greece

George Samaras
University of Cyprus, Cyprus

ABSTRACT

Advances in wireless communications and information technology have made the Mobile Web a reality. The Mobile Web is the response to the need for anytime, anywhere access to information and services. Many wireless applications have already been deployed and are available to customers via their mobile phones and wirelessly-connected PDAs. However, as communications and other IT usage becomes an integral part of many people's lives and the available products and services become more varied and capable, users expect to be able to personalize a service to meet their individual needs and preferences. The involved sectors have to meet these challenges by reengineering their front-end and back-end office. This article will examine the interaction requirements regarding a friendlier, personalized and more effective multi-channel services environment. It will present the mobility challenges and constraints implemented into the business sector, investigating the current m-commerce situation and the extended user characteristics presenting a high level user-centric m-commerce architecture.

INTRODUCTION

The Internet revolution has brought about a new wave of conducting business and proved to be an important marketing tool for all sorts of business fields that found this new means as convenient as it is creative. With the emergence of wireless and mobile technologies, new communication platforms and devices, apart from PC-based Internet access, are now emerging, making the delivery of services available through a variety of multi-channel mediums without loosing their integrity or quality of their content (Germanakos et al., 2005a). Inevitably, this increases user requirements, which are now focused upon an "*anytime, anywhere and anyhow*" basis. Moreover, the explosive growth in the size and use of the World Wide Web as well as the complicated nature of most Web structures may lead in orientation difficulties, as users often lose sight of the goal of their inquiry, look for stimulating rather than informative material, or even use the navigational features unwisely. To alleviate such navigational difficulties, researchers have put huge amounts of effort to identify the peculiarities of each user group and design methodologies and systems that could deliver an adapted and personalized Web content. Challenges therefore range not only on adapting to the heterogeneous user needs and user environment issues, such as current location and time (Panayiotou, & Samaras, 2004), but also on a number of other considerations with respect to multi-channel delivery of the applications concerning multimedia, services, entertainment, commerce, and so forth. That is why, when the next big thing in technological gadgets, the mobile phone flooded the market, offering cheap SMS services using the GSM technology and enabling the extremely successful i-Mode and mobile Internet, various business institutions felt they couldn't stay away from this new rising opportunity.

Relevant channel and distribution strategies are critical for future advancement of m-commerce services to achieve accessible, customer-focused and responsive services. Following the growing user demands and requirements as well as the rapid development of the technological advancements and infrastructure capabilities, the development of m-commerce services should not only focus on making the service available on the Internet, but also examine the different delivery platforms. Indisputably, this is the vision of an interoperable, transparent and secure continent whereby multi-channel service delivery integration is considered fundamental.

This article emphasizes the proliferation of m-commerce services delivery starting with a reference to multi-channel delivery characteristics of user-centric services and the investigation of the m-commerce status and dimensions in various application areas (market fields). The adaptation and personalization considerations with regards to new user requirements and demands is also analyzed emphasizing the significance and peculiarities of user profiling for providing a more personalized m-commerce services result. It further presents a high level architecture for personalizing m-commerce services, introducing a comprehensive user profiling that incorporates intrinsic user characteristics such as user perceptual preferences (visual, cognitive and emotional processing parameters), on top of the "traditional" ones (such as name, age, education, etc.).

SERVICE REQUIREMENTS AND DELIVERY

To struggle against the amplification of the digital divide and therefore to think 'user interaction' whatever the age, income, education, experience, and the social condition of the citizen. (Europe's Information Society, 2004)

The specific theme above reveals exactly the need for user-centric m-government services development and personalized content delivery. In many ways, the new technology provides greater opportunities for access. However, there are important problems in determining precisely what users want and need, and how to provide Web-based services content in a user-friendly and effective way. User needs are always conditioned by what they already get, or imagine they can get. A channel can change the user perception of an m-government application: when users have a free choice between different channels to access an application, they will choose the channel that realizes the highest relative value for them. However, separate development of different channels for a single service (multi-channel delivery) can lead to inconsistencies such as different data formats or interfaces. To overcome the drawbacks of multiple-channel content delivery, the different channels should be integrated and coordinated.

Since successful m-government service delivery depends on a vast range of parameters, there is not a single formula to fit all situations. However, there have been reported particular steps (IDA, 2004) that could guide a provider throughout the channel selection process. Moreover, it should be mentioned that the suitability and usefulness of channels depends on a range of factors, out of which technology is only one element. Additional features that could affect the service channels assessment could be: directness, accessibility and inclusion, speed, security and privacy and availability. To realize their potential value though, channels need also to be properly implemented and operated.

The design and implementation complexity is rising significantly with the many channels and their varying capabilities and limitations. Network issues include low bandwidth, unreliable connectivity, lack of processing power, limited interface of wireless devices and user mobility. On the other hand, mobile devices issues include small size, limited processing power, limited memory and storage space, small screens, high latency, and restricted data entry (Germanakos et al., 2005a).

AN OVERVIEW OF M-COMMERCE/ M-MARKETING THEORETICAL FRAMEWORK

It is generally agreed that m-commerce is the use of mobile or handheld devices for communication and commerce transactions without any formal conceptualization of the term currently in existence, due to the continuous technological evolution. Therefore, there is no clear, inclusive and broad definition of all the aspects it might include (Balasubramanian et al., 2002).

What needs to be taken into account is what m-commerce means for an organisation belonging to a specific market field and which definition is more close to the needs of these institutions. Therefore, Easton's (2002) approach could be more descriptive, simple and close to what these organisations need to know about the m-commerce term. Easton claims that: (a) m-commerce is the purchase of a product or a service through wireless devices; (b) there is a number of physical limitations to the above devices indicating to the end-user that it is more convenient to sign up to mobile services by using the traditional interface of the PCs; and (c) the models used by the organisations in the case of m-commerce are more close to "click and mortar" paradigm. For example, in a theatre context the users will receive information about the performance and the box office contact details through mobile commerce and then they will purchase the ticket at the box office, or payment of the ticket will be done wirelessly with various methods, but still the transaction will be fulfilled when the purchaser attends the performance at the theatre.

Furthermore, what someone should be interested in is not the mobile Internet itself, "the representation of the physical world" in the wireless devices, but rather the mobile marketplace in which different kinds of transactions take place: communication services and wireless purchase. The mobile marketplace covers three dimensions (see *Figure 1*): personalisation according to local

Figure 1. M-marketplace dimensions

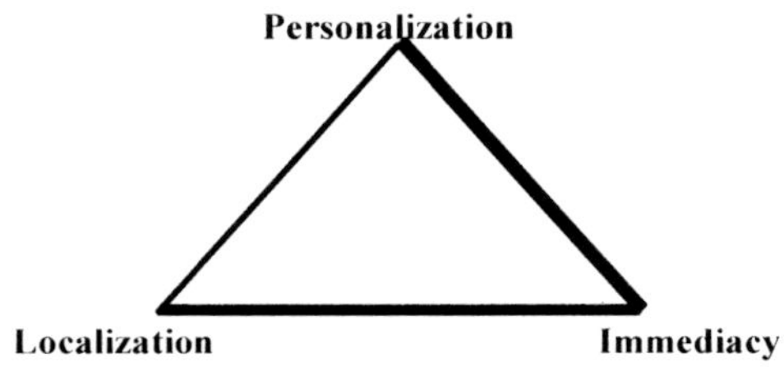

position of the holder and the relevance of information to his preferences, localization through the local based services, and immediacy (Lindgren, Jedbratt, & Svebsson, 2002).

As literature indicates, mobile phones are a communication medium that keeps people connected at any place, wherever they go and any time. Therefore, it is accounted as the most effective tool of direct marketing, while wireless marketing involves reaching and servicing customers and developing relationships with them through premium services (Bayne, 2002). Towse & Elgew (2003) describe marketing as "the process whereby the organisation's goals for its products and services are strategically developed and implemented to meet carefully reached price, product, place and promotion preferences of targeted customer groups." In this mentioned marketing mix (5Ps: product, place, promotion, price and, for arts organisations, people) it is clear that the mobile phone sets a new spectrum of marketing applications.

The infrastructures to support great financial transaction through mobile devices are still under development and consequently the mobile phone is used more closely to marketing applications. Therefore, the greater part of analysis in this article mainly concerns the modern m-marketing services era and dedicates a narrower part to the transactional nature of the medium. However, m-marketing is included by definition in the term of "m-commerce," and in the article the terms are used according to each case.

Under the term "m-marketing," two main methods are analyzed: push and pull marketing. Pull marketing method involves the consumers pulling what interests them toward themselves, either the product or information itself. Here the messages are sent instantly after a personal request of the recipient. In the case of push marketing method the marketer sends—or better—pushes the message to the consumers, but in this case the message is sent massively to the subscribers of such service. Another term that is under the umbrella of push marketing is "location-based services" (LBS) or "mobile location services" (MLS). These are "applications that utilize information related to the geographical position of their users in order to provide value-added services to them" (Giaglis et al., 2003).

Mobile Related Technological Aspects

The present technologies used in telecommunication are GSM (générale système mobile/global system for mobile communications), WAP (wireless application protocol), and GPRS (general packet radio service), which, accompanied by the third generation mobile devices (3G), offer to the end-user a broad variety of capabilities. Mobiles can hardly replace computers due to device limitations such as the screen and the low memory. Even processing capabilities were quite low in the beginning of the 3G development. The first services offered were mostly data services like text alerts for news and entertainment. These alerts are closely related to the wider used form of communication, text messages or else SMS. Text messages are enriched in the later version of the 3G devices with graphics, videos or picture launched as the MMS (multimedia message services) services (Paavilainen, 2002). The services that are slowly launched in the telecommunication markets are the location-based services, which are closely related to the global positioning system (GPS), cell identification and Bluetooth. With these technologies each mobile is detected according to its exact location (Giaglis et al., 2003). With Bluetooth it will be easier to receive text alerts customized to the geographical location, for example, information about the nearest theatre, or receive a promotion alert when the user is near to a theatre house.

The Role of New Media in Modern Society

It would be inadequate to examine the use of mobile commerce and marketing by the various business institutions without presenting the general context of the modern era and what has led to the decision of following this innovative technology. Subsequent to the suggestions that technologies should not be seen without being connected to other processes and factors that set the frameworks, we present an overview to the social and economic tendencies that lead to the spread of m-commerce (Hesmondhalgh, 2002).

Table 1. Challenges of marketing in digital era

Frameworks	Trends
Globalisation	1. global market
	2. global brands, i.e., Tate
	3. Global organizations, i.e., Guggenheim
Society	1. cash rich, time poor → materialistic hedonism
	2. increased leisure time, "the hurried leisure class"
	3. increased technology ownership (mobile, PCs, PDAs)
	4. Individualism
Technology	1. digitalistion of cultural products and data compression, i.e., musical pieces in Mp3 formats, films in DVD
	2. Internet → development of peer-to-peer transactions, e-commerce
	3. rise of Internet → growth of commerce
	4. Information, easy access o data
Virtualisation of digital channels	1. development of new channels, i.e., mobile phones and virtualization of products/services
Virtualisation of payments	1. plastic cards with magnetic stripe/ chip & pin (debit, credit, loyalty cards)
	2. Smart cards
	3. electronic cash/ electronic wallet
	4. Micropayments
Media trends	1. from mass messages to individual
	2. From generic to specific
	3. from print to electronic
	4. from passive to interactive
	5. from plenty of time to fastest way of communication
	6. 24 hours a day/ 7 days a week, global presence
Law	1. Issues of privacy and security
	2. Data Protection Act
	3. Intellectual Property issues
	4. fraud

The use of mobile is not just a technological trend that needs to be followed. Practice has shown that society accepts technological developments not just for the sake of fashion but when they offer an advanced solution to the needs of each era. The social and economical index change continuously and the market itself decides which technology will spread and survive. The next step is the high penetration of video phones with high resolution screens or i-mode services whose music portals and ringtone downloads have created havoc to the networking marketplace. Therefore, it is the market that allows ideas and technological visions to become part of the everyday consuming life (Knell, 2003). In *Table 1* that follows, we refer to the main issues rising in the modern digital era (Hesmondhalgh, 2002; O'Connor, 2001; Postma, 1999).

Taking into account all these factors, the business institution has to face a number of challenges in modern life. As consumers live in a multimedia and cross-cultural environment, they demand combinations of forms and new delivery methods, the marketers should keep in mind that they have to satisfy these needs in order to keep their audiences close to their institution.

EVALUATING M-COMMERCE IN RELATION TO VARIOUS MARKET FIELDS

It is quite clear so far that the literature offered on the subject of m-commerce is quite narrow and even sometimes inadequate. In the past five years there were a number of managerial handbooks analyzing the new mobile era either for commerce or marketing. The initial stimuli derived from the book of Turban et al. (2002) where the authors dedicate the last chapter to mobile commerce as a natural evolution of e-commerce through a new portable medium. The m-commerce is perceived as a follower of e-commerce (Easton, 2002) as long as mobile Internet is developed and so the consumers are able to use portable devices to purchase items. Mobile phone is seen by many authors as the physical transformation of the Internet and the leader of the new media (Stafford et al., 2004; Candace, 2005).

Paavilainen (2002) introduces the mobile Internet as the alternative of desktop Internet. Mobile advertising, text alerts and mobile commerce models are described through different case studies from the commercial sector. Mobile entertainment, that is, ringtones, downloads and text alerts, is rich with cases from music and film industries, as sector journals like the New Media Age show. Although someone would think that this marketing trend would hardly expand in other forms of the business institutions, like culture, there is a release by the DigiCULT (2004) scheme that proves the opposite. Mobile devices applying location-based services (LBS) can be integrated in a museum or gallery exhibition giving the option to be another communication channel after the audio guides and the information kiosks. Moreover, scenarios like booking tickets through mobile, receiving text alerts for the latest exhibition or information how to get to an art gallery are becoming more obvious examples in the cultural sectors. The referred applications are included according to Mahatanankoon et al. (2005) in two modes of m-commerce: (i) the content delivery mode and (ii) the transaction mode.

Mobile Marketing

As part of the new strategies, the use of the mobile as a marketing and communication tool plays mainly an important role. With the introduction of text messaging boom marketers were totally influenced and seized the SMS as a new advertising chance. Haig (2002) describes the history of SMS and introduces various applications of mobile devices in marketing like the text alerts. With the introduction of the third generation mobile phones (3G) the capabilities of the devices to support image, sounds and movie clips grow, increasing the number of applications. Marketers find themselves holding a powerful interactive and totally personal marketing tool. However, there are still a great number of barriers which limit the widespread of the medium and its success according to Scharl et al. (2005), and they have to do either with the technological features of the devices, the transmission process or customer behaviour itself.

Customer Behaviour

Kleijnen et al. (2004) used the Technology Acceptance Model (TAM) to examine a number of factors that influence the quicker penetration of mobile technologies in the market in order to explain this trend of text messaging. Factors like cost of the devices and services, the value they offer to the end-user as well as the easiness of use are important in order to make special mobile services apart from voice mail more acceptable. Moreover, Brumer & Kumar (2005) also introduced the influential factors of utility, convenience and fun, which are the most interesting point, for example, for an arts organisation. An innovative and entertaining way of advertising, promotion or ticket booking could be critical in order to catch the customers' attention and make them act. This last factor mainly influences the mobile advertisements (m-advertisement), toward which the attitude of consumers is more positive than toward traditional advertising, as Tsang et al. (2004) claim. What influences the attitudes of the end-users apart from the entertaining way of presentation is the content of the provided advertising services and its amount of information.

A key concept that should be looked into is the limitation of the irritation caused by interfering and unwanted messages. Therefore, it is important for organisation using the mobile services to ask for permission to deliver them to customer. Permission marketing is the term that should be closely related to any action of direct marketing.

Permission Marketing and Trust

Godin's book "Permission Marketing" (2002) is the most common reference in the literature under the subject of mobile marketing (Kavassalis et al., 2003) and there is no critical point suggested by other authors toward what it claims. The introduction of permission in marketing tactics, especially in direct promotions, comes in contraposition to the traditional way of advertising—interruptive marketing, as explained later.

Additionally, trust, security and privacy are the emerging issues for m-commerce. The fear of manipulation of personal data is the main limit in the consumer's shift to positive attitude. Good sources of information about privacy policies are the independent protecting bodies set to protect the end-users [for example, the Data Protection

Figure 2. An m-commerce personalization architecture

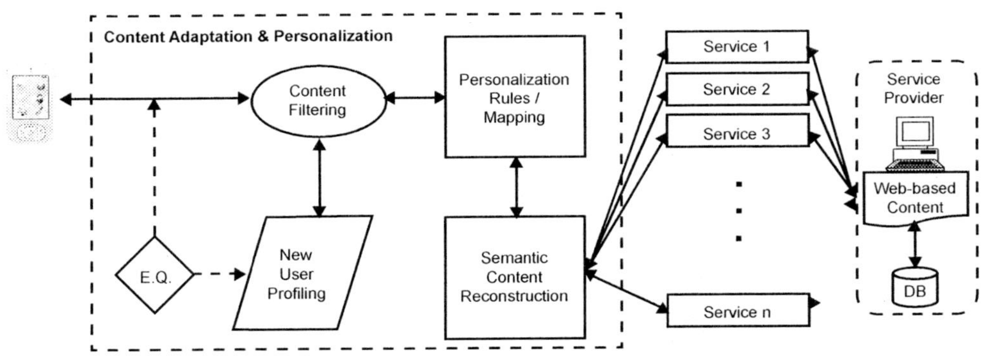

Act of 1998 is followed by the majority of the cases and the organisations show great respect to its lines, the Wireless Location Industry Association (WLIA) declares an adopted privacy policy by its members and so did the Mobile Marketing Association (MMA) with a relative declaration]. It must also be said though, that in cases of transactions through mobile phones, the risk of hacking is lower than that of the Internet due to the evolving technologies and the use of high standard encryption.

Effectiveness

How could effectiveness be defined in a very unstable context? This is the main question underlying this subject. It is quite difficult to find research on the effectiveness of the new medium due to the very recent introduction of mobile commerce in the market. The given difficulty still rises given the peculiarities of the various market fields. More specifically, for an arts organisation, for example, it is even more difficult to define effectiveness, as this can be meant under totally different aspects depending on what aims each cultural institution has set. Drucker (1995) and Heman & Renz (1999) also briefly referred to the effectiveness in non-profit organisations.

The factors that can make mobile an effective new media for organizations were examined in depth. A good attempt to measure the wireless marketing results by introducing the Mobile Return of Investment (ROI) was also made by Bayne (2002) and that is a useful guide for any organisation which wishes to monitor its investment on this new technology. Moreover, Kavassalis et al. (2003) introduced more indicators of effectiveness in their empirical studies, while Frolick & Chen (2004) assess the mobile commerce applications.

The majority of the research converges to the same conclusion. No matter what the effects of the mobile practices are, which in each case can be totally different, what is important is the benefit that someone can gain by being the first in the market for using this new technology.

PRINCIPLE DRIVERS OF AN M-COMMERCE OPEN SERVICE INFRASTRUCTURE

The deployment of a m-commerce open service platform that could be shared by networked private and public institutions could be a promising approach with further insights on maintaining wireless service provision sustainability in a long-term perspective. Wireless technology is about extending the availability of an e-commerce infrastructure to mobile and wireless channels. It becomes more fully developed and, as bandwidth increases with the availability of "always on" connectivity, next generation applications and entirely new practices will arise different from those delivered over existing static networks (Caldow, 2001).

The large array of new communication technology opportunities and the rapid emergence and change of standards, as well as the variety of mobile channels, offer different technical capabilities for sustainable architectures and technology frameworks in order to meet critical requirements like broadband, interoperability, scalability, transparency, personalization, privacy and security (Germanakos et al., 2005a).

Extended User Requirements and the Personalization Problem

The user population is not homogeneous, nor should be treated as such. To be able to deliver quality services, m-commerce systems should be tailored to the needs of individual users providing them with personalized and adapted information based on their perceptions, reactions, and demands. Therefore, a serious analysis of user requirements has to be undertaken, documented and examined taking into consideration their multi-application to the various delivery channels and devices. Some of the user (customer) requirements and arguments anticipated could be clearly distinguished into (Germanakos et al., 2005c): (a) general user service requirements (flexibility: anyhow, anytime, anywhere; accessibility; quality; and security), and (b) requirements for a friendly and effective user interaction (information acquisition; system controllability; navigation; versatility; errors handling; and personalization).

Although one-to-one Web-based service provision may be a functionality of the distant future, user segmentation is a very valuable step in the right direction. User segmentation means that the user population is subdivided into more or less homogeneous, mutually exclusive subsets of users who share common user profile characteristics. The subdivisions could be based on: demographic characteristics (i.e., age, gender, urban or rural-based, region); socio-economic characteristics (i.e., income, class, sector, channel access); psychographic characteristics (i.e., lifestyle, values, sensitivity to new trends); individual physical and psychological characteristics (i.e., disabilities, attitude, loyalty).

The issue of personalization is a complex one with many aspects and viewpoints that need to be analyzed and resolved. Some of these issues become even more complicated once viewed from a moving user's perspective; in other words, when constraints of mobile channels and devices are involved. Such issues include, but are not limited to: what content to present to the user, how to show the content to the user, how to ensure the user's privacy, how to create a global personalization scheme. As clearly viewed, user characteristics and needs, determining user segmentation and thus provision of the adjustable information delivery, differ according to the circumstances and they change over time (Panayiotou & Samaras, 2004).

User Profiling Characteristics: A More Comprehensive Approach

One of the key technical issues in developing personalization applications is the problem of how to construct accurate and comprehensive profiles of individual users and how these can be used to identify a user and describe the user behaviour, especially if they are moving (Adomavicious & Tuzhilin, 1999). According to the Merriam-Webster Dictionary, the term profile means "a representation of something in outline." User profile can be thought of as being a set of data representing the significant features of the user, like preferences, characteristics, and activities, derived from a set of keywords that are compared against information items.

User profiling can either be *static*, when it contains information that rarely or never changes (e.g., demographic information), or *dynamic*, when the data change frequently. Such information is obtained either *explicitly*, using online registration forms and questionnaires resulting in static user profiles, or *implicitly*, by recording the navigational behaviour and/or the preferences of each user. In the case of implicit acquisition of user data, each user can either be regarded as a member of group and take up an aggregate user profile or be addressed individually and take up an individual user profile. The data used for constructing a user profile could be distinguished into: (a) the data model, which could be classified into the demographic model (which describes who the user is), and the transactional model (which describes what the user does); and (b) the profile model, which could be further classified into the factual profile (containing specific facts about the user derived from transactional data, including the demographic data), and the behavioural profile (modeling the behaviour of the user using conjunctive rules, such as association or classification rules). The use of rules in profiles provides an intuitive, declarative and modular way to describe user behaviour (Adomavicious & Tuzhilin, 1999)). Additionally, in the case of a mobile user, by user needs it is implied both the thematic preferences (i.e., the traditional notion of profile) as well as the characteristics of their personal device called "device profile." Therefore, here, adaptive personalization is concerned with the negotiation of user requirements and device abilities. As Web developers regard personalization as the best way to filter out unnecessary or irrelevant information for their users, some argue on issues like it may restrict the extent and the variety of information users receive, that people often do not have well-defined preferences, they need to answer detailed questions to personalize their Web pages, that the recommendation process is a black box for end-users and so on (Wang & Lin, 2002).

But, could the user profiling be considered complete incorporating only these dimensions? Do the designers and developers of m-commerce applications take into consideration the real users' preferences in order to provide them with really personalized Web-based service content? Many times this is not the case. How can a user profiling be considered complete, and the preferences derived optimized, if it does not contain parameters related to the user perceptual preference characteristics? We could define *user perceptual preference characteristics* as all the critical factors that influence the visual, mental and emotional processes liable of manipulating the new information received and building upon prior knowledge that is different for each user or user group. These characteristics, which have been primarily discussed in Germanakos et al. (2006), determine the visual attention, cognitive and emotional processing taking place throughout the whole process of accepting an object of perception (stimulus) until the comprehensive response to it (Germanakos et al., 2005c). The proposed comprehensive user profiling could be considered as the main raw content filtering module of an m-commerce personalization architecture. To our knowledge, nowadays, there is not research that moves towards the consideration of user profiling incorporating optimized parameters taken from these research areas in combination.

AN OVERVIEW OF M-COMMERCE PERSONALIZATION ARCHITECTURE

Based on the above-mentioned considerations, an m-commerce personalization architecture is overviewed trying to convey the essence and the peculiarities encapsulated. The current system, depicted in *Figure 2*, is composed of a number of interrelated components, as detailed below:

Entry Point

It is the user access interface of the system. It accepts multi-device (enables the attachment of various devices on the infrastructure, such as mobile phones, PDAs, desktop devices, etc.) identifying the characteristics of the device and the preferences as well as the location of the user (personalization/location-based) and multi-channel (due to the variety of multi-channel delivery, for example, over the Web, telephone, interactive kiosks and so on, this module will identify the different characteristics of the channels) requests. It is directly communicating with the "content filtering" component exchanging multi-purpose data.

Content Filtering

This component is considered the main link of the "entry point" with the "'new' user profiling" and the "personalization rules/mapping" components of the architecture. It actually transmits the data accumulated both directions and it makes the filtering of the content, according to the personalization processing characteristics, delivering the adapted and personalized service. The whole processing varies from security, authentication, user segmentation, service content identification, "new" user profiling characteristics and so forth.

"New" User Profiling

This is the main component of the architecture and it is called "'new' user profiling" component. At this component all the requests are processed. It is responsible for the custom tailoring of information to be delivered to the users, taking into consideration their habits and preferences, as well as, for mobile users mostly, their location ("location-based") and time ("time-based") of access (Panayiotou & Samaras, 2006). It also keeps the user logs and retrieves data according to user login to the system. This component accepts requests from the "entry point" component and, after the necessary processing and further communication with the "content filtering" component, either sends information back or communicates with the next component ("personalization rules/mapping") accordingly. This component is comprised of the following two elements:

The "Traditional" User Profile

It contains all the information related to the user necessary for the Web personalization processing. It is composed of two elements: the (a) user characteristics (the so-called "traditional" characteristics of a user: knowledge, goals, background, experience, preferences, activities, etc.), and (b) the device/channel characteristics (contains characteristics referring to the device or channel the user is using and contains information like: bandwidth, size, connectivity, power processing, interface and data entry, memory and storage space, battery lifetime, etc. These characteristics are mostly referred to mobile users and are considered important for the formulation of a more integrated user profile, since it determines the technical aspects of it). Both elements are completing the user profiling from the user's point of view.

User Perceptual Preference Characteristics

This is the new component/dimension of the user profiling. It contains all the visual attention, cognitive and emotional processing parameters that completes the user preferences and fulfils the user profile. User perceptual preference characteristics could be described as a continuous mental processing starting with the perception of an object in the user's attentional visual field and going through a number of cognitive, learning and

Figure 3. User perceptual preference characteristics—three-dimensional approach

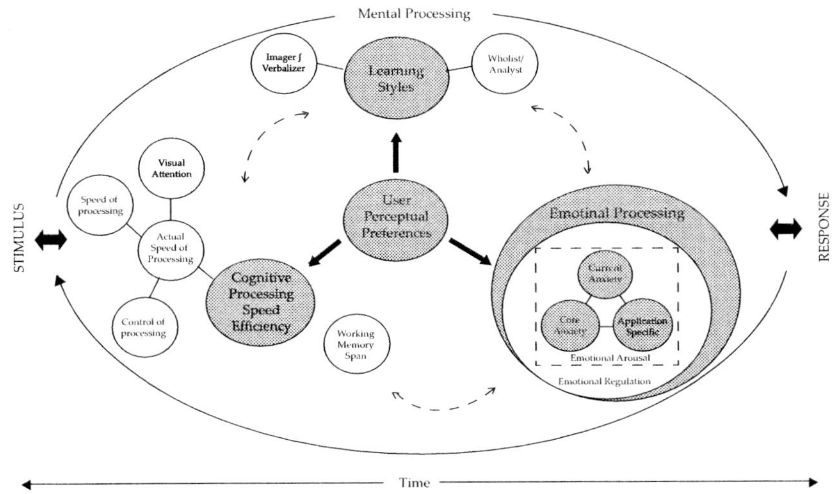

emotional processes giving the actual response to that stimulus, as depicted in *Figure 3*. These characteristics have been primarily discussed in Germanakos et al. (2006) and Germanakos et al. (2005a).

As can be observed, its primary parameters formulate a three-dimensional approach to the problem.

The first dimension investigates the visual and cognitive processing speed efficiency of the user, the second his/her learning style, while the third captures his/her emotional processing during the interaction process with the information space. All the above dimensions have specific characteristics and implications into the information space based on which the personalization rules in the "personalization rules/mapping" component will be constructed. Suggestively, we present in *Figure 4* the data implications diagram of the learning styles chosen.

For a better understanding of the three dimensions' implications and their relation with the information space a diagram that presents a high level correlation of these implications with selected tags of the information space (a code used in Web languages to define a format change or hypertext link) is depicted in *Figure 5*. These tags (images, text, information quantity, links – learner control, navigation support, additional navigation support, and aesthetics) have gone through an extensive optimization representing group of data affected after the mapping with the implications.

The particular mapping is based on specific rules created, liable for the combination of these tags and the variation of their value in order to better filter the raw content and deliver the most personalized Web-based result to the user. As it can be observed from the diagram, each dimension has primary (solid line) and secondary (dashed line) implications on the information space altering dynamically the weight of the tags.

Figure 4. Riding's learning styles characteristics and implications

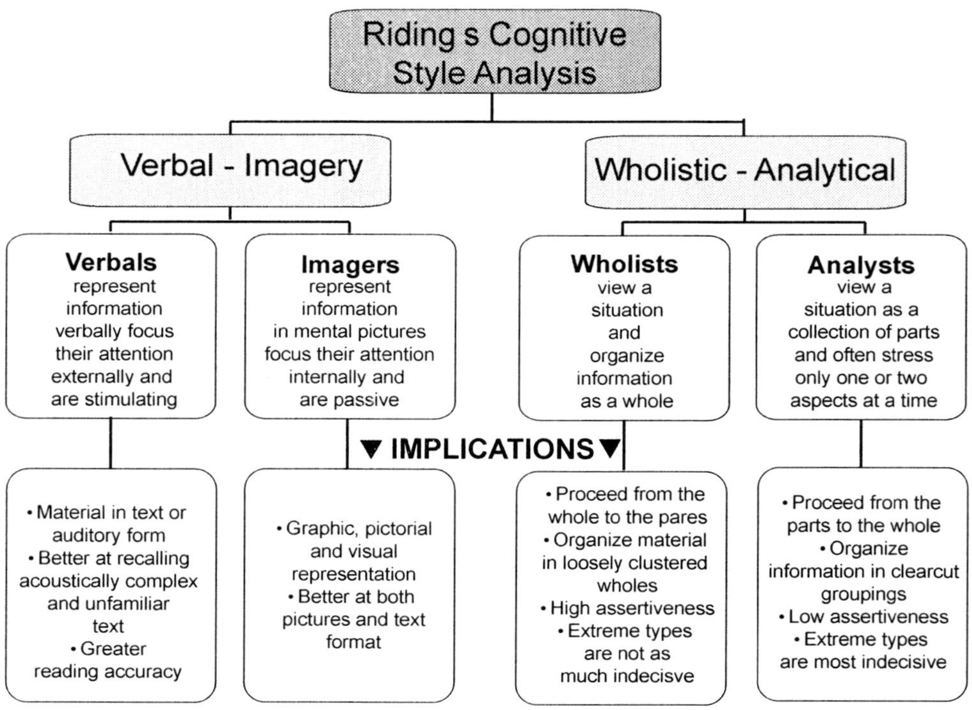

Figure 5. Data: Implications correlation diagram

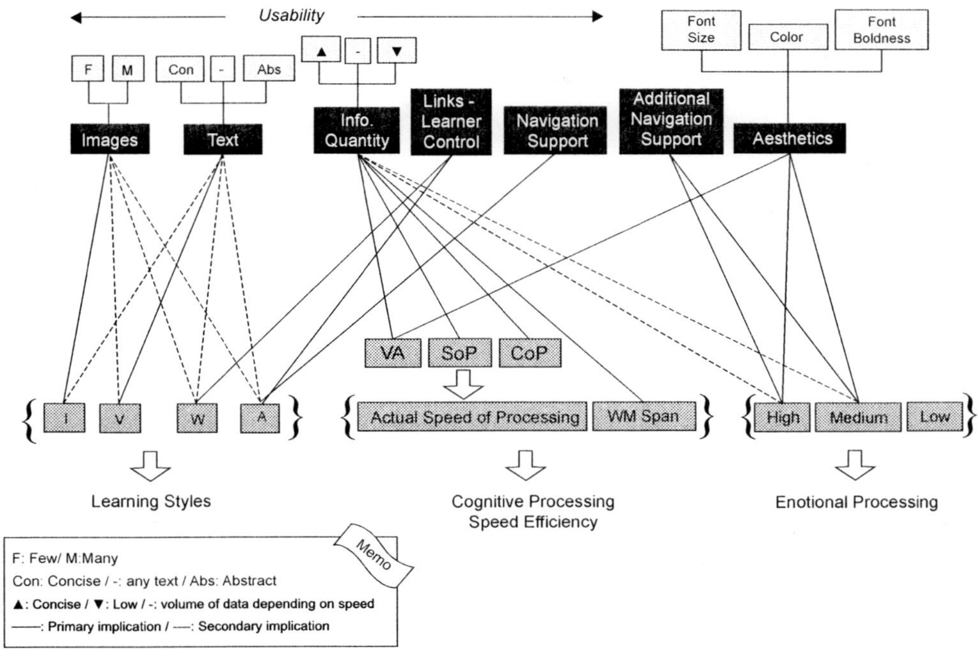

Therefore, we include in the learning styles dimension Riding's Cognitive Style Analysis, which applies in a greater number of information distribution circumstances, since it deals rather with cognitive than learning style. Henceforth, for example, the number of images (few or many) to be displayed has a primary implication on imagers, while text (more concise or abstract) has a secondary implication. An analyst may affect primarily the links – learner control and navigation support tag, which in turn is secondary affected by high and medium emotional processing, while might secondary affect the number of images or kind of text to be displayed, consequently. Actual speed of processing parameters (visual attention, speed of processing, and control of processing), as well as working memory span, is primarily affecting information quantity. Eventually, emotional processing is primarily affecting additional navigation support and aesthetics, as visual attention does, while secondary affects information quantity.

A practical example of the Data – Implications Correlation Diagram could be as follows: a user might be identified that is: Verbalizer (V) – Wholist (W) with regards to the learning style, has an actual cognitive processing speed efficiency of 1000 msec, and a fair working memory span (weighting 5/7), with regards to his/her cognitive processing speed efficiency, and (s)he has a high emotional processing. The tags affected, according to the rules created and the Data – Implications Correlation Diagram, for this particular instance are the: images (few images displayed), text (any text could be delivered), information quantity (less info since his/her cognitive speed is moderate), links – learner control (less learner control because (s)he is Wholist), additional navigation support (significant because (s)he has high emotional processing), and high aesthetics (to give more structured and well defined information, with more colors, larger fonts, more bold text, since (s)he has high emotional processing).

At this point it should be mentioned that in case of internal correlation conflicts primary implications take over secondary ones. Additionally, since emotional processing is the most dynamic parameter compared to the others, any changes occurring at any given time are directly affecting the yielded value of the adaptation and personalization rules and henceforth the format of the content delivered.

Personalization Rules/Mapping

At this component all the calculations such as the user categorization and mapping, content reconstruction and content adaptation takes place. In order for the current component to run properly, Web-based semantic services content is conveyed in the form of metadata from the "semantic content reconstruction" component. Once the provided content is adjusted based on the developed rules to the user characteristics it returns the corresponding adapted and personalized result to the "content filtering" component.

Semantic Content Reconstruction

This component is based on metadata and it is responsible for describing the content (data) available from the "service provider" (service 1, service 2, ...service n). In this way a common understanding of the data, that is, semantic interoperability, openness, is achieved. The data manipulated by the system is described using metadata that comprises all needed information to unambiguously describe each piece of data and collections of data. This provides semantic interoperability and a human-friendly description of data. This component is also directly related to the "personalization content reconstruction" component providing the altered Web-based metadata service content. It is consisted of two elements:

- Perceptual provider characteristics: It identifies the provider characteristics assigned to the Web-based services content. They are involving all these perceptual elements that the provider has been based for the design of the content.
- Semantic content properties: This element performs the identification and metadata description of Web-based services content based on predetermined ontologies. It is implemented in a transparent manner removing data duplication and the problem of data consistency.

Service Provider

This is the last component of the architecture and is directly connected to the "semantic content reconstruction" component. It contains transition mechanisms and the databases of Web-based services content as supplied by the provider without being through any further manipulation or alteration.

CONCLUSION

Indisputably, by implementing mobile commerce and mobile marketing a business institution attains a leading image of an organisation close to the technological trends and the needs of modern audiences. It differentiates itself in an era when each institution competes for a share of market and a share of heart not only through its work but also its marketing tactics. The healthy image of an innovative and avant-garde organisation can only work for the benefit of its long-term viability. It is hard to predict what the future will bring. Surely there will be an increasing number of organisations adopting the mobile services. However, the future is nearer than generally estimated while the circumstances rapidly change. There can be futuristic scenarios which one day might become reality, but what is of more importance is always to know what the end-users finally need. The various institutions need to keep close and open to any kind of evolvement, not only to technological developments but also to how their audiences respond to them.

On these grounds, this article made a reference to the m-commerce content delivery investigating multi-channel delivery characteristics of user-centric m-commerce services and the adaptation and personalization considerations with regards to new user requirements and demands. It analyzed the current situation of m-commerce services, ranging from theoretical considerations to technological parameters, and it underpinned the significance of the user profiling, introducing the comprehensive user profiling that incorporates intrinsic user characteristics such as user perceptual preferences. It has finally presented an overviewed m-commerce personalization architecture that considers cognitive learning styles as its main personalization filter, in an attempt to provide the user with the most personalized m-commerce content result.

REFERENCES

Adomavicious, G., & Tuzhilin, A. (1999). User profiling in personalization applications through rule discovery and validation. In *Proceedings of the ACM Fifth International Conference on Data Mining and Knowledge Discovery (KDD'99)* (pp. 377-381).

Balasubramanian, S., Peterson, R.A., & Jarvnpaa, S.L. (2002). Exploring the implications of m-commerce for markets and marketing. *Journal of the Academy of Marketing Science*, *30*(4), 348-361.

Bayne, K.M. (2002). *Marketing without wires*. Wiley.

Caldow, J. (2001). *E-gov goes wireless: From Palm to shining Palm*. Institute of Electronic Government, IBM Corporation.

Candace, D.P. (2005). *E-commerce and m-commerce technologies*. IRM Press.

Drucker, P.F. (1995). *Managing the non-profit organisation: Practices and principles*. Butterworth-Heinemann.

Easton, J. (2002). *Going wireless: Transform your business with mobile technology*. Harper Business.

Europe's Information Society. (2004a). *User interaction*. Retrieved from http://europa.eu.int/information_ society

Frolick, M.N., Chen, L.-D. (2004, Spring). Assessing m-commerce opportunities. *Information Systems Management*, pp. 53 -61.

Germanakos, P., Tsianos, N., Lekkas, Z., Mourlas, C., & Samaras, G. (2006). Capturing essential intrinsic user behaviour values for the design of comprehensive Web-based personalized environments. *Computers in Human Behavior Journal*, Special Issue on Integration of Human Factors in Networked Computing. (accepted)

Germanakos, P., Samaras, G., & Christodoulou, E. (2005a). Multi-channel delivery of services—the road from eGovernment to mGovernment: Further technological challenges and implications. In *Proceedings of the First European Conference on Mobile Government* (Euro mGov 2005) (pp. 210-220).

Germanakos, P., Tsianos, N., Mourlas, C., & Samaras, G. (2005b). New fundamental profiling characteristics for designing adaptive Web-based educational systems. In *Proceeding of the IADIS International Conference on Cognition and Exploratory Learning in Digital Age (CELDA2005)* (pp. 10-17).

Germanakos, P., Mourlas, C., & Samaras, G. (2005c). Considering the new user requirements for apt mobile Internet services delivery. In *Proceedings of the IADIS International Conference on WWW/Internet 2005* (pp. 148-152).

Giaglis, G.M., Kourouthanassis, P., & Tsamakos, A. (2003). Towards a classification framework for mobile location services. In B.E. Mennecke & T.J. Strader (Eds.), *Mobile Commerce: Technology, Theory, and Applications*. Hershey, PA: Idea Group Publishing. Retrieved from http://www.eltrun.gr/papers/mBook-classification.pdf..

Haig, M. (2002). *Mobile marketing: The message revolution*. Kogan Page.

Heman, R.D., & Renz, D.O. (1999). Theses on nonprofit organisational effectiveness. *Nonprofit and Voluntary Sector Quarterly, 28*(2), 107-126.

Hesmondhalgh, D. (2002). *The cultural industries*. SAGE.

Interchange of Data Between Administrations. (2004). *Multi-channel delivery of eGovernment services*. Retrieved from http://europa.eu.int/idabc/.

Kleijnen, M., Wetzels, M., & de Ruyter, K (2004). Consumer acceptance of wireless finance. *Journal of Financial Services Marketing, 8*(3), 206-217.

Knell, S. (2003). The shape of things to come: Museums in the technological landscape. *Museum and Society, 1*(3), 132-146.

Mahatanankoon, P., Wen, J.H., & Lim, B. (2005). Consumer-based m-commerce: Exploring consumer perception of mobile applications. *Computer Standards and Interfaces, 27*, 347-357.

O'Connor, J., & Galvin, E. (2001). *Marketing in the digital age*. Prentice Hall.

Panayiotou, C., & Samaras, G. (2006). Mobile user personalization with dynamic profiles: Time and activity. In *Proceedings of On the Move to Meaningful Internet Systems 2006: OTM 2006 Workshops* (PerSys 2006). Part II (pp. 1295-1304).

Panayiotou, C., & Samaras, G. (2004). mPERSONA: Personalized portals for the wireless user: An agent approach. *Mobile Networks and Applications (MONET), Special Issue on Mobile and Pervasive Commerce, 9*(6).

Paavilainen, J. (2002). *Mobile business strategies: Understanding the technologies and opportunities*. Wireless Press.

Postma, P. (1999). *The new marketing era: Marketing to the imagination in a technology-driven world*. McGraw-Hill.

Stafford, M.R., & Faber, R.J. (2004). *Advertising, promotion and new media*. M.E. Sharpe.

Towse, R. (Ed.). (2003). *A handbook of cultural economics*. Elgew Edward.

Tsang, M.M., Ho, S.-C., & Liang, T.-P, (2004). Consumer attitudes toward mobile advertising: An empirical study. *International Journal of Electronic Commerce, 8*(3), 65-78.

Turban, E., King, D., Lee, J., Warkentin, M., Chung, H.M. (2002). *Electronic commerce 2002: A managerial perspective*. Prentice Hall.

Wang, J., & Lin, J. (2003). Are personalization systems really personal? – Effects of conformity in reducing information overload. In *Proceedings of the 36th Hawaii International Conference on Systems Sciences* (HICSS'03) (p. 222c). IEEE Computer Society.

This work was previously published in Journal of Electronic Commerce in Organizations, Vol. 6, Issue 1, edited by M. Khosrow-Pour, pp. 1-18, copyright 2008 by IGI Publishing (an imprint of IGI Global).

Chapter 10
The Evolution of Online Relationships in Business to Consumer E-Commerce

Pauline Ratnasingam
University of Central Missouri, USA

ABSTRACT

The growth of business-to-consumer (B2C) e-commerce has gained a lot of attention among SMEs. Most B2C firms are turning their attention on how to retain new customers and are left in a situation to compete with larger firms. This chapter aims to examine the evolution and impact of online relationships in B2C firms. Drawing upon the theories of customer relationship management, e-commerce, trust and loyalty, the author develops an integrated model to illustrate the stages customers go through when shopping online and its impact on trust, hard and soft factors that reflect the level of transactional and relational components of communication thereby impacting the customers shopping time lifecycle experiences. This chapter develop a number of hypotheses to facilitate testing of the model. The author then provides a discussion of the model leading to the theoretical and practical implications of this study and directions for future research.

INTRODUCTION

Business to consumer (B2C) e-commerce, a significant subset of total Internet business, can provide significant benefits to customers and Small-Medium-Enterprises (SMEs) including the ability to market their products and services online thereby achieving global connectivity, high accessibility, scalability, interoperability, interactivity, and greater information richness (Turban, Lee, King & Chung, 2006). Online relationships evolve through stages of loyalty and trust when consumers engaged in B2C shopping.

DOI: 10.4018/978-1-60566-964-9.ch010

Information Data Center predicts that e-commerce spending will reach $496.7 billion in the U.S. and $1.3 trillion globally by the year 2009. In recent years we have witnessed an increased growth of small businesses turning towards the Internet to market their products. SMEs are an important and integral part of every country's economy and have been long recognized as different from larger businesses (Street & Meister, 2004). The Internet helps SMEs to manage customer relationships effectively, as it provides an easy and inexpensive way to advertise, lowering the barriers to entry for SMEs. Despite that customers are now even more demanding than ever before and B2C e-commerce firms are challenged to keep pace with the changing customers needs in maintaining a competitive edge. Unlike traditional means of communication such as newspapers or television, there are many Web revenue models that assist customers in gaining control when choosing and processing informing about the firm. Web tools such as; online feedback forms, 1800 telephone numbers, help desk operators assist customers in obtaining information and clarifying questions quickly. Products such as LikeMinds from Macromedia uses complex algorithm to create affinity groups for each customer, so that B2C firms can better understand their customers engaged in buying from their web site. The Web overturns the traditional hierarchical system of distribution channels by allowing integration of systems and information between the marketing, sales and servicing activities thereby forcing SMEs to view online relationships seriously. Beyond blogs, pod casts and wikis, social media comprises social networks of all strips, social tagging that provides organized customer generated content for small businesses and customers to share views. The breadth of the medium allows wider availability, accessibility and selection of hard-to-find products and services. Web tools thus serve two purposes namely; to provide web content and to manage customers' interactions.

This study aims to examine the evolution and impact of online relationships in the context of B2C e-commerce among SMEs. Drawing upon the theories of customer relationship management, e-commerce, trust and loyalty, we develop an integrated model to illustrate the stages customers go through when shopping online and its impact on trust, hard and soft factors that reflect the level of transactional and relational components of communication thereby impacting the customers shopping time lifecycle experiences leading to four modes namely; learning, monitoring, collaborating and distancing modes. The rest of the paper is organized as follows. The next section discusses the hard and soft factors in online relationships. While hard factors reflect the transactional component of communication including web site's performance, user friendliness, quality of web content, customer acquisition/search costs and perceived security. Soft factors reflect on the relational component of communication including the quality of products/services, satisfaction, trust, loyalty and reputation of the firm. We develop a number of hypotheses to facilitate testing of the model. We then report and discuss the model leading to theoretical and practical implications of the study and directions for future research. This study contributes to theory as we provide a holistic framework that incorporates both hard and soft factors and its impact on customers shopping time lifecycle experiences. The study also contributes to practice as B2C e-commerce practitioners will be aware of the mechanisms used to achieve timely, cost-effective, scalable, manageable and reliable feedback on their customer relationship performance.

STAGES OF TRUST IN B2C E-COMMERCE

Trust may be even more critical in the context of B2C e-commerce because Internet purchase is based on consumers' confidence in the processes that are not transparent online, in contrast to that of the traditional brick and mortar businesses where trust is based on personal relationships and face to face interactions between the customer and the merchant (Kim, Ferrin & Rao, 2008).

Schneider (2009) suggests that customer loyalty with a merchant evolves in five stages. First, is the awareness stage where customers recognize the company's name and product via advertisements, word of mouth and during this stage no purchases have been made. Second, is the exploration stage where customers learn more about company and their products. There is frequent communication between the customer and the merchant. This leads to the third stage which is familiarity where customers complete several transactions with the merchant and customers experience competence trust in their ability to successfully complete the business transactions online and predictability trust as they are aware of the firm's returns, credits policies, and pricing flexibility. The fourth stage is commitment where customers experience considerable number of highly satisfactory encounters. They develop increased trust and would prefer to only purchase from this merchant. This trust called the goodwill trust is developed over time. Finally, in the fifth stage customer encounter a negative experience with the merchant and decide to disband the relationship. The relationship terminates and declines at this stage due to the lack of trust. In the next section we discussion the importance of hard and soft factors and how it facilitates transactional versus relational components of communication.

HARD FACTORS

B2C e-commerce is socio-technical system used to advertise, sell and share business information and products by conducting business transactions over the Internet. We argue that it is critical to consider the role of the Internet and other web applications that facilitate the *transactional component* of communication thereby contributing to the hard factors such as; website's performance as in increased sales, reduced complaints, user friendliness, quality of web content, customer acquisition/search costs and perceived security. The transactional components include both the transactions of information and money pertaining to the purchasing process. The transactional component of communication is reflected in the more formal and precise reporting of outcomes and quantitative results such as sales figures. Transactional attributes of communication also refers to the accuracy, timeliness, adequacy, correctness, and credibility of information exchanged between the firm and their customers (Daft & Lengel, 1986; Huber & Daft, 1987; Mohr & Spekman, 1994).

Hard factors thus refer to the technological mechanisms that facilitate customers to use the B2C firm's web site to engage in online shopping. They form the foundation for B2C e-commerce transactions to occur and are derived from the Technology Acceptance Model (TAM) which is one of the most widely accepted models used for predicting e-commerce adoption. Drawing from the TAM, which has its origins in the Theory of Planned Behavior (TPB) (Davis, Bagozzi, & Warshaw, 1981), and Theory of Reasoned Action (TRA) (Ajzen & Fisbein, 1980), we propose that technical solutions and security solutions provide impersonal assurances that contribute to positive expectations, intentions, and behaviors in e-commerce relationships. This is consistent with Zucker (1986), who claims that trust is created in impersonal economic environment (without familiarity) and similarity (communality). Trust forms, not because people know each other personally, but because institutional structures that are akin to policies, auditing, and recourse embedded in the e-commerce technology. McKnight & Chervany (2001, p.13) advance this argument in the context of B2C e-commerce, arguing *"beliefs that the Internet has legal or regulatory protection for consumers (institution-based trust) should influence relationship trust of a particular e-vendor (interpersonal trust)."*

TAM thus posits that e-commerce adoption is affected by prior use related beliefs. It identifies two beliefs namely; perceived usefulness (as a response to the B2C firm's extrinsic value – accuracy, quality, timeliness of business transactions); and perceived ease of use (as a response from the customers pertaining to the intrinsic characteristics of the e-commerce system (ease of learning, flexibility, security solutions). E-commerce solutions and other third party application providers inform the B2C firm's expectations of the transactional capabilities of the technology platform. Customers form beliefs regarding the accuracy and efficiencies of the system thereby reducing search costs. The e-commerce system is also used to support a wide range of applications including standing purchase orders, just-in-time sourcing, invoicing and payments, as well as collaborative design and development. The technology also provides a mechanism to monitor the customers' interactions and communications with the firm and ultimately the firm's economic performance. Further, best business practices as in IT solutions and proven problem solving methods are used to attract and retain customers. Best business practices also include; enforcing governance mechanisms in the form of contracts, regular audit policies, top management commitment, high quality standards, adequate and complete training of the employees, and enforcing risks management procedures that impact and control the quality of the web content, security and web site performance.

SOFT FACTORS

Soft factors are derived from the *interpersonal or relational component* of communication between the firm and their customers. They include; quality of products/services, satisfaction, trust, loyalty and reputation of the firm. McKnight & Chervany (2001, p.13) advance this argument in the context of B2C e-commerce arguing *"beliefs that the Internet has legal or regulatory protection for consumers should influence relationship trust of a particular e-vendor."* Trust is frequently defined in the literature as (1) the confidence or predictability in one's expectations (Zucker, 1986), and (2) the confidence in others goodwill (Ring & Van de Ven, 1994). The benefits of trust in e-commerce relationships seem wide ranging; including lowering transaction costs (Gulati, 1995), promoting desirable behaviors, reducing the extent of formal contracts and facilitating dispute resolutions (Ring & Van de Ven, 1994). The behaviors reflect integrity, and dependability as well as the customers' knowledge of the products and satisfaction of the firm increases. Previous research suggests that customer loyalty increases profit and growth in many ways to the extent that increasing the percentage of loyal customers by as much as 30% to 85% depending on the industry (Gefen, 2002). This is evidenced by Keen (2000:1) who noted: *"We are moving from an IT economy to a trust economy."* Similarly, a significant number of studies support the underlying importance of trust in e-commerce relationships (Dyer & Chu, 2000; Jarvenpaa, Tractinsky & Vitale, 2000; Keen, 2000; Sako & Helper, 1998).

Based on the hard and soft factors that impact the customers' interactions and communications using the firm's web site we identify four modes derived from the customers shopping time lifecycle experiences leading to the four modes namely; learning, monitoring, collaborating and distancing modes. A mode refers to the characteristics of a particular e-commerce customer relationship at any given moment in the life of the relationship and it represents a pattern of behavior reflecting the customers' desire to improve predictability and reduce uncertainty in their relationship with the firm. The firm also monitors behaviors of their customers, assesses performance, identifies ways to improve their marketing strategies and adjusts their behaviors in order to retain and attract new customers.

THE EVOLUTION OF B2C SHOPPING TIME LIFECYCLE EXPERIENCES

The customers shopping time lifecycle experiences based on the hard and soft factors reflected on the transactional and relational components of communication leading to the four modes namely; learning, monitoring, collaborating and distancing modes. The firm monitors behaviors of their customers, assesses their shopping behaviors and identifies room for where the firm can improve, and adjusts in order to align it with their future intentions of their customers. The firm also focuses on the operations of their e-commerce system, to potentially reveal additional information about their needs and expectations of their customers. The communications the firm has with their customers will reflect their satisfaction and intentions to continue the e-commerce relationship. Previous research suggests that the quality of communication is related to satisfaction of the customers, intentions to continue a relationship, and a willingness to provide referrals (Morgan & Hunt, 1994; Sawhney & Zabin, 2002). MacNeil (1981) acknowledges that honest and open lines of communication encourage continued growth and close ties between trading partners. The customers shopping time lifecycle experiences impacts both the hard and soft factors that are reflected in the transactional and relational components leading to the four modes discussed below.

Learning Mode

The learning mode of the online B2C relationships occurs when the impact of the hard factors is low and the soft factors are high. Novice customers using the B2C firm's web site experience a positive relationship with the firm (Vatanasombut et al., 2004). During the initial stages of an e-commerce relationship, the firm encourages their customers to contact them via 1800 call numbers and email. The emphasis is on the relational component of communication and experiences the customers have with the firm and to what extent the firm will satisfy the customers' learning curve in using their web site. Because the B2C firm values the customers they will go out of their way to make additional investment in assuring that their customers are satisfied. The firm is tolerant of the initial mistakes made by their customers. The firm focuses on how to attract new customers and build customer relationships. By properly integrating the e-commerce systems the customers experience reduction in consumers search costs. Typically, the B2C firm will concentrate on negotiating and implementing technical security solutions so that their e-commerce system can provide speed, accuracy, privacy, accountability thereby reducing the customer acquisition/search costs. Furthermore, certain web technologies such as search engines and intelligent agents directly reduce the customers' search costs by providing relevant information for consumer decision making at a much lower cost and time (Lager, 2005). The B2C firm will continue to attract sophisticated customers as the firm provides user friendly technical solutions in the form of security, privacy policy information, functionality of the web site, warranty details and increase in both the ease and depth of navigation (Vatanasombut et al., 2004). It is essential for firms to offer online attractive promotions on their web site backed with high quality products and services in order to convert potential customers to actual sales. Therefore, we propose that:

RH-1: *The customers in a learning mode are more likely to experience a positive relationship with the firm.*

Monitoring Mode

The monitoring mode of the online B2C relationships occurs when the impact of the hard factors is high and the soft factors are low. The situation occurs when the hard factors are not used in a proper manner by the customers in order to create a positive relationship. The emphasis is on the economic returns from the transactional components. Because the firm has confidence in the functionality of their e-commerce system, they increase the monitoring measures to their benefit and assess the performance of their customers. Novice customers who do not have sufficient experience with the Internet may feel comfortable and tend to use the Internet for simple tasks (Lager, 2005). Their concerns relate to the compatibility of the browsers, hardware/software, cookies, user friendliness of the web site, and sometimes being unable to remember how to log onto the B2C web site. Sophisticated customers on the other hand have a higher understanding of the Internet technology demand speed and timely information. Their concerns relate to search costs, timeliness and accuracy of information/transaction updates, system and transaction reliability, new features, web site functionality and accurate, timely product information (Vatanasombut et al., 2004). The sophisticated customers who have become the firm's regular customers' may not purchase the firm's products and services if their concerns were not met. Therefore, we propose that:

RH-2: *The customers in a monitoring mode are more likely to experience a negative relationship with the firm.*

Collaborating Mode

The collaborating mode of the online B2C relationships occurs when the impact of both the hard and soft factors are high. The emphasis is on the relational component of the relationship between the customers and the firm. The customers experience cooperation, increased positive communications, trust satisfaction, and loyalty when interacting with the firm. Furthermore, the customers' positive experiences with the firm are associated with a willingness to recommend new customers derived from the reputation and image of the firm. Previous research suggests that there is a positive relationship between trust and commitment (Doney & Cannon, 1997; Ganesan, 1994; Gulati, 1995; Lewicki & Bunker, 1996; Rousseau, Sitkin, Burt & Camerer, 1998). This if reflected in current research which suggests that customer relationship management significantly impacts the level of trust of B2C e-commerce firms and their customers (Bunduchi, 2005). Trust can be achieved by providing customers with valuable information in a timely manner. According to Ganesan (1994) trust is the extent to which the customer believes that the vendor has intentions and motives that are beneficial to the customer. When customers trust a B2C e-commerce firm, they will share and communicate with the firm. Loyal customers are more tolerant when something goes wrong, and are easier to satisfy because the B2C firms are aware of their customers' expectations (Gefen, 2002). Thus, loyal customers are satisfied and are more likely to continue shopping with the B2C firm thereby increasing both the number and/or frequency of purchases (Kim et al., 2008). Satisfaction with delivered products and services have been suggested and empirically documented as affecting the customers' intention to continue, thereby increasing the B2C firm's reputation. Therefore, we propose that:

RH-3: *The customers in a collaborating mode are more likely to continue purchasing with the firm.*

Distancing Mode

The distancing mode of the online B2C relationships occurs when both the hard and soft factors are low. The customers perceive that the firm is not meeting their needs and are not keen to shop on their web site. Because the customers lack confidence in the firm and their e-commerce system they are more likely to encounter difficulties in trying to resolve issues and spending more time searching for their products and services. The customers lack trust and experience negative interactions with the firm. Most SMEs who conduct business over the Internet do not follow the robust technical, and security standards needed as they are challenged by financial resources, lower technical expertise, and limited management skills. They often do not have the time to gather customer information, which sometimes leads to poor decision mak-

Figure 1. The evolution of B2C relationships model

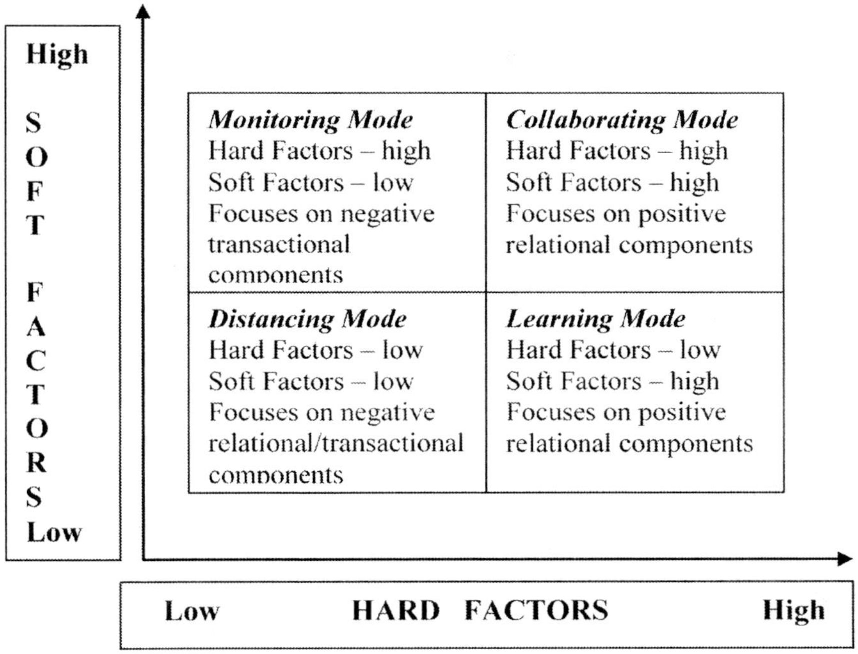

ing. The customers perceived the B2C firm to be untrustworthy, lacking quality service thereby leading to a negative image and poor reputation. The employees in a distancing mode wind up with reduced productivity and limited job satisfaction. The customers are inclined to distance themselves from this situation and are not willing to return to the web site. The dissatisfaction of the customers' contributes to an inclination to withdraw from the firm. Previous studies suggest that customers are unlikely to purchase from a B2C e-commerce firm if they perceive their online experience is not sufficiently secured (Freed, 2006; Vatanasombut et al., 2004). Therefore, we propose that:

RH-4: *The customers in a distancing mode are unlikely to continue shopping with the firm.*

THE EVOLUTION OF B2C RELATIONSHIP MODEL

The evolution of B2C relationship model Figure 1 below, illustrates the four modes and their impact on the both the hard and soft factors.

DISCUSSION

The model paved the way to some insights on the customers behaviors based on the impact of the hard and soft factors that were reflected in the transactional and relational components of communication leading to the four modes of the customers shopping time lifecycle experiences. The hard factors that reflected the emphasis on transactional components of communications were seen in the learning and monitoring modes, whereas the impact of the soft factors that reflected the emphasis on relational components of communications were seen in the collaborating and distancing modes. For example, while it normally takes time for new customers to evolve from the learning mode to the collaboration mode based on previous research, we suggest that it may take only a couple of attempts for a customer to experience the distancing mode. We believe that the monitoring mode is the deciding stage for the customers whether to pursue the relationship with the firm leading to collaborating mode or to disband from the relationship leading to the distancing mode. Thus, each mode represents a broad reflection of the effect of both the transactional and relational components based on the hard and soft factors leading to relationship continuity. Although shopping online can be a simple process, firms must periodically survey their customers to determine the impact of their online solutions on customers attributes and behaviors. Based on these insights firms will be able to identify their regular, profitable customers and customers who stopped shopping on their web sites. Further the frequent communications with their customers via email will enable firms to identify the needs, attitudes and behaviors of their customers thereby taking actions to increase their profit by meeting the needs of their customers.

CONCLUSION

This study expands upon the existing theory of the evolution of online relationships by explicitly considering the role of a B2C e-commerce firm. The study contributes to theory as we suggest that the important characteristics of the hard and soft factors that are in turn reflected on the transactional and relational components of communication leading to the four modes of the customers shopping time lifecycle experiences. The hard factors that emphasized the transactional components focused on the outcomes of the exchange between the customers and their firms. Alternatively, soft factors that emphasized on the interpersonal relational components were characterized by conversations via the telephone and email between the customers and the firm. At a simple level, we recognize that technology affects the transactional component and interactions between the customers and the firm. Technology provides a distinct basis for retaining customers that is separate from an assessment of the customers as it has an effect on profit. More importantly, we draw upon this expanded understanding over the *customers shopping time lifecycle experiences* based on the impact of both the hard and soft factors that is reflected on the transactional and relational components of communication leading to the four different modes in the B2C e-commerce relationships namely; learning, monitoring, collaborating and distancing modes.

This study contributes to theory as the evolution of B2C relationships model provides a holistic view of the firm by simultaneously examining the customers shopping time lifecycle experiences and performances leading to the four modes. Further, the study contributes to practice as we discuss how firms that have adopted B2C e-commerce are likely to experience customers shopping time lifecycle experiences. Future research should attempt to examine the evolution of B2C relationship model via multiple case studies on how the hard and soft factors can impact the customers shopping time lifecycle experiences thereby contributing to meaningful generalizations. In conclusion, customer attraction and retention for B2C e-commerce firms remains a complex, elusive yet extremely important phenomenon. The proliferation of the Internet has significantly threatened the ability of SMEs engaged in B2C e-commerce to retain their customers.

REFERENCES

Ajzen, I., & Fishbein, M. (1980). *Understanding Attitudes and Predicting Social Behavior*. Englewood Cliffs, NJ: Prentice-Hall.

Bunduchi, R. (2005). Business relationships in Internet-based Electronic Markets: The role of goodwill trust and transaction costs. *Information Systems Journal, 15*, 321–341. doi:10.1111/j.1365-2575.2005.00199.x

Daft, R., & Lengel, R. (1986). Organizational information requirements, media richness, and structural design. *Management Science, 32*(5), 554–571. doi:10.1287/mnsc.32.5.554

Davis, F. D. (1989). Perceived usefulness, perceived ease of use, and user acceptance of information technology. *MIS Quarterly, 13*(3), 319–340. doi:10.2307/249008

Doney, P. M., & Cannon, J. P. (1997). An Examination of the Nature of Trust in Buyer-Seller Relationships. *Journal of Marketing*, (April): 35–51. doi:10.2307/1251829

Dyer, J. H., & Chu, W. C. (2000). The Determinants of trust in Supplier Auto-maker relationships in the U.S., Japan and Korea. *Journal of International Business Studies, 31*(2), 259–185. doi:10.1057/palgrave.jibs.8490905

Freed, L. (2006). *Best Practices for Measuring the Website Contribution to Multi-Channel Purchase Behavior – Introducing the Multi-Channel Value Index* (Forsee Results White Paper, Winter, 1-10).

Ganesan, S. (1994). Determinants of long-term orientation in buyer-seller relationships. *Journal of Marketing, 58*, 1–19. doi:10.2307/1252265

Gefen, D. (2002). Customer loyalty in e-commerce. *Journal of the Association for Information Systems, 3*, 27–51.

Gulati, R. (1995). Does Familiarity Breed Trust? The Implications of Repeated Ties for Contractual Choice in Alliances. *Academy of Management Journal, 38*(1), 85–112. doi:10.2307/256729

Huber, G., & Daft, R. 1987. The information environment of organizations. In F.Jablin et al. (Ed.), *Handbook of Organizational Communication* (pp. 130-164). Thousand Oaks, CA: Sage Publications.

Jarvenpaa, S. L., Tractinsky, N., & Vitale, M. (2000). Consumer trust in an Internet store. *Information Technology and Management, 1*, 45–71. doi:10.1023/A:1019104520776

Keen, P.G.W. (2000, March 13). Ensuring E-trust. *Computerworld*.

Kim, D. J. Ferrin, D.L. & Rao, H.R. (2008). Trust and Satisfaction, Two Stepping Stones for Successful E-Commerce Relationships. *Information Systems Research, Articles in Advance* (pp. 1-21). Informs.

Lager, M. (2005). *Breaking Down the Silos, Customer Relationship Management*. Retrieved from http://www.destinationCRM.com

Lewicki, R. J., & Bunker, B. B. (1996). Developing and Maintaining Trust in Work Relationships. In R.M. Kramer, & T.R. Tyler (Eds), *Trust in Organizations: Frontiers of Theory and Research* (pp. 114-139). Thousand Oaks, CA: Sage Publications.

Macneil, R. (1981). *The new Social Contract*. New Haven, CT: Yale University Press.

McKnight, D. H., & Chervany, N. L. (2001). What trust means in e-commerce customer relationships: an interdisciplinary conceptual typology. *International Journal of Electronic Commerce, 6*(2).

Mohr, J., & Spekman, R. (1994). Characteristics of Partnership Success: Partnership attributes, Communication Behavior, and Conflict Resolution Techniques. *Strategic Management Journal, 15*, 135–152. doi:10.1002/smj.4250150205

Morgan, R. M., & Hunt, S. D. (1994). The commitment-trust theory of relationship marketing. *Journal of Marketing, 58*, 20–38. doi:10.2307/1252308

Ring, P. S., & Van de Ven, A. H. (1994). Developing processes of cooperative inter-organizational relationships. *Academy of Management Review, 19*, 90–118. doi:10.2307/258836

Rousseau, D. M., Sitkin, S. B., Burt, R. S., & Camerer, C. (1998). Not so different after all: A cross-discipline view of trust. *Academy of Management Review, 23*(3), 393–404.

Sako, M., & Helper, S. (1998). Determinants of trust in supplier relations: Evidence from the automotive industry in Japan and the United States. *Journal of Economic Behavior & Organization, 34*,387–417. doi:10.1016/S0167-2681(97)00082-6

Sawhney, M., & Zabin, J. (2002). Managing and measuring relational equity in the network economy. *Journal of the Academy of Marketing Science, 30*(4), 313–332. doi:10.1177/009207002236908

Schneider, G. P. (2009). *Electronic Commerce* (8th ed.). Course Technology, Cengage Learning

Street, C. T., & Meister, D. B. (2004). Small Business Growth and Internal Transparency, The Role of Information Systems. *MIS Quarterly, 28*, 473–506.

Turban, E. Lee, J. King, D. & Chung, H.M. (2006). *Electronic Commerce: A Managerial Perspective*. Prentice Hall Inc.

Vatanasombut, B., Stylianou, A. C., & Igbaria, M. (2004). How to Retain Online Customers . *Communications of the ACM, 47*(6), 65–69. doi:10.1145/990680.990682

Zucker, L. G. (1986). Production of trust: Institutional sources of economic structure: 1840-1920. In B.Staw & L.Cummings (Eds.), *Research in organizational behaviour*, 8, 53-111.

Chapter 11
Business Models for M-Services:
Exploring the E-Newspaper Case from a Consumer View

Carina Ihlström Eriksson
Halmstad University, Sweden

Thomas Kalling
Lund University, Sweden

Maria Åkesson
Halmstad University, Sweden

Tobias Fredberg
Chalmers University of Technology, Sweden

ABSTRACT

This article presents the consumer view of an innovative m-service, that is, the e-newspaper, which is published for a mobile reading device equipped with an e-paper display. The research question of this article is: What are the implications of the consumer view of future m-service innovation on business models? In this article, we present empirical results from an online survey with 3,626 respondents representing the consumer view. The factor analysis revealed three aspects of consumer preferences, that is, Ubiquitous access, Prestige of news source and Local anchorage and advertising. These were then correlated with media behavior and e-newspaper preferences to indicate three possible market segments. We then discuss the implications that these consumer views could have on developing new business models, followed by a comparison to earlier research. Finally, we propose an integrated e-newspaper business model framework consisting of three models, ubiquitous, local, and prestige.

INTRODUCTION

Today, the market for mobile e-services (m-services) has expanded in a wide range of areas, including the newspaper industry. There are two ongoing trends that influence the newspaper industry and related m-service initiatives.

First, the media industry is converging and boundaries between different media domains are blurring (Fidler, 1997). For example, newspapers have become media houses, offering content and services in multiple channels such as the Web, radio, mobile phones, PDAs, and TV (Åkesson & Ihlström, 2006a). Companies often merge their businesses with free newspapers (Chan, 2004). *Second*, the newspaper industry faces structural change due to declining subscription and advertising revenues as well as new competition from companies attracting advertisers to digital channels, for example, Google and Yahoo. This is forcing the newspaper industry to explore new revenue opportunities (The Economist, 2006).

M-services have attracted growing interest from the industry over the last couple of years due largely to the high penetration of mobile phones and the possibility of reaching new and younger audiences (The Swedish Newspaper Publisher's Association, 2007; IFRA 2006). The mobile phone penetration in the UK, Sweden and Italy already passed 110% in 2006 (Wallace, 2006), as teenagers and young adults in Sweden are in general more frequent mobile users than elderly (Westlund, 2006).

In Japan, the mobile channel has been the most common point of access to online services for years (Ratcliff, 2002). However, the adoption of m-services has not taken off as expected in Europe (Hammond, 2001; Carlsson et al., 2006). A growing body of research into m-services and m-commerce has sought to understand the reasons for this disappointing development (Amit & Zott, 2001; Pedersen et al., 2002; Samtini et al., 2003; Sarker & Wells, 2003; Vrechopoulos et al., 2003; Constantinou et al., 2005; Mallat et al., 2006). Among the explanations are *business model related factors*, *usability factors* and *technological factors*.

The newspaper industry has experienced problems finding successful business models for mobile content, mainly due to the revenue split with mobile operators, the difficulty of attracting advertisers and the low adoption rate of m-services among readers (Ihlström, 2005). Moreover, it has been challenging to offer a quality news reading experience on small screens. However, a technological innovation, e-paper technology (further described in section 2), has the potential of offering a quality news reading experience because this technology gives the same reading experience as on print. Incorporated in a mobile reading device, e-paper has the potential of revolutionizing the m-service market for this industry, with its core business of the "written word," but the challenge of finding suitable business models remains. If planned carefully, the e-newspaper introduction could also address the challenges of revenue splitting and attract advertisers in new and innovative ways.

In addressing this challenge, we present empirically the consumer views on this particular future m-service. The core research question of this article is: *What are the implications of the consumer view of future m-service innovation on business models?*

The purpose of this article is twofold, theoretical and practical. The first objective is to contribute to existing literature on business models for m-services by illustrating the consumer view on a possible future m-service innovation. The second objective is to provide knowledge useful for the content providers to develop business models for the introduction of the e-newspaper.

The article is structured as follows; section 2 outlines the importance of e-newspapers, followed by the theory used in section 3. The methodology and data collection is presented in section 4, and the consumer view is presented in section 5.

Section 6 discusses the findings and concludes the article.

THE IMPORTANCE OF E-NEWSPAPERS

An e-newspaper is a newspaper m-service published in a mobile reading device with an e-paper display. The e-newspaper should not be confused with electronic newspapers offered online, such as PDF versions of newspapers, which are digital replicas of the printed edition in 72 dpi resolution.

E-paper technology is reflecting, giving the same reading experience as on paper (such as high contrast and the possibility to read in sunlight), and is thin, flexible and nonsensitive. The resolution is 160 dpi, which is the same resolution as on printed newspapers, compared to the 72 dpi of a LCD screen. The electronically made print also enables e-paper the same viewing angles as printed material because it is a reflecting display technology in contrast to projecting technologies such as LCD screens. Another advantage of the technology is the low power consumption; power is only needed when the display is updated. Taken together, these properties make e-paper technology especiallly suitable for content traditionally offered in print such as newspapers, books, and magazines.

There are three e-paper based products, that is, eReaders, on the market today, iRex iLiad with an 8.1" screen (iRex Technologies, 2006), Sony Reader with a 6.9" screen (Sony Styles, 2007), and Amazon Kindle with a 6" screen (Amazon, 2008), which makes them highly mobile but also allows larger presentation space than mobile phones and PDAs. E-paper technology is developing rapidly and many are investing in "plastic electronics" (E Ink Corporation, 2007; Polymer Vision, 2007), which is predicted to become a multi million dollar industry (Plastic Logic, 2007).

The optimal future e-newspaper combines the portability, readability and overview from the printed newspaper with the possibilities of online media such as constant updates, interactivity and video, offering a high quality news reading experience anytime and anywhere (Ihlström, Åkesson, & Nordqist, 2004). However, at this point some of the most essential attributes like color, multimedia and interactivity are not supported in the available eReaders. At this moment, due to the fact that these devices do not support GSM, 3G or HSDPA, they do not support updated news anytime and anywhere if there is no WLAN coverage.

De TIJD, a belgium financial newspaper, made a test trial publishing an e-newspaper on the iRex iLiad in the spring of 2006 (The Editors Weblog, 2006a) followed by a test trial with an e-newspaper from Sundsvalls Tidning in Sweden (Ihlström Eriksson & Svensson, 2007). Yantai Media Group started to publish e-newspapers on iRex iLiads in China during 2006 (iRex Technologies, 2006) and in May 2007 the French daily economic paper Les Echos started to print their e-newspaper (The Editors Weblog, 2006b). More recently, New York Times, Wall Street Journal, Washington Post and Le Monde among others, have started to publish on the Amazon Kindle (Amazon, 2008). In In Figure 1, an e-newspaper from Sundsvalls Tidning is presented in the iRex iLiad.

The European project, DigiNews (ITEA 03015), was a 2-year project including partners from Belgium, Spain, Netherlands, France and Sweden, including several major technology firms, media houses and universities. Several studies regarding e-newspapers have been conducted by DigiNews. The overall goal of this project was to explore research and development issues regarding technology, design and business models for the future e-newspaper. The DigiNews project ended in June 2006, but this line of research has continued within a project called Designing Ubiquitous Media Services through Action Research (UbiMedia), which is a Swedish project with partners from nine Swedish newspapers,

Figure 1. An example of an e-newspaper in an iRex iLiad

the Swedish Newspaper Publishers' Association and Stampen. This 2-year project targets the challenge of designing ubiquitous media services for a multitude of devices and contexts to be consumed anytime and anywhere. The survey from which results are presented in this article was conducted as a joint activity between the two projects. In the following we refer to publications within these two projects.

The e-newspaper concept has attracted the interest of the newspaper industry mainly because of the print quality and readablity. Publishing news on e-paper raises hopes in the newspaper industry as a future replacement of print, which would heavily reduce production and distribution cost while preserving the qualities of news reading. Another reason for their interest is the possibility of expanding the market reach, for example reaching subscribers located far away (Ihlström, 2005). Moreover, due to the possibility of uniqely identifying each e-paper device to the database that handles the distribution and subscription, it is possible to target individuals with advertising, which in turn could attract more advertisers (Ihlström Eriksson, Akesson, & Hakeröd, 2007).

The newspaper industry has for centuries built its business on selling news to consumers and exposure to advertisers. In this business model, the newspaper companies have control over production, distribution and customer relations. As the introduction of the e-newspaper involves both new technology and new content at the same time, the introduction involves multiple stakeholders in the value chain. Besides content providers, consumers and advertisers, device producers and infrastructure providers need to be involved. Due to the unique readability, the eReaders have the potential of becoming "the iPod of reading," including content like books, manuals, maps, magazines, newspapers and personal documents, bringing together several content providers as well.

To test the e-newspaper concept, three e-newspaper prototypes (Figure 2) were developed within the DigiNews project in collaboration with three newspaper partners, that is, Aftonbladet, Göteborgs-Posten and Sundsvalls Tidning (Åkesson & Ihlström, 2006b; Ihlström Eriksson & Svensson, 2007). These prototypes were also presented in conjunction to the questionnaire used in this study (see section 4).

The newspapers are prepared to take a more aggressive role in the e-newspaper introduction to avoid the competition from other stakeholders, such as mobile operators, (Ihlström, 2005). As stated in the introduction, the challenge of finding initial business models still remains. In order to address this challenge, the content provider views of potential business models for the e-newspaper introduction have been studied (Ihlström Eriksson & Kalling, 2007), but in this article we focus on the consumer view and how their preferences can be of use for content providers in developing business models, and compare the results with the content provider view.

Figure 2. E-newspaper prototypes

THEORY

In this study the business model construct has been used to capture the consumer view of the future e-newspaper. The business model concept is used by practitioners as well as academics. It has been discussed as a substitute for strategy (e.g., Normann, 1977; McGrath & MacMillan, 2000; Normann, 1977; Schumpeter, 1934), and, more importantly has been used to discuss electronic business strategies (e.g., Amit & Zott, 2000; Rappa, 2002; Timmers, 1998).

We have chosen to use Hedman and Kalling's (2002, 2003) generic business model construct, which is based on established business strategy theory. This business model construct suggest that business models include six causally related firm-internal and firm-external components. Internally, firms consist of *resources* (e.g., Barney, 1991; Peteraf, 1993) which are *activated and organized* (e.g., Porter, 1985) to generate *offerings* (Porter, 1980). This requires interaction with the *factor market* (e.g., Barney, 1986; Porter, 1991; Wernerfelt, 1984), and, of course, that *customers* (Porter, 1980) are made to prefer the offering over those of *competitors* (Porter, 1980). Figure 3 depicts the construct.

The construct can be used for the analysis of the relation between a business and its environment as well as between internal factors. It has a causal

Figure 3. The business model construct (Hedman & Kalling, 2002, 2003)

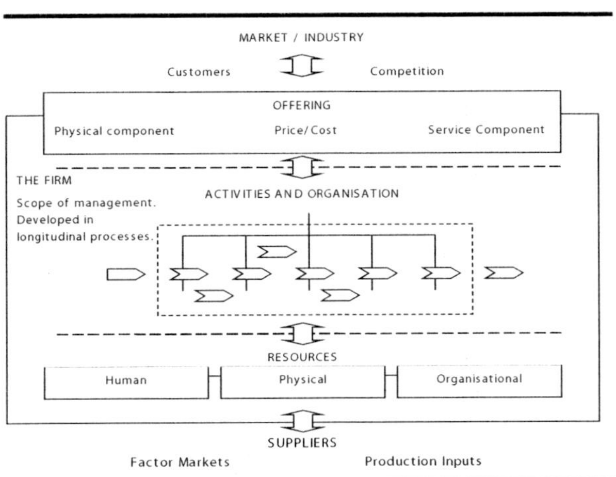

logic linking input factors through its conversion into offerings demanded at profitable levels in relation to competing offerings. It provides plausible explanations to firm performance variation and competitive advantage. Below we discuss the business model components and link these to current literature on e-news services offered online.

Customers

Customers put pressure on suppliers to increase the quality/price ratio (Porter, 1980), and understanding customers includes segmentation and individual customer knowledge in terms of preferences, price sensitivity, demographics, quality requirements, switching costs, locations, market position, and so on. Within online media, readers and advertisers are the most central customers (Clemons et al., 2002; Walter & Stolarova-Ornek, 2001), although here we concentrate on readers. Within research, the focus is on lock-in aspects (Sääksjärvi & Santonen, 2003), niche-marketing possibilities (Krueger et al., 2004) and the tendency to demand commoditized media products for free (Chyi, 2002).

Competition

Features of competition that make a difference and need be understood generically include scale and size, product range, cost structure, customer relations, competencies, value chain configurations, organization structure, locations, barriers to entry/exit, switching costs, and growth rate. Yet competition has received scarce attention in online media literature. Clemons et al. (2002) claim advertising has low barriers to entry whereas news has higher barriers due to credibility and reputation gained over years of news publishing. Punie et al. (2001) anticipate increasing consolidation in the sector, due the fragmentation of actors.

Offering

Any offering has a cost and a price, and tangible and intangible features that customers compare with those of competitors. Understanding customer needs and the features of the offering is central to perceived quality and uniqueness. Online offering aspects are well covered in current literature (e.g., Chyi. 2002; Punie et al., 2002; van der Beek et al., 2005; Ihlström & Henfridsson, 2005). Online news is frequently free, and alternative sources of revenue include differentiating certain news for price-flexible customers, loyalty programs, and advertising. Personalization, archiving and versioning become important (Ihlström & Palmer, 2002). With respect to features, online media must consider content, advertisements, design, branding, and complementary products, in addition to several technical challenges. Screens, batteries, security, usability, technology ownership, hardware and software stability, speed and capacity are important factors in the functionality of mobile services (Tsalgatidou & Pitoura, 2001).

Organization and Activities

A value-chain analysis includes sequences of activities, links to outside stakeholders and between internal activities, as well as structure and means of control. In addition, it requires an analysis of *drivers* of cost and differentiation. Possible metrics include productivity, effectiveness, and perceived quality of output. In the literature on online media, the organization of activities is typically assumed to resemble a traditional newspaper. Clemons et al. (2002) say it is principally about creating, producing, and authenticating news and that confidence is a central feature of online news providers. Furthermore, advertising will become more important, given that it is not the typical core competence of newspapers.

Resources and Factor Market Interaction

Competencies, brands, image, and proprietary technology are typical firm resources. Resources that matter are valuable but also unique, and analyzing under which circumstances they are valuable and unique is imperative. This includes how individual resources are activated and organized and how this links to critical features of the offering. Resources also need to be acquired and sourced cost-effectively on the factor market. In online media literature, strong customer and advertiser relations are highlighted (Walter & Stolarova-Ornek, 2001), as are reputation, brand name and advertising (Clemons et al., 2002). Reputation and recognition are important (because readers can turn anywhere to find news) and often characterized by time compression diseconomy (Dierickx & Cool, 1989).

In a recent study, this business model construct (Figure 3) was applied to address content provider views on possible business models for the e-newspaper introduction (Ihlström Eriksson & Kalling, 2007). This study was based on interviews and workshops with newspaper representatives. The empirical findings revealed 16 factors as the most important aspects when the e-newspaper market is being established.

Four typical business models (Figure 4) were suggested based on the input side of news distribution, that is, the degree of bundling, and the output side, that is, geographical coverage. This four business models are: *The Interpress News Agent*, *International Newspaper*, *Local Newspaper* and *Community Micronews*. The "International Newspaper" refers to the global actor who provides its own news, without bundling news from multiple sources. An example here includes the New York Times and companies with a global brand name and fully controlled editorial workflows. The "Interpress Newsagent" has a global coverage yet sources its content from a wide array of local providers. The "Community Micronews" actor has a very strong local approach, and covers local news in great detail, using a community journalism approach, not necessarily using their own journalists. The "Local Newspaper" resembles very much the ordinary local, regional or national newspapers of today.

Based on these four typical business models (Figure 4), a business model framework was suggested to represent the content provider view for the e-newspaper introduction (Table 1).

The business model framework includes the 16 empirically identified aspects, as well as business model components for all four typical business models, that are relevant in a discussion

Figure 4. Four typical business models

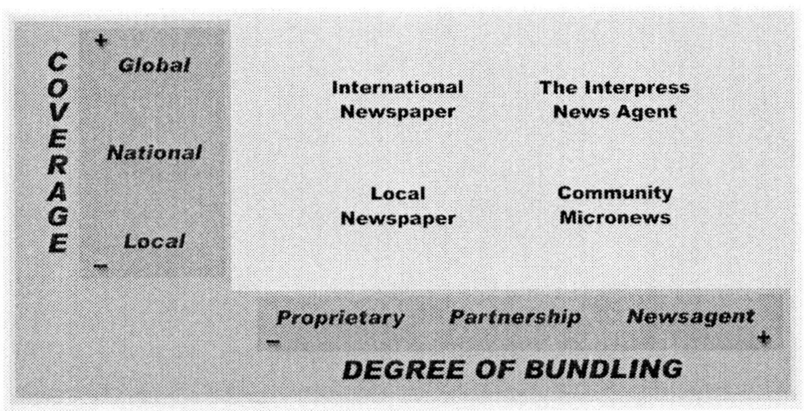

Table 1. An e-newspaper business model framework (Ihlström Eriksson & Kalling, 2007)

	Empirical findings	The Interpress News Agent	Inter-national Newspaper	Local Newspaper	Community Micronews
Customers	Niche marketing – readers Niche marketing – advertisers Customer reading habits	Global readers Global and local advertisers Source awareness Anytime, anywhere	Global brand-aware readers Global advertisers "Global readers" Anytime, anywhere	Regional readers Regional advertisers Dedicated readers	Community readers Community advertisers Local awareness Traditional, routine readers
Competition	Content Time Distribution	Local + global newspapers 24/7 Distribution network	Other internatio-nal media houses 24/7 Distribution ownership	Other local newspapers Regularity Distribution ownership	Morning Distribution network
Offering	Price model Device features Content features	Reader price pressure Advertising price premium News wholesale Standard device Up-to-date world news	Price premium Prestigious devices Up-to-date world news	Sub-scription rate Branded device Local news and ads	Low price Standard device Extremely local news and ads
Activities and organization	Distribution Editorial workflow Advertising workflow New activities	Provider relations 24/7 Differen-tiated advertising Umbrella branding Legal functions	Supply chain manage-ment 24/7 Global advertisers	Owner of distribution Several editions Local and national advertisers New mediator roles	Local provider relations One edition Micro advertising New mediator roles Umbrella branding
Resources and factor market interaction	Journalism Advertising skills Perceived credibility	Source reach Global and local brands Advertising networks Recognized wholesaler	Well recognized journalists Global brand name Additional channel Serious journalism	Multimedia journalists Local and national advertisers Additional channel Established brand	Community journalists Multimedia journalists Local brands Local relations, small firms Authenti-city

of e-newspaper strategies from a content provider view. This earlier work suggests that the business model construct (Hedman & Kalling, 2002, 2003) is applicable to e-services, such as mobile e-news services.

METHODOLOGY AND DATA COLLECTION

In addressing consumer views on mobile e-services we have chosen to conduct a Web-based survey. This was done in the spring of 2006 at three Swedish online newspapers. The Swedish market is characterized by high newspaper readership (Lund, 2007) and high penetration of mobile phones (Wallace, 2006) suitable for a survey on a future innovative mobile e-newsservice. We choose to use online questionnaires because that allowed us to provide information for the respondents to obtain an understanding of the e-newspaper concept. Because we assumed that the e-newspaper concept was not known to most of the potential respondents we provided them with the possibilities to:

- read more about e-paper technology,
- watch three concept videos (see below), and
- explore the e-newspaper prototype of the newspaper in question (see below).

In nonprobability sampling there is a risk that the sample does not accurately represent the population. In the case of voluntary sampling, the sample tends to be overrepresented by individuals who have strong opinions. However, surveys on the Internet, as in other approaches, are limited by the willingness of the respondents to answer the questionnaire. Ritter et al. (2004) made a comparative study using Internet vs. mailed questionnaires and argue that: "*among a convenience sample recruited via the Internet, results from those randomly assigned to Internet participation were at least as good as, if not better than, among those assigned mailed questionnaires, with less recruitment effort required. The instruments administered via the Internet appear to be reliable and to be answered similarly to the way they are answered when they are administered via traditional mailed paper questionnaires.*" Furthermore, Buchanan and Smith (1999) argue that Web samples can be as representative as traditionally collected samples because of the heterogeneity of the online population.

Admittedly there are inherent problems in controlling who responds to online questionnaires. Control for cases with multiple submissions from the same IP number was handled in the data collection. Furthermore, the respondents that had given an age under 15, those who did not complete or answered the questions included in this study inconsistently were excluded form the data set. We are aware that we might have captured an audience that is more interested in media or technology than the average consumer as the questionnaire was placed on newspaper sites and regarded new technology as well as being a voluntary sample. On the other hand, it would not have been possible in this early stage of development to receive the preferences of a larger sample of consumers, without the properties of the Internet as we needed to show the prototypes and concept videos (discussed below) in order for the respondents to understand the concept of an e-newspaper.

We presented the questionnaires at the news sites of the three Swedish newspapers that we have collaborated with in developing e-newspaper prototypes (Figure 2), that is, Aftonbladet, Göteborgs-Posten and Sundsvalls Tidning. Aftonbladet is a tabloid with the most visited news site in Sweden, Göteborgs-Posten is a regional morning paper covering Göteborg (the second largest city in Sweden) and its surroundings, and Sundsvalls Tidning is a local morning paper in the north of Sweden.

Three short concept movies of future e-newspaper use were developed to further illustrate the potential of the e-newspaper. These movies were created in collaboration with SVID (Stiftelsen Svensk Industridesign), the design company Propeller AB and the Swedish Newspaper Publishers' Association. The movies envisioned the benefit of the e-newspaper for three different personas:

the business women, the student and the senior citizen. Close-ups on the designed user interface together with examples of functions showed the future e-newspaper in detail. The scenarios were based on preferences articulated by respondents in workshops and interviews and showed how a future e-newspaper could support the media consumption of the three personas in different contexts. These movies were presented to the respondents to further illustrate the e-newspaper concept.

Three approaches for presenting the questionnaire were used in our study, a) a newspaper article explaining the e-newspaper concept at Aftonbladet's www.aftonbladet.se, b) a banner ad at Göteborgs-Posten's www.gp.se, and c) a pop-up window at Sundsvalls Tidning's www.st.nu.

a. In conjunction to the article at the site, links to the three concept videos and to the Aftonbladet e-newspaper prototype was presented. There was also a written invitation to participate in the questionnaire presented together with a link to the starting page of the questionnaire. The article was placed on the front page of www.aftonbladet.se on April 7th (11:23 AM) and was moved to the IT section on April 10th and was removed on April 11th (5:00 PM). At www.aftonbladet.se, 3757 respondents answered the questionnaire.
b. The banner read *"Do you want to influence the future e-newspaper? Click here!"* Clicking the banner directed the visitor to the starting page of the questionnaire. The banner was placed on www.gp.se between April 7th (10:23 AM) to April 12th (07:54 AM).
c. A full size pop-up window was presented each time a visitor entered the front page at www.st.nu between April 7th (12:10 AM) to April 12th (09:15 AM) resulting in 447 respondents. The pop-up window contained the starting page of the questionnaire.

In total, 4339 respondents answered the questionnaire (Table 2).

After reducing the sample by removing inconsistencies and so forth, as discussed above, the dataset used in the analysis contained 3626 respondents. In Table 3, an overview of the demographics of the respondents is presented.

As can be seen in Table 3, there are more men than women and the average age of men is slightly higher. Most of the respondents are in the medium income segment, have secondary or university education, and work full time or study.

The questionnaire was divided in four different parts, 127 questions in total. The first part contained 20 background questions. The second part consisted of 39 questions regarding business models for media services, the third part 26 questions about media behavior and e-newspaper preferences, and finally, the fourth part 41 questions concerning mobile services. Many of the questions where of multiple choice and several open questions were included as well. In this article we report from part one, two and three.

The empirical ambition of this study is to track key concerns and considerations for media consumers with regard to e-newspaper business models. This includes, naturally, the objective of identifying patterns of preference and behavior, groups of factors that explain consumption of the

Table 2. Newspapers hosts for questionnaires and number of respondents

Newspaper	*URL*	*Unique visitors/day*	*Total no. of respondents*
Aftonbladet	aftonbladet.se	1.200.000	3757
Göteborgs-Posten	gp.se	41.500	135
Sundsvalls Tidning	st.nu	14.500	447

offerings of e-newspaper suppliers, and possible ways to link these factors to optional strategies and business models, all with the ambition to say something empirically and theoretically informed about the emerging e-newspaper industry.

- **Background questions (part one):** The background questions included questions of demographic nature (5), media consumption (3), technology use (5) and the use of mobile services (7). In this paper only the demographic questions are used, that is, gender, age, occupation, education and income level.
- **Business model questions (part two)***:* Our questions to users were designed in a way that enabled respondents to signal the importance of matters related to the different components of e-newspaper business models. Respondents were asked to indicate, by means of 1-7 Likert scale questions, the importance of certain features related to preferences, e-newspaper offerings, competing offerings, organization and activities, and resources. The questions were also designed to enable particular consideration of geographical coverage and degree of bundling (see Ihlström Eriksson & Kalling, 2007). Because there is relatively little written on e-newspapers, and even less using the five components of the business model construct, we operationalized the model partly openly and inductively, partly by reference to our past experience and knowledge of the e-newspaper business, and partly based on relevant indicators as proposed by Hedman and Kalling (2002).

The questions on "customers" were to capture consumer behavior, priorities and preferences with regards to the functionality offered (Hedman & Kalling, 2002). Q25, Q28, Q38, and Q52 are examples of questions related to these general aspects. Lock-in aspects (see Sääksjärvi & Santonen, 2003) are covered by Q34, Q49, and Q58. Niche-

Table 3. Demographics of respondents

Respondents		***All***		***Men***		***Women***	
No of respondents		3626	100%	2216	61,1%	1410	38,9%
Age range		15-86		15-79		15-86	
Average age		37,1		38,0		35,8	
Std. dev		13,8		14,4		12,7	
Income level	low	1088	30,0%	561	25,3%	527	37,4%
	medium	1520	41,9%	853	38,5%	667	47,3%
	high	1018	28,1%	802	36,2%	216	15,3%
Education (54 missing)	elementary	331	9,3%	212	9,7%	119	8,6%
	secondary	1646	46,1%	1002	45,8%	644	46,6%
	university	1595	44,7%	975	44,0%	620	44,8%
Occupation	full time	1995	55,0%	1353	61,1%	642	45,5%
	part time	281	7,7%	107	4,8%	174	12,3%
	unemployed	210	5,8%	115	5,2%	95	6,7%
	retired	227	6,3%	164	7,4%	63	4,5%
	student	688	19,0%	387	17,5%	301	21,3%
	sick leave	132	3,6%	50	2,3%	82	5,8%
	other	93	2,6%	40	1,8%	53	3,8%

marketing possibilities (e.g., Krueger et al., 2004) are operationalized by Q33 and Q47. The demand for more commoditized media products (see Chyi, 2002) is tested through Q22 and Q32.

"Competition" questions are aimed here at understanding how readers view alternative offerings. Relatively little has been written on competition in the electronic news sector, and so the questions are based on general considerations of competing offerings in a fragmented sector (e.g., Punie et al., 2001), and particularly if and how competing alternatives can be integrated. The local vs. national vs. international dimensions (Ihlström Eriksson & Kalling, 2007) are covered by Q33, Q51 and Q59. Q23 and Q29 relate to the view on integrating competing offerings.

The "offering" questions are meant to indicate price sensitivity and the actual functionalities and quality properties that customers are interested in (Hedman & Kalling, 2002). Pricing (e.g., Ihlström & Palmer, 2002) is captured in Q27 and Q49. The importance of brand and confidence is covered in Q56, the real-time access preference in Q24 and Q43, the local vs. world news priorities in Q53, and finally the preference for electronic vs. physical papers in Q40 (see Chyi, 2002; Punie et al., 2002).

Looking at "activities and organization," this business model component is meant to cover aspects related to the infrastructure of e-newspaper supply. We have looked in particular at the timing dimension of news reporting (e.g., Clemons et al., 2002), operationalized in Q36, Q50, Q54 and Q55; and also in to the view on the importance and relevance of networking externally outside the media house itself. Q45 and Q48 indicate how readers view their own role in news reporting (e.g., community journalism) and how they rate the interaction with advertisers.

Finally, the "resources and factor market interaction" questions aim at describing readers' view on the requirements on assets and their management (Kalling & Hedman, 2002). Customer relations, including the ability to show local and international news (Q30, Q35, Q37) and the ability to interact and make use of advertising (Q26 and Q46) are covered (see Walter & Stolarova-Ornek, 2001), as are the importance of credibility and recognition, particularly brands, journalists and editors, authenticity and swiftness (Q42, Q41, Q21, Q44 and Q57) (see also Clemons et al., 2002).

- **Media behavior and e-newspaper preference questions (part three):** Of the 26 questions in part three, six regarded general e-newspaper preferences, based on earlier research (Ihlström et al., 2004; Ihlström, Svensson, & Akesson, 2005, Åkesson & Ihlström, 2006b). In the correlation analysis in this article we use three of these (Q61, Q64 and Q67), that used a 7-grade Likert scale. Part two of the questionnaire also included four questions (Q62, Q63, Q68 and Q69) about acceptance of the e-newspaper concept as described in this study. These questions were given fixed options for answers. The questions regarding e-newspaper preferences could be found in Appendix B. The remaining 20 questions concerned media behavior.

To research media behavior, we first constructed questions that regarded the social and interactive nature of the media (Q70-71). Seven questions (Q73-79) were adopted and adjusted from the Technology Readiness Index (Parasuraman, 2000). We then looked at the need for news and focused on the different character of news items. Q80 refers to the attention creation mechanisms used by media companies (Fredberg, 2003), Q81 serves as an indicator of the extent to which readers prioritize localness or simple headline news (as is the business of content aggregators such as Yahoo News). Media markets are becoming increasingly differentiated where the local nature of news (geographic or topical) becomes increasingly important (Fredberg, 2003), and Q82-83 refers to the tendency of readers to seek value such news. The last two questions relate to the relation between the different media outlets of a news organization and aims at creating an understanding of readers' opinion about the integration/competition between outlets. The questions used in the analysis of this article are presented in Appendix C.

The responses to the questionnaire were analyzed using SPSS v14.0. To analyze the structure of the interrelationships among the variables in part two we conducted an exploratory factor analysis. The sample met the necessary thresholds for conducting a factor analysis (Hair et al., 2005), KMO Measure of Sampling Adequacy = 0,88. We used principal component factor analysis with principal component extraction method and Varimax rotation method. After 5 rotations that accounted for 44% of the variance, this analysis resulted in a set of three factors representing common underlying dimensions of the consumers media consumption patterns and preferences. To examine if these three dimensions are related to different consumer profiles we compared them against background factors such as age, level of education and income. This analysis focused on calculation of mean scores for each background factor, and the mean differences were examined with ANOVA analysis (Hair et al., 2005). Thereafter, we correlated these three factors to the questions about media behavior and preferences of a future e-newspaper. Thereby, we could relate the consumer's perceptions of what is attractive for a future e-newspaper to the dimensions of business models.

CONSUMER VIEWS ON MOBILE E-NEWS SERVICES

In this section we present the data and analysis from the survey among newspaper consumers. First, the result of the factor analysis of the business models questions is outlined and second, the mean scores of the factors between different groups of respondents based on demographics are compared. Finally, the correlations between the factors and the question about preferences for a future e-newspaper and media behavior are presented.

Factor Analysis: Questions About Future Mobile E-News Services

In order to examine the consumer view on future mobile e-news services, to inform the development of business models, we performed a principal component factor analysis on the questions in Appendix A. Initially all items were included and in an iterative process the items that scored lower than 0,3 or loaded on more than one component by 0,3 or higher were eliminated from the solution. For sample sizes 350 and larger the significance level for a factor loading is 0,30 or above (Hair et al., 1995). In the factor analysis 11 items with low factor loadings (>0,3) or cross-factor loadings were eliminated to ensure the factors to be unidimensional and distinct. The final principal components factor analysis (varimax rotation) suggests a three-factor solution (see Table 4) based on 28 items that after five iterations accounted for 44% of the total variance.

The three-factor solution presented in Table 4 gives three consumers views on mobile e-news services. The items in the first factor are related to how consumers access news content and to the independence of time and place. We have labeled this factor *Ubiquitous access* (1). The items forming the second factor concern the brand and the resources related to the news source and we have labeled this factor *Prestige of news source* (2). Finally, the third factor is related to items referring to local anchorage and advertising, thereby labeled *Local anchorage and advertising* (3). These three factors can be considered to be three viewpoints on mobile e-news services from a consumer perspective that inform the construction of business models for the introduction of the e-newspaper.

Table 4. Factor solution of consumer views on future mobile e-news services

Q no		*Component* *F1*	*F2*	*F3*
32	I want access to news all day	0,736		
55	My news source must update its news several times every day	0,720		
22	I read the news on the Internet every day	0,667		
43	I must have access to news in real time	0,669		
54	It is of utmost importance that I have access to news instantly	0,648		
25	I want access to news also when I am traveling or commuting to work	0,616		
24	I want access to several news sources simultaneously	0,601		
39	I am happy to pay for technical equipment giving me access to news (e.g., computers, Internet connection)	0,558		
40	Within a few years I will not read the physical paper regularly	0,534		
29	My demands for news is not satisfied by one individual paper	0,488		
50	I expect my paper to be more active in reporting news	0,492		
26	I want to be able to interact more with journalists and editors (e.g., to ask questions)	0,473		
23	I would appreciate if several newspapers could share one portal/Website	0,465		
33	I combine local papers with national or foreign	0,403		
44	My news source must be well known		0,814	
42	My paper must have well known journalists and editors		0,780	
56	For me, the brand of the paper is important		0,775	
41	The brand of the news source is important		0,746	
45	My paper or my news sources must have their own editorial board and journalists		0,687	
34	It is important that the news source is well known		0,684	
59	I prefer local newspapers to national			0,731
47	Newspaper advertisements are an important source of information to me			0,670
38	I am primarily interested in local news			0,679
46	I find the advertisements in my paper useful			0,658
33	I prefer local advertisements to national			0,580
49	I have strong confidence in my local paper			0,588
48	I want to be able to communicate with the companies that advertise in my paper			0,545
58	My primary news source is the morning paper			0,520

Extraction Method: Principal Factor Component Analysis. Rotation Method: Varimax with Kaiser Normalization. Rotation converged in 5 iterations

To compare the three viewpoints against consumer demographics we split the dataset into the different categories given by the background questions. We calculated the mean score for each of the three factors (from a seven grade scale) in relation to each of the demographic categories. Differences of mean scores were tested using ANOVA analysis. The results are presented in Table 5.

As demonstrated in Table 5, the mean score differences are significant (p<0,05) with the exception of *Ubiquitous access*/Occupation and *Local anchorage and advertising*/Income level. There are some interesting differences to be noted. *Ubiquitous access* and *Prestige of news source* are rated higher by men than women while women rate *Local anchorage and advertising* higher. The

Table 5. Mean scores of the three factors per demographic category

Factor		*Ubiq-uitous access*	*Std dev*	*Prestige of news source*	*Std dev*	*Local anchorage and advert-ising*	*Std dev*
Gender	Men	4,7	1,1	4,1	1,5	3,3	1,2
	Women	4,5	1,1	3,9	1,4	3,6	1,2
	ANOVA F/sig	19,451/0,000		12,436/0,000		33,193/0,000	
Age	15-29	4,6	1,1	3,8	1,4	3,3	1,1
	30-44	4,8	1,1	4,0	1,5	3,5	1,2
	45-59	4,5	1,2	4,1	1,5	3,5	1,3
	60-	4,6	1,1	4,7	1,4	3,6	1,2
	ANOVA F/sig	1,479/0,008		2,074/0,000		1,336/0,037	
Income level	low	4,6	1,1	3,9	1,4	3,4	1,1
	medium	4,7	1,1	4,0	1,4	3,5	1,2
	high	4,9	1,2	4,3	1,5	3,4	1,2
	ANOVA F/sig	10,149/0,000		8,095/0,000		1,709/0,163	
Education (54 missing)	elementary	4,5	1,2	3,9	1,5	3,6	1,3
	secondary	4,6	1,1	3,9	1,4	3,5	1,3
	university	4,7	1,1	4,1	1,4	3,3	1,2
	ANOVA F/sig	7,248/0,001		11,517/0,000		16,978/0,000	
Occu-pation	full time	4,7	1,1	4,0	1,4	3,4	1,2
	part time	4,5	1,2	3,9	1,5	3,4	1,2
	unem-ployed	4,6	1,1	3,9	1,5	3,4	1,1
	retired	4,6	1,2	4,6	1,4	3,7	1,2
	student	4,6	1,0	3,8	1,3	3,3	1,1
	sick leave	4,5	1,2	3,9	1,6	3,5	1,3
	other	4,6	1,2	3,8	1,4	3,4	1,3
	ANOVA F/sig	1,491/0,177		9,003/0,000		3,270/0,003	

age group that rate *Ubiquitous access* the highest is 30-44, and the ones that rate *Prestige of news source* and *Local anchorage and advertising* is 60-. The age group 15-29 generally rate *Prestige of news source* and *Local anchorage and advertising* lower. *Ubiquitous access* and *Prestige of news source* are rated slightly higher by the higher educational and income level whereas *Local anchorage and advertising* is about the same for all income levels. Noticeable is that the higher educational level, the lower rating for *Local anchorage and advertising.* As a final point, *Ubiquitous access* and *Prestige of news source* are rated somewhat higher by full time occupied respondents whereas *Local anchorage and advertising* is rated higher by retired respondents.

Correlation against E-Newspaper and Media Behavior Questions

The three factors representing consumer viewpoints were correlated against the questions about how consumers would prefer a future e-newspaper. The correlation coefficients are presented in Table 6.

The three factors/viewpoints were then correlated against the questions asked about their media behavior. The correlation coefficients are presented in Table 7.

As indicated by the statistics in Table 6 and Table 7, there is a pattern of how newspaper preferences and media behavior correlate to the three viewpoints of future mobile e-news services. These patterns will be discussed below.

In the factor analysis, *Ubiquitous access* centered around the notion of being able to access news (including international news) ubiquitously. In correspondence, the correlation patterns related to Ubiquitous access segment centre upon five groups of features: *all possible functionality* (Q 61, 64, 67), *networking* (Q 70, 71), *advanced use of media and technology* (Q 73-76, 78), *constant updates and relevance* (Q 80, 82, 83), and *portal consumption* (Q 81, 84, 85).

The *Prestige of news source* rests on questions related to well-known and recognized media brands, editors, journalists and news sources. The correlation patterns include preferences partly similar to those of the *Ubiquitous access* and *Local anchorage and advertising* segments: *new and advanced technology* (Q 61E, 67A, 67H, 73, 74, 75, 77) is important (cf. the Ubiquitous access segment), while the perception that the e-newspaper is seen as a *differentiated e-news service* (Q 61F, 84, 85) resembles the preferences of the *Local anchorage and advertising* patterns.

The *Local anchorage and advertising* viewpoint emphasizes factors related to local news and local advertisements. The analysis also revealed that this viewpoint correlates to four groups of features: *low degree of use of media and technology* (Q 73, 75, 76, 78), *differentiated media* (Q 61F, 81, 84) where the e-newspaper should add value in relation to the paper version, *basics only* (Q 61A, 61B, 64H, 64F) meaning e-newspaper features such as environmental-friendliness and cost-effectiveness are important, and *passive character* (Q 67B, 67F, 70). We have summarized the features correlating to the three viewpoints in Table 8.

As indicated by these statistics, we suggested that the market for the e-newspaper is a segmented market. There are three emergent segments; the *Ubiquitous access* segment, *Prestige of news source* segment, and the *Local anchorage and advertising* segment. All of these segments have certain features (see Table 8), that can be used to describe the preferences and consumption behavior of these segments.

Finally, to examine the consumer's acceptance of the concept of an e-newspapers four questions were asked concerning accepted cost level, the financing of an e-paper device, their willingness to exchange the printed newspaper to an e-newspaper in the future, and finally, within which timeframe they would consider exchanging. The results are presented in Table 9.

Table 6. Correlations between the three viewpoints and e-newspaper preferences

Q no	*Preferences for a future e-newspaper* (Alternatives on 7-grade Likert scale)	*Correlation coefficient* *F1*	*F2*	*F3*
61	*What reasons are critical for you to exchange the traditional newspaper to the e-newspaper in the future?*			
A	Environmental reasons, that is, you are concerned about future environment	0,204**	0,026	0,068**
B	Cost savings, that is, you see economical advantages for the exchange	0,323**	-0,007	0,046*
C	Time savings, for example you do not have to go to a news stand to buy the newspaper	0,418**	-0,029	-0,015
D	Accessibility, that is, the newspaper is accessible wherever I am in the world	0,419**	0,022	0,014
E	Satisfaction, for example, satisfy interest in new technology	0,428**	0,037*	0,029
F	Added value, that is, the new technology offers new services that the traditional newspaper does not	0,444**	0,056**	0,044*
64	*Apart from reading the news, what added services do you think should be available in a future e-newspaper?*			
A	Personalization, for example, possibility to choose the parts of the content that you are willing to pay for	0,163**	-0,019	-0,015
B	Community information, for example, time tables, maps, traffic information	0.156**	-0.003	0,003
C	Personal information, for example, phone book, schedule	0,177**	-0,021	0,010
D	Common information, for example, dictionaries, manuals	0,172**	0,022	0,002
E	Fiction and factional literature, magazines	0,177**	0,025	-0,029
F	Possibility to access yesterdays papers	0,154**	0,025	-0,049**
G	Transactions, for example, e-commerce, orders, bookings	0,187**	0,030	0,030
H	Entertainment, for example, games, horoscopes, movie tips	0,133**	-0,024	0,049**
67	*How important are the following attributes/factors for you choosing to read your newspaper on e-paper?*			
A	The appearance of the e-paper and its contents	0,129**	0,077**	-0,021
B	The news are constantly updated	0,182**	0,015	-0,049**
C	Offer of additional functions such as chat, motion pictures, interactivity	0,197**	0,001	0,031
D	Ease of use and ease of learning	0,068**	0,024	0,013
E	Trust in technology performance	0,114**	0,032	-0,031
F	Observation of people in my surrounding using the technology	0,024	0,067**	0,067**
G	Positive effect on the environment for example, less paper production and tree fell	0,099**	-0,20	0,000
H	Importance of take part in new technology	0,249**	0,049**	-0,12
I	Ease of finding what I am looking for/wish to read	0,145**	-0,002	-0,21

* Correlation is significant at the 0.05 level (2-tailed). ** Correlation is significant at the 0.01 level (2-tailed).

As can be seen in Table 9, the majority of consumers expect an e-newspaper to be cheaper than the traditional printed newspaper and prefer the device to be included in the subscription. As many as 67.1% would consider exchanging the printed newspaper for an e-newspaper and the majority are willing to do so today or within 5 years. This indicates that there is an acceptance among consumers for the concept of an e-newspaper as described in this study.

Table 7. Correlations between the three viewpoints and media behavior

Qno	Media behavior (Alternatives on 7-grade Likert scale)	Correlation coefficient F1	F2	F3
70	I actively contribute with letters to the editor, debates or comments about papers on the Internet	0,156**	0,030	-0,56**
71	I would like to communicate more with other readers	0,196**	0,012	0,007
73	I am often better informed than people in my surrounding about media content	0,212**	0,114**	-0,39*
74	I am often asked about media content	0,200**	0,110**	-0,35
75	I am often well to the front with the latest technical products, for example, mobile phones	0,238**	0,045*	-0,046*
76	I am often asked for advice about new technology	0,222*	0,021	-0,64**
77	I prefer simple technology to advanced functionality	-0,23	0,042*	0,011
78	According to my opinion it is safe to use credit cards for payment on the Internet	0,165*	0,004	-0,091**
79	It worries me that other people can see what information I send on the Internet	-0,006	0,019	0,020
80	I have a desire to check for updates on the online newspapers I read	0,293**	0,030	-0,035
81	I benefit from reading news on sites that summarize news, for example, Yahoo News	0,225**	-0,007	-0,049**
82	My online newspapers address local issues and debates that are of importance to me	0,164**	0,031	0,052
83	My online newspapers address topics that are of importance for my leisure interests	0,157**	0,029	-0,004
84	I follow the same topics and news online as I do in other media	0,155**	0,041*	-0,037*
85	It irritates me if I find the same paper in the printed newspaper as in the online newspaper	0,047*	0,042*	0,003

Table 8. Summarized features of the different viewpoints

Ubiquitous access	*Prestige of news source*	*Local anchorage and advertising*
all possible functionality	new and advanced technology	low degree of use of media and technology
networking	differentiated e-news service	differentiated media
advanced use of media and technology		basics only
constant updates and relevance		passive character
portal consumption		

Table 9. Consumers acceptance of an e-newspaper

Question and alternatives	***Results***	
	No.	***%***
What cost level is acceptable for you to change to an e-newspaper?		
Cheaper than the traditional printed newspaper	2119	58,4
Same price	371	10,2
Can be more expensive if there is added value	281	7,7
Price is unessential	171	4,7
Missing	684	18,9
Total	3626	100
How do you think the e-paper device should be financed/ paid for?		
Hire-purchase	281	7,7
Buy the device	711	19,6
Included in subscription	1708	47,1
Other	592	16,3
Missing	334	9,2
Total	3626	100
Would you consider to, some time in the future, exchange your traditional printed newspaper for the e-newspaper?		
Yes	2431	67,1
No	1070	29,5
Missing	125	3,4
Total	3626	100
Within which time frame are you ready to read your newspaper on e-paper?		
Today	1521	41,9
Within 5 years	683	18,8
Within 10 years	209	5,8
Within 20 years	88	2,4
Never	223	6,2
Missing	902	24,8
Total	3626	100

DISCUSSION AND CONCLUSION

In this article we have addressed the challenge of finding suitable business models for future innovative m-services, in this case the e-newspaper, by empirically illustrating the consumer view of such an innovation. We have done so by conducting a broad survey analyzed by a) a factor analysis of consumer views on an innovative e-news service, b) comparing the results with demographics, and c) correlating against media preferences and behavior. The analysis revealed three consumer views, that is, *Ubiquitous access*, *Prestige of news source* and *Local anchorage and advertising*. Further, the analysis indicate that these three viewpoints represent three possible market segments. These insights provide useful knowledge to content providers when developing initial business models. In the following we discuss the practical implications of addressing the research question of *what are the implications of the consumer view on a future mobile m-service innovation to initial business models*. Thereafter, we discuss how these findings add to the literature on business models for m-services.

The findings have some practical implications for content providers to develop initial business models for the e-newspaper. The results from this article can inform management in their efforts to target market segments. As shown by the comparison to demographic background data, the differences between different demographic groups regarding their viewpoints on future mobile e-news services are not very large but, with few exceptions, they proved to be significant. Together with the patterns given by the correlations to preferences and media behavior, this study outlines possible strategies for market segmentation. The results from the factor analysis challenge the present notion of suitable business models held by content providers (Ihlstöm Eriksson & Kalling, 2007). While content providers notions form four different business models, that is, *The Interpress News Agent, International Newspaper, Local Newspaper,* and *Community Micronews*, the consumer viewpoint centers around three factors, that is, *Ubiquitous access*, *Prestige of news source* and *Local anchorage and advertising*. In the following, we will discuss the three possible market segments and how they relate to the content provider views on business models for the e-newspaper introduction.

Taken together, we argue that the three possible segments can be described as two main positions *Ubiquitous access* and *Local anchorage and advertising* with *Prestige of news source* as a mixture of properties from both. In other words, we argue that the *Ubiquitous access* segment expects a high degree of ubiquity of media, news and communication and interaction, prefers more or less any new possible feature that follows with the e-newspaper, uses the latest technology, prefers all news in one place (one integrator preferably), and who probably will not read paper versions in the future. In sharp contrast, the *Local anchorage and advertising* segment representative is not as interested in technology, networking, or the ubiquity of news and media, but rather sees the e-newspaper as a complement to the paper version, and who typically appreciates a strong local character with regard to advertisements and news content. Sharing preferences with both, the *Prestige of news source* segment is yet more aware and updated consumers, anxious to be part of the world of media and news. They are interested in new technology but also prefer their media sources to be well-known and trustworthy, and recognition appears important. Like the *Local anchorage and advertising* segment, they still see the e-newspaper as a complement to the paper version.

In comparison to the business model options suggested by Ihlström Eriksson and Kalling (2007), we suggest that the *Ubiquitous access* segment is probably best served by the Interpress News Agent business model. The Interpress News Agent business model includes customer preferences such as "news anytime, anywhere" customers and a multiplicity of sources, and it includes offering components such as low prices, "news wholesale" product scope, up-to-date world news and standard (nonbranded) e-reader devices. Furthermore, the activities of this business model include 24/7 operations, provider relations and differentiated advertising. Among resources required by the business model are source reach, global and local media brands, advertising networks, and to be perceived as a recognized wholesaler. This business model appears to fit hand in glove the preferences and consumer patterns of the *Ubiquitous access* segment.

Likewise, we think the *Prestige of news source* segment fits the International Newspaper business model (Ihlström Eriksson & Kalling, 2007), which includes brand aware customers concerned about the credibility and image of the media provider. The price model of the offering addresses the need to charge for news, which appears plausible for this segment. The offering is also based on prestigious and branded devices and updated world news. Activities include 24/7 service, and global advertisement sourcing, and resources required are well-recognized journalists perceived as serious, and a global brand name.

The *Local anchorage and advertising* segment is probably well served by both the Community Micronews and the Local Newspaper business models. The question is whether it is important that the provider has a "local wholesaler" strategy with multiple sources or an "inhouse production" strategy. If this local customer is able to appreciate community journalism, is only interested in extremely local news and ads, and does not care too much about branded media and prestigious devices, then the Community Micronews model that probably suits the segment best. If customers are more tied to their local newspaper and media provider, and if they can accept a slightly higher price, then it might be the Local Newspaper model that fits the segment better.

As highlighted by the discussion above, the business model framework suggested for the content provider view corresponds to the three possible market segments suggested in this article. We therefore suggest a new integrated business model framework consisting of three models, that is, *ubiquitous*, *local* and *prestige* (Table 10), that could be helpful to content providers when developing business models for the introduction of the e-newspaper. The implications of these three models are further discussed below.

The possibilities and future potential for these business models vary, but the primary speculation is that the ***Ubiquitous model*** will need to run a cost-effective operation and, at the same time, be able to source news and ads from all over the world. Infrastructure will become central, as will marketing in order to be perceived as trustworthy. Perhaps the most important tactic is to become *the* e-newspaper quickly in order to set standards and develop customer groups. It is likely, we believe, that this business model, more than the others, is subject to increasing returns, network externalities and first-mover advantages (see Amit & Zott, 2001). Cheaper standard devices, possibly distributed for free or included in the subscription, could be a way for this strategy to be implemented. But it must be complemented by a commitment to adjust the technology as the market matures.

Furthermore, the ***Local model*** might face challenges due to technological development. The strong feelings for local news and advertisers will be matched by the global and international scope of news and advertising, but also lower prices, more frequent updates, and more. A device such as the eReader may not be able to compete with the physical paper unless its functionality is used to the maximum, including news and ads sourced from other places than the local town, and more frequent updates. Possibly, this business model will last as long as there are readers primarily interested in local matters.

The ***Prestige model*** has one central feature that will impact the opportunities, and that is the ability and will of customers to pay. How long this preference persists is another question. Crucial here is carefully nurturing of the brand, including potentially expensive things such as promoting trustworthiness, including journalists, authenticity, in-house productions, global networks of advertisers, marketing and more. One central task is to manage the internal infrastructure, which is bound to be global. There will be constant discussions about being "bureaucratic," and the question of insourcing vs. outsourcing will probably surface frequently. For e-newsreaders, this business model is likely to include expensive and branded devices.

Adding to the literature on business models for m-services, we present this integrated framework including content provider as well as consumer views on m-services. This framework may serve as guidelines for future introductions of an e-newspaper, but may also be of guidance

Table 10. Integrated e-newspaper business model framework

	The Ubiquitous Model	The Local Model	The Prestige Model
Customers	"Anytime, anywhere;" multiple sources and portal consumption; price sensitivity; all functionality; networking	Local interest; "basics only;" paper and e-paper seen as complements	New and advanced technology; e-newspaper a differentiated media; image awareness; global brand recognition
Competition	Bundling; distribution network; and low costs	Local papers and e-papers; ownership; outsource or insource?	Global news sources and media houses; distribution ownership; brand vs. "wholesale" strategies
Offering	Price pressure; advertising price premium; "wholesale;" standard devices	Informative, local news and advertisement; regular and few updates of news; lower price sensitivity	Informative; global news and advertisements; 24/7 updated news; lower price sensitivity; prestigious and branded devices
Activities and organization	24/7, provider relations; sourcing to be prominent; umbrella trading and differentiated advertising	Local relations to advertisers; local journalists; possibly community journalism; distribution ownership	24/7 service; in-house journalists; distribution ownership; global advertisement sourcing
Resources and factor market interaction	Sourcing; global and local brands; orchestration and efficiency; strong on international advertisement	Both internal and external local journalists; local relations; authenticity; recognised brand locally	Globally recognised brand; well-known and serious journalists

to other types of m-service content. As of yet, no major market analysis with the e-newspaper have been done, and it is therefore difficult to foresee patterns of usage, or where, when and how the e-newspaper will be used. More research is needed to understand in depth the possible role of the e-newspaper in people's lives and in relation to other media forms such as the Internet, the printed newspaper, television and already existing electronic mobile media. Also, there are several questions regarding the functionality of e-newspapers that remain unanswered. The preferences of the segments identified here may perhaps serve as guides for such functionality.

To conclude, the e-newspaper innovation attracts interest among the consumers. Two thirds of the consumers in our study are willing to read their newspaper on e-paper today or within 5 years. Matching this consumer interest with the content providers' hopes for the e-newspaper, we believe that the e-newspaper introduction in Sweden is not far away, making it an important issue to consider at the managerial level. We believe that the practical implications based on the proposed integrated business model framework could be helpful for newspaper management when developing initial business models for this introduction. The article also contributes to existing literature on business models for m-services by creating a further understanding of customer preferences in this market and by further developing the conclusions one may draw of these preferences. Although this research has been conducted with consumers in the newspaper industry, we believe that the concept of these models is relevant to other m-services as well. Further research will address the advertiser view, also relevant in building business models for mobile news services. The proposed integrated business model framework must also be empirically validated, both in Sweden and internationally.

ACKNOWLEDGMENT

We would like to thank VINNOVA for the funding our participation in the DigiNews project, and the KK Foundation for funding the UbiMedia project during which the analysis was done. We would also like to thank the three newspapers that allowed us to publish our questionnaire. Furthermore, Tobias Fredberg is grateful for the general support from the Jan Wallander and Tom Hedelius foundation.

REFERENCES

Åkesson, M., & Ihlström, C. (2006a). Towards an ubiquitous media environment-adding the e-newspaper channel. In *Proceedings of TAGA 2006*, Vancouver, Canada.

Åkesson, M., & Ihlström, C. (2006b). Designing and evaluating the calm e-newspaper. In *Proceedings of ECIS 2006*, Göteborg, Sweden.

Amazon (2008). *Kindle: Amazon's new wireless reading device*. Retrieved January 14th, 2008, from http://www.amazon.com/Kindle-Amazons-Wireless-Reading-Device/dp/B000FI73MA

Amit, R., & Zott, C. (2001). Value creation in e-business. *Strategic Management Journal, 22*, 493-520.

Barney, J. (1991). Firm resources and sustained competitive advantage. *Journal of Management, 17*, 99-120.

Buchanan, T., & Smith, J. L. (1999). Using the Internet for psychological research: Personality testing on the World-Wide Web. *British Journal of Psychology, 90*, 125-144.

Carlsson, C., Carlsson, J., Hyvönen, K., Puhakainen, J., & Walden, P. (2006). Adoption of mobile devices/services – searching for answers with the UTAUT. In *Proceedings of the 39th Hawaii International Conference on System Sciences*, Hawaii. (CD-ROM). Computer Society Press.

Chan, A. (2004). *Mediacorp and Singapore press holdings merge their TV and free newspaper operations*. Retrieved November 15, 2007, from Singapore press holdings, from http://www.sph.com.sg/news/archive/press_040917_003.html

Chyi, H. I. (2002). No one would pay for it? Web content as inferior goods. In *Proceedings of the Fifth World Media Economics Conference*, Turku, Finland.

Clemons, E. K., Gu, B., & Lang, K. R. (2002, January 7-10). Newly vulnerable markets in an age of pure information products: An analysis of online music and online news. In *Proceedings of the 35th Annual Hawaii International Conference on System Sciences* (CD-ROM). Computer Society Press.

Constantiou, I., Damsgaard, J., & Knutsen, L. (2004). Investigating mobile services' adoption and diffusion in the Danish market. In *Proceedings of the 27th Information Systems Research Seminar in Scandinavia*, Sweden.

Dierickx, I., & Cool, K. (1989). Asset stock accumulation and sustainability of competitive advantage. *Management Science*, *35*(12), 1504-1511.

E-ink Corporation. (2007). *Electronic displays*. Retrieved November 15, 2007, from http://www.eink.com/products/customers.html

Fidler, R. (1997). *Mediamorphosis: Understanding new media*. Thousand Oaks, CA: Pine Forge Press.

Fredberg, T. (2003). *Interface strategies: Internet and the business of large Swedish daily newspapers*. Gothenburg: Institute for Management of Innovation and Technology.

Hair, J. F., Black, B., Babin, B., Anderson, R. E., & Tatham, R. L. (2005). *Multivariate data analysis*. Upper Saddle River, NJ: Prentice Hall.

Hammond, K. (2001). B2C E-commerce 2000-2010: What experts predict. *Business Strategy Review*, *12*, 43-50.

Hedman, J., & Kalling, T. (2002). *IT and business models: Theories and concepts*. Copenhagen: CBS Press.

Hedman, J., & Kalling, T. (2003). The business model concept: Theoretical underpinnings and empirical illustrations. *European Journal of Information Systems*, *12*, 49-59.

IFRA. (2006). *eNews initiative*. Retrieved November 15, 2007, from http://www.ifra.com/website/website.nsf/html/CONT_ENEWS?OpenDocument&ENW&

Ihlström, C. (2005). The e-newspaper innovation-converging print and online. In *Proceedings of the International Workshop on Innovation and Media: Managing changes in Technology, Products and Processes*, Stockholm, Sweden.

Ihlström Eriksson, C., Åkesson, M., & Hakeröd, J. (2007, August 11-14). Advertising in ubiquitous media environments. In *Proceeding of IRIS'30*, Tampere, Finland.

Ihlström, C., Åkesson, M., & Nordqvist, S. (2004). From print to Web to e-paper-the challenge of designing the e-newspaper. In *Proceedings of ICCC 8th International Conference on Electronic Publishing, ELPUB 2004*, Brasilia, (pp. 249-260).

Ihlström, C., & Henfridsson, O. (2005). Online newspapers in Scandinavia: A longitudinal study of genre change and interdependency. *IT & People*, *18*(2), 172-192.

Ihlström Eriksson, C., & Kalling, T. (2007, June 7-9). Proposing a business model framework for the e-newspaper. In *Proceedings of the 15th European Conference on Information Systems, ECIS 2007*, St. Gallen.

Ihlström, C., & Palmer, J. (2002). Revenues for online newspapers: Owner and user perceptions. *Electronic Markets: The International Journal of Electronic Commerce & Business Media*, *12*(4), 228-236.

Ihlström Eriksson, C., & Svensson, J. (2007, May 31-June 2). The mobile e-newspaper innovation-examining the pre-adoption phase. *LA Global Mobility Round Table*, Los Angeles.

Ihlström, C., Svensson, J., & Åkesson, M. (2005). How would you like your e-newspaper? Converging the best of two worlds. In *Proceedings of HCI International 2005*, Las Vegas. CD-ROM.

iRex Technologies. (2006). *The iLiad*. Retrieved November 15, 2007, from http://www.irextechnologies.com

Jungk, R., & Müllert, N. (1996). *Future workshops-how to create desirable futures.* London: Institute for Social Inventions.

Krueger, C. C., van der Beek, K., & Swatman, P. (2004). New and emerging business models for online news: A survey of 10 European countries. In *Proceedings of the 17th Bled eCommerce Conference*, Slovenia.

Lund, A.B. (2007). Media markets in Scandinavia: Political economy aspects of convergence and divergence. *Nordicom Review, Jubilee Issue 2007*, 121-134.

Mallat, N., Rossi, M., Tuunainen, V. K., & Öörni, A. (2006). The impact of use situation and mobility on the acceptance of mobile ticketing services. In *Proceedings of the 39th Hawaii International Conference on System Sciences*, Hawaii. (CD-ROM). Computer Society Press.

McGrath, R. G., & MacMillan, I. C. (2000). *The entrepreneurial mindset. Strategies for continuously creating opportunity in an age of uncertainty.* Cambridge: Harvard Business School Press.

Miles, M. B., & Huberman, A. M. (1994). *Qualitative data analysis.* Thousand Oaks, CA:Sage Publications.

Normann, R. (1977). *Management for growth.* Chichester: John Wiley.

Osborn, A. (1953). *Applied imagination.* New York: Scribner.

Parasuraman, A. (2000). Technology readiness index (TRI): A multiple item scale to measure readiness to embrace new technologies. *Journal of Service Research, 2*(4), 307-320.

Patton, M. Q. (2002). *Qualitative research & evaluation methods (3rd ed.).* Thousand Oaks, CA: Sage Publications.

Pedersen, P E., Methlie, L B., & Thorbjørnsen, H. (2002). Understanding mobile commerce end-user adoption: A triangulation perspective. In *Proceedings of the 35th Hawaii International Conference on System Sciences, Hawaii.* (CD-ROM). Computer Society Press.

Peteraf, M. A. (1993). The cornerstones of competitive advantage: A resource-based view. *Strategic Management Journal, 14*(3), 179-191.

Plastic logic. (2007). *Plastic Logic raises $100 million to enable the first "take anywhere, read anywhere" electronic reader products.* Retrieved November 15, 2007, from http://www.plasticlogic.com/news-detail.php?id=300

Polymer Vision. (2007). *Press releases.* Retrieved November 15, 2007, from http://www.polymervision.com/News-Center/Press-Releases/Index.html

Porter, M. E. (1980). *Competitive strategy.* New York: Free Press.

Porter, M. E. (1985). *Competitive advantage.* New York: Free Press.

Porter, M. E. (1991). Towards a dynamic theory of strategy. *Strategic Management Journal, 12*, 95-119.

Punie, Y., Burgelman, J-C., & Bogdanowicz, M. (2002). The future of online media industries. *Scenarios for 2005 and Beyond. IPTS, 64*, 35-42.

Rappa, M. (2002). *Business models on the Web.* Retrieved November 15, 2007, from http://digitalenterprise.org/models/models.html

Ratliff, J. M. (2002). NTT DoCoMo and its i-mode success: Origins and implications. *California Management Review, 44*, 55-70.

Ritter, P., Lorig, K., Laurent, D., & Matthews, K. (2004, July-September). Internet vs. mailed questionnaires: A randomized comparison. *Journal of Medical Internet Research, 6*(3), e29.

Samtani, A., Leow, T.T., Lim, H.M., & Goh, P.G.J. (2003). Overcoming barriers to the successful adoption of mobile commerce in Singapore. *International Journal of Mobile Communications, 1*(1/2), 194-231.

Sarker, S., & Wells, J. D. (2003). Understanding mobile handheld device use and adoption. *Communications of the ACM, Special Issue on mobile commerce opportunities and challenges, 46*(12), 35-40.

Schumpeter, J. A. (1934). *The theory of economic development.* Cambridge: Harvard University Press.

Sony Styles. (2007). *Sony reader.* Retrieved November 15, 2007, from http://www.sonystyle.com/webapp/wcs/stores/servlet/CategoryDisplay?catalogId=10551&storeId=10151&langId=-1&categoryId=16184&XID=O:ebook%20reader:corp_read_gglsrch

Sääksjärvi, M., & Santonen, T. (2003). Customization as business model for online newspapers. In *Proceedings of the 16th Bled eCommerce Conference*, Bled, Slovenia.

Tsalgatidou, A., & Pitoura, E. (2001). Business models and transactions in mobile electronic commerce: Requirements and properties. *Computer Networks, 37*(2), 221-236.

The Editors Weblog. (2006a). Belgium*: E-paper test launch.* Retrieved November 15, 2007, from http://www.editorsweblog.org/news/2006/02/belgium_epaper_test_launch.php

The Editors Weblog. (2006b). *France: Les Echos publishes e-paper.* Retrieved November 15, 2007, from http://www.editorsweblog.org/print_newspapers/2006/03/france_les_echos_publishes_epaper.php

The Swedish Newspaper Publisher's Association. (2007). Retrieved November 15, 2007, from http://www.tu.se/

Timmers, P. (1998). Business models for electronic markets. *Electronic Markets, 8*(2), 2-8.

van der Beek, K., Swatman, P., & Krueger, C. (2005. January 3-6). Creating value from digital content: E-business model evolution in online news and music. In *Proceedings of the 38th Annual Hawaii International Conference on System Sciences* (CD-ROM). Computer Society Press.

Vrechopoulos, A., Constantiou, I., & Sideris, G. (2003). The critical role of consumer behavior research in mobile commerce. *International Journal of Mobile Communications, 1*(3), 329-340.

Wallace, B. (2006). *30 countries passed 100% mobile phone penetration in Q1.* Retrieved November 15, 2007, from http://www.telecommagazine.com/newsglobe/paper.asp?HH_ID=AR_2148

Walter, Z., & Stolariva-Ornek, J. (2001). Strategies for local newspapers in the digital economy. In *Proceedings of the Seventh Americas Conference on Information Systems*, Boston, M.A.

Wernerfelt, B. (1984). A resource-based view of the firm. *Strategic Management Journal, 5*(2), 171-180.

Westlund, O. (2006, September 27-30). Beyond time and space Sweden and their mobile Internet news users (MINU).In *Proceedings of the AOIR 7.0 Conference "Internet and convergence,"* Brisbane.

Who killed the newspaper?. (2006, August 24). *The Economist.*

This work was previously published inJournal of Electronic Commerce in Organizations, Vol. 6, Issue 2, edited by M. Khosrow-Pour, pp. 29-57, copyright 2008 by IGI Publishing (an imprint of IGI Global).

Chapter 12
The Evolving Portfolio of Business-to-Business E-Services:
Service and Channel Innovation

Christine Legner
European Business School (EBS), Germany

ABSTRACT

E-services have been recognized for their contribution to strengthening customer relationships and fostering customer loyalty. While prior research has focused on the role of e-services in business-to-consumer interactions, this chapter studies e-services in the business-to-business domain. More specifically, it explores how the preferences for e-services and electronic channels change over time as a result of developing technology and evolving customer requirements. Based on the case study of ETA SA, a Swiss manufacturer of watch movements and components, three stages of increasing electronic interaction with customers have been identified. The company pioneered online ordering of watch spare parts based on a first generation e-commerce solution and subsequently introduced a comprehensive customer service portal. This research suggests that moving from the e-commerce solution to the customer process portal involved a significant extension of the e-service portfolio (service innovation), whereas the subsequent stage introduced an additional electronic channel (channel innovation). Every stage in this evolution path forces companies to substantially re-architect their interorganizational process and system linkages. From the experiences of ETA SA, we conclude that a well-designed B2B architecture is required to cope with the growing number of e-services and the complexity of serving multiple electronic channels.

INTRODUCTION

Progress in information technology (IT), and more specifically the emergence of the Internet, is considered a major accelerator in realizing closer forms of collaboration between business partners. On the customer side, the electronic channel has proved valuable in offering supplementary self-services that are information-based and need not be co-located with the product (e.g. Day & Hubbard, 2003; Piccoli, Brohman, Watson, & Parasuraman, 2004). In fact, many companies report closer customer relation-

DOI: 10.4018/978-1-60566-964-9.ch012

ships and improved operational efficiencies in their customer-facing or downstream processes (Chen & Chen, 2004). While the literature on electronic services is heavily skewed toward the business-to-consumer (B2C) field and focuses on online retail channels (e.g. Cenfetelli, Benbasat, & Al-Natour, 2008; Madlberger, 2006; Palmer, 2000; Piccoli, et al., 2004), the specificities of business-to-business (B2B) interactions have attracted less attention so far. These interactions tend to be more complex in nature since they often imply trading sophisticated bundles of goods and services in mid- to long-term business relationships. Thus, traditional mechanisms for customer service in industrial markets such as large (and expensive) field forces and customer support centers will tend to be complemented rather than replaced by self-service strategies (e.g. Archer & Yufei, 2000; Bhappu & Schultze, 2006; Merrilees & Fenech, 2007; Wilson & Daniel, 2007).

In the literature, the term 'e-services' is sometimes used as a synonym for Web sites (van Riel, Liljander, & Jurriëns, 2001), and many studies assess e-service quality based on factors like Web site esthetics, accessibility or reliability, typically using the SERVQUAL service quality framework. Recently, a more differentiated view has been adopted which introduces a functional perspective of Web-based services (Cenfetelli, et al., 2008; Otim & Grover, 2006; Piccoli, et al., 2004). In the context of our research, e-services denote services which are delivered using electronic channels, most frequently the Internet. Interestingly, many e-services that once were a source of differentiation, like online product configuration or parcel tracking, become state-of-the-art functionality over time. Whereas the emergence of such a dominant design has been described by Piccoli et al. (2004), it is not well understood how the preferences for e-services and electronic channels change over time as a result of developing technology and evolving customer requirements. Previous studies in this field, e.g. by Yeung & Lu (2004) and Piccoli et al. (2004), focus on the evolution of commercial Web sites. They do not take other electronic interaction channels into account which are of particular relevance in industrial markets (Cullen & Webster, 2007). This paper takes on the particular challenges of business-to-business e-services and explores the following questions:

(1) How does the e-service portfolio as well as the nature of electronic interaction evolve over time?
(2) In which way does the evolving e-services portfolio impact the process and integration architecture of an enterprise?

Since there has been little conceptual and empirical research on the evolution of e-services and the implications for organizational and IS design, our study is exploratory in nature. We adopted a case study research method which is particularly suitable for understanding phenomena within their organizational context (Benbasat, Goldstein, & Mead, 1987; Eisenhardt, 1989; Yin, 2002). As case site, we chose ETA SA, a manufacturer of watch movements and components, that launched one of the first e-shops targeted at B2B customers in Switzerland and today is successfully interacting through electronic channels with 80% of its global customer base. Based on a longitudinal case study, we explore different stages of electronic interaction with customers and deduce the future need for offering electronic services using both direct and portal-based electronic channels.

This paper is structured as follows: Following an introduction to e-services and the role of electronic channels, this article goes on to outline typical issues in implementing higher levels of process integration with external partners and the convergence of internal and external integration technologies. Then, the research approach and the research methodology are discussed. Next, the case of ETA SA is introduced: The firm is successfully offering e-services through a customer process portal, but is confronted with some class of customers demanding higher levels of process integration. In order to offer electronic services via different channels and to satisfy the requirements of each customer segment, ETA conceives a multi-channel architecture, which is presented in detail. The final part of the article identifies three distinct stages in the evolution of B2B e-services. It concludes with a discussion of the associated challenges of interorganizational process and system architecture design.

LITERATURE REVIEW

E-Services and the Role of Electronic Channels

Since the advent of the Internet, the role and the adoption of e-services have gained attention and recognition from academia and practitioners. Different streams of research are exploring the phenomenon of e-services: While some studies emphasize channel choices between physical and online channels and the associated threat of disintermediation (e.g. Madlberger, 2006; Walker, Craig-Lees, Hecker, & Francis, 2002), a large body of research aims at assessing e-service quality and user acceptance (e.g. Cristobal, Flavian, & Guinaliu, 2007; Lu & Zhang, 2003; Ranganathan & Ganapathy, 2002; van Riel, et al., 2001; Yeung & Lu, 2004). Many of the existing research focuses on consumer Web sites and measures e-service quality based on factors derived from the SERVQUAL model (Parasuraman, Zeithaml, & Berry, 1988). However, these studies neither consider composite online service offers nor the fact that e-service categories might evolve over time. With regard to the role and impact of e-services on customer relationships, a more differentiated view has evolved (e.g. Barnes, Hinton, & Mieczkowska, 2005; Bitner, Brown, & Meuter, 2000; Day & Hubbard, 2003; Otim & Grover, 2006; Piccoli, et al., 2004; Salomann, Kolbe, & Brenner, 2006). Based on a survey of 352 senior managers, Day & Hubbard (2003) report that the electronic channel is judged to "offer opportunities to reduce customer service costs while tightening customer relationships by encouraging dialogue, linking more parts of customer contact and enabling the personalisation of communications". Otim & Grover (2006) differentiate Web-based service categories, i.e. pre-purchase, transaction-related, and post-purchase services, and examine their impact on customer loyalty. Their empirical study demonstrates that post-purchase services consisting of support of order tracking, on-time delivery, and customer support positively influence customer retention. (Cenfetelli, et al., 2008) introduce the concept of supporting services functionality (SSF), i.e. "the extent to which a website uses IT to provide services that support a core product or service transaction". They empirically assess SSF by adapting 14 individual service dimensions of the Customer Service Life-Cycle, and find that it is an important contributor to an online customer's perceptions of usefulness, the customer's satisfaction with the website, and ultimately their continued usage of that site. Consequently, and despite the emotional and privacy concerns of some customers (Barnes, et al., 2005), many companies offer a wide range of customer self-services over the Internet with the goal to strengthen their customer relationships (Salomann, et al., 2006).

Although most scholars explore consumer rather than industrial markets, there is broad consensus that electronic channels create enormous opportunities for improving the scope and strength of B2B customer relationships (e.g. Archer & Yufei, 2000; Merrilees & Fenech, 2007). In industrial markets, e-services are often used to support the acquisition and the use of a physical product by offering self-services which are information-based and need not be co-located with the product. Because industrial markets are characterized by a higher complexity of customer interactions, they impose additional challenges when designing and managing electronic channels in a successful and cost-effective manner. In a recent study, Wilson &

Daniel (2007) identify seven dynamic capabilities for channel transformation. They conclude that "organisations must not only be able to find how to make use of new channels, but also bring those channels into their standard way of operating, if the organization is going to deliver an integrated customer experience and make the most of synergies between channels."

With the introduction and extension of e-services, companies face the challenges of multi-channel management, more specially the coordination of electronic and physical channels (Payne & Frow, 2004; Stone, Hobbs, & Khaleeli, 2002). From a firm-level perspective, interaction with other individuals or organizations usually occurs via several (a) institutional or organizational channels (e.g. subsidiaries, sales agents or call centers) and (b) communication channels (e.g. phone, fax, Web portal or XML-based message exchange). According to Stone, Hobbs & Khaleeli (2002) and Bucklin et al. (1996), maintaining channels (including marketing, advertising and managing the channels themselves) can typically account for around 40 per cent of the overall costs of a company and between 15 and 50% of the retail price of goods and services in an industry. As Day & Hubbard (2003) and Payne & Frow (2004) report, the coordination of this channel mix requires careful planning in order to avoid competition and cannibalization between channels. Schierholz et al. (2007) distinguish two facets of multi-channel management, (a) the internal alignment and coordination of different organizational channels (*channel management*) and (b) the support of the interaction between customer process activities and core process services by utilizing the appropriate communication channels (*interaction management*). Both facets are interrelated, since one goal of multi-channel management is to guide each interaction into that channel which is best suited to it (e.g. because of customer preferences or operational cost savings). It is argued that previous experiences, perceived benefits and risks, product characteristics and market environment influence channel choice across multi-channel options (Subramaniam, Shaw, & Gardner, 2000).

Electronic Interaction

The nature of the electronic interaction can be described according to their interaction type (Löwer, 2005; McAfee, 2005; Reimers, 2001) which is depending on the communication channel. Electronic human-human-interaction describes traditional forms of interaction between humans which are supported by electronic means, e.g. groupware or e-mail. In the case of human-machine-interaction, external users are getting direct access to internal data and applications. This is typically realized by Web front-ends or portals bundling data and applications on the basis of users and roles. Machine-machine interaction finally describes the direct communication between two information systems without human intervention. It leads to consistently automated processes and requires a linkage of applications across company borders. Electronic Data Interchange (EDI) is the standard example for machine-machine interaction, but is being succeeded by XML and Web Services.

Due to the semiotic structure of communication (Kubicek, 1992), a large number of agreements on different levels need to be made explicit and to be formalized in order to allow IS-mediated interaction (McAfee, 2005; Reimers, 2001): At the lowest level, information systems have to share agreements about how data is to be transported over a network. Once agreements at this basic transport level, e.g. through standard Internet protocols such as HTTP, are in place, human-human and human-machine interactions can take place. Since information systems are not as flexible as humans in interpreting documents, further ex ante agreements have to be made for machine-machine interactions. According to McAfee (2005) they encompass the payload, i.e. the content and structure of electronic messages, and the processes which make use of these messages.

Figure 1. Convergence of interorganizational information systems {adapted from \Alt, 2008 #6}

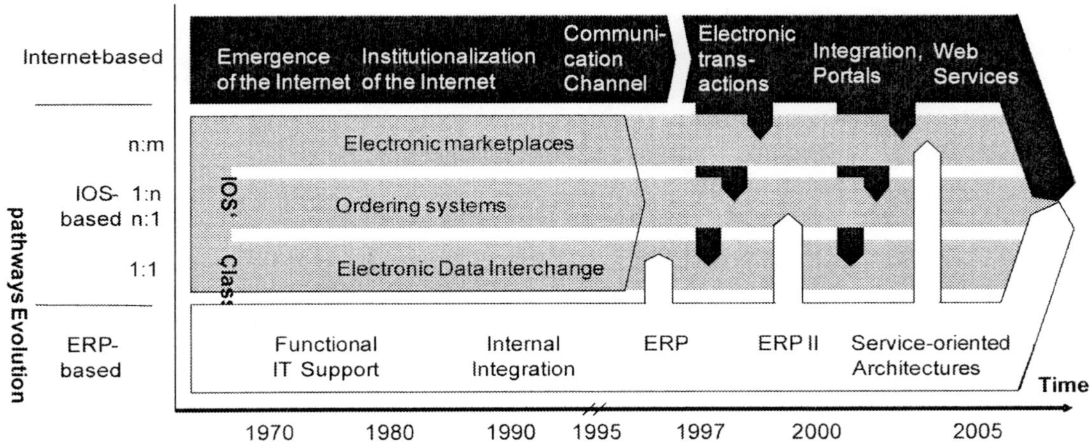

Technological Evolution

With the more intensive electronic interaction, infrastructure and technology requirements increase and the need for channel integration rises (Payne & Frow, 2004). Starting in the 1970s, the evolution of information systems for interorganizational integration is characterized by the convergence of three parallel technological evolution pathways (Figure 1):

- *Interorganizational Systems (IOS)* denote network-based information systems that transcend organizational boundaries (Hong, 2002; Johnston & Vitale, 1988). Since the early 1970s, EDI connections have been used to link one or more firms to their customers and suppliers through value-added networks. They were supplemented by ordering systems with higher-order data integration, shared databases and functionality, like AMADEUS in the travel industry or SWIFT in the financial industry. Electronic marketplaces finally were established in order to provide many-to-many (m:n) integration (Bakos, 1991; Kaplan & Sawhney, 2000; Mahadevan, 2003).
- Parallel to the emergence of IOS, *ERP-systems* introduced internal information integration into the company by providing cross-functional data integration and process support. The reach of ERP-systems gradually expanded to support collaboration with external partners (Davenport & Brooks, 2004): ERP II (Weston, 2003) or Business Networking Systems (Alt & Fleisch, 2000) provide functionality for both, customer relationship management and supply chain management.

Since the mid 1990s, the commercial diffusion of the Internet and especially the World Wide Web has completely overhauled interorganizational integration. Following a phase of rapid growth of enterprise Web sites since 1994/1995, companies gradually integrated Internet technologies with their existing information systems by connecting their internal applications directly to the Web front-end. This was supported by enhanced capabilities of portal solutions for integrating content and functionality from heterogeneous data sources (Dias, 2001). Web sites and portals are widely used in industrial practice today (Daniel & White, 2005), while the exchange of XML messages is taking off slower than expected, and EDI still remains the most frequent form of two-way machine-to-machine integration in most industries.

These three parallel technological evolution pathways increasingly converge towards a technological platform for interorganizational integration, reproducing the development of integrated ERP-systems in an interorganizational environment. Web services and service-oriented architectures (SOA) allow for application-to-application integration across heterogeneous platforms using open Internet standards, and are expected to stimulate this development (Daniel & White, 2005; Hagel & Brown, 2001).

RESEARCH METHODOLOGY

In view of our research objective, we were particularly interested in understanding how technology development and changing customer requirements impact the e-services portfolio and the nature of electronic B2B interactions. Thus, our research can be classified as "Theory for Explaining (Type II)" according to Gregor (2006). It meets the criteria for case study research established by Benbasat, Goldstein & Mead (1987) and Yin (2002) in several aspects: First, the e-services phenomenon has to be understood in the organizational context, taking into account the complex interplay between organizational and IS design of several business partners. Second, a case study research method is particularly appropriate for understanding the dynamics in the adoption of new technologies. Finally, case studies support our long-term research goal, i.e. the development of a conceptual model for e-services evolution which considers organizational and technical contingencies. As a first step in filling the existing research gap related to B2B e-services evolution, this research contributes to establishing a theoretical base by using inductive principles.

Case studies have long been used as a teaching method, especially in the area of managerial sciences, but have also been criticized for theory building due to their lack of rigor and their limited empirical basis. However, qualitative case study research is increasingly accepted as a valid research strategy within the IS research community (Gregor, 2006). In order to ensure the validity of this study, we have closely followed the process of building theory from case study research by Eisenhardt (1989) and Yin (2002), and adhered to the methodology established for researchers who participate in organizational change by Miles & Huberman (1994).

We chose a longitudinal single case study design over a multiple case design owing to the demands of gaining access for intensive data collection over a period of several years. The main criterion for case selection was the successful track record with deploying B2B e-services to a heterogeneous international customer base. In this regard, ETA SA is a particularly appropriate setting for our study: In 1999, the company launched a first generation e-commerce solution and, unlike many other e-commerce pioneers, established electronic interactions with the majority of its global customer base within the space of two years. Although this particular industry context might appear to be a niche industry at first sight, the watch industry has very similar characteristics to other markets for technical components. The case selection was also driven by the accessibility of data and the company's willingness to participate in our research. We began our research in 2002, when ETA SA sought academic advice for the further evolution of their Customer Service Portal, and completed it in 2006 with the conceptualization of a multi-channel architecture. Throughout the project, we conducted regular meetings with ETA's managers, process and system experts. Also significant was our participation in workshops with both ETA project members and customers. Thus, data collection consisted in the review of qualitative data, mainly internal project documentation, and workshop and meeting reports. Additionally, we had access to detailed statistical data related to ETA's electronic channels for the entire period from 2001 to 2006. In order to analyze and consolidate our rich data, we described ETA's evolving process and system architecture using the documentation model and notation suggested by (Österle 1995) for business engineering projects. In the course of this research, we developed comprehensive documentation comprising ETA's business network configuration, process models (activity chains), and the application landscape and integration mechanisms. This documentation was re-checked with ETA experts and linked to the empirical data and additional qualitative customer feedback. On this basis, we have been able to identify three stages of e-service evolution which can be characterized by a specific set of e-services and the related level of external process and application integration.

THE CASE OF ETA SA

Company Background and Business Network

Founded in 1793, ETA SA Manufacture Horlogère Suisse is today one of the world's largest manufacturers of finished watch movements and components. It is a member of 'The Swatch Group', a Swiss-based group that covers the entire value chain of the watch industry, including component manufacturers, well-known brands (e.g. Omega, Rado, Tissot, Certina and Swatch) and local sales organizations. Among ETA's approximately 1,500 business customers worldwide are large brands, which in some cases belong to the same parent company, as well as spare parts dealers, watchmakers and training organizations for the watch industry. The heterogeneous customer base imposes very different demands on ETA's customer-facing processes. At one end of the spectrum there are large customers placing high volume orders and using sophisticated information systems in their procurement. At the other are watchmakers with occasional orders and no IT support except for a personal computer. The business network further comprises external logistics and financial service providers which handle out-tasked activities in the order fulfillment process (cf. Figure 2).

E-Service Portfolio

The desire for greater customer orientation and better servicing its global customer base was the starting point for e-Business activities of ETA SA in 1998. The activities focused mainly on streamlining order fulfillment in ETA's customer service (ETA-CS), which is responsible for the distribution of spare parts, repair of watch movements and technical support. The ETA Online Shop (EOS), a first generation e-commerce solution, was launched in 1999 with an initial functional range of a typical electronic catalog (search, transaction workflow, and shopping basket) (Alt, Reichmayr, & Zurmühlen, 2000). ETA gradually upgraded the EOS by introducing additional transaction-related services such as retainable delivery addresses and online payment. However, it did not yet support some critical parts of the customer process, namely spare parts identification as well as the repair and maintenance processes.

Consistently pursuing the ultimate customer support, ETA decided in 2002 to extend the range of electronic services offered to their customers and introduce the Customer Process Portal (CSP). Traditionally, a portal denotes an Internet-based system that provides link libraries and guides users according to their interests (Dias, 2001). Extending this concept to incorporate the more transaction-oriented electronic commerce vision, process portals are comprehensive portal systems that bundle electronic services to sup-

Figure 2. Business network of ETA SA

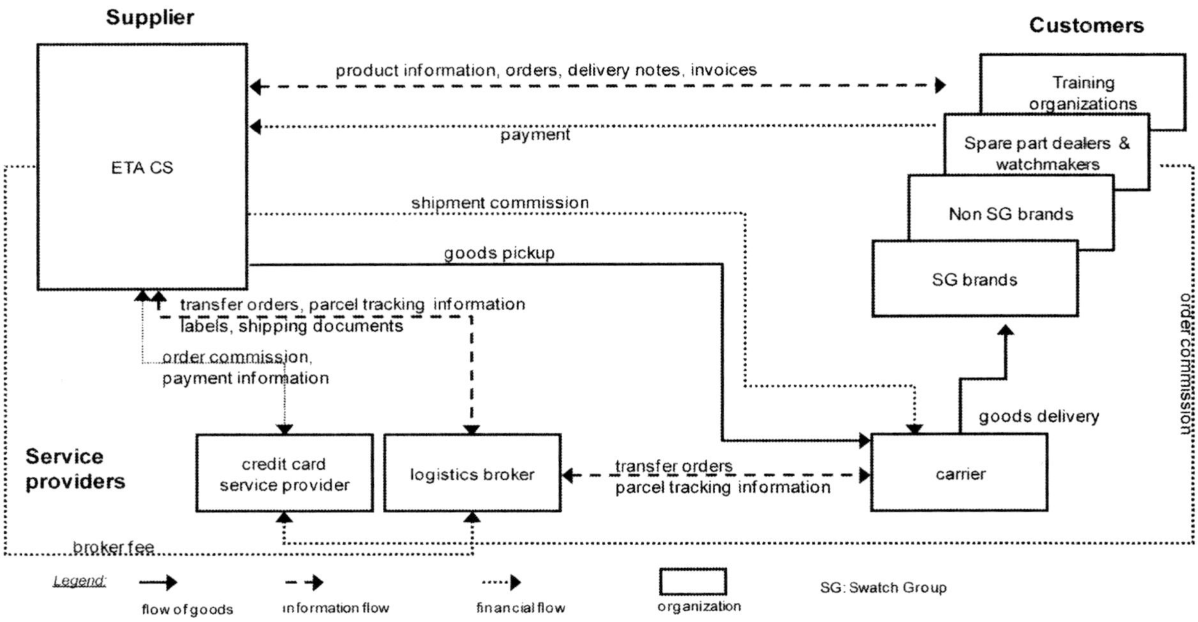

port customer-facing processes and thus provide customers with a single point of contact (Österle, 2001; Puschmann & Alt, 2005).

During the design phase, ETA identified around 50 potential electronic services and bundled them into customer process components reflecting their support for the customer's activities (Alt, Reichmayr, Cäsar, & Zurmühlen, 2002): Portal services within the process component *information* can be considered pre-purchase services in the taxonomy developed by Otim & Grover (2006). They support requirements specification and product evaluation by providing product documentation, sales conditions and organizational information. The *spare parts* component subsumes a variety of transaction-related portal services for spare parts ordering ranging from availability check, to pricing, order capture and order as well as parcel tracking. Additional process components cover post-purchase services, such as technical instructions and related notifications. In 2004, the Customer Service Portal (CSP) went online, implementing a first set of services and integrating the existing EOS functions. Figure *3* illustrates an extract of the portal services along with the implementation status as of 2006.

Integration Architecture

Although virtually integrated from the customer's perspective, the Customer Service Portal relies on two separate systems: The ETA Online Shop provides the electronic catalog and order entry functionality whereas the portal system itself provides additional services such as authorization and authentication or tracking functions. User master data is managed by an active directory database which is queried online.

Since electronic services usually originate from a multitude of applications and data sources, an appropriate integration architecture will reduce the need to build custom interfaces and speed up application integration and deployment (Puschmann & Alt, 2005). For ETA, the key challenge has been to integrate electronic product information from multiple sources, in order to relate spare

Figure 3. Electronic services in the Customer Service Portal and implementation status

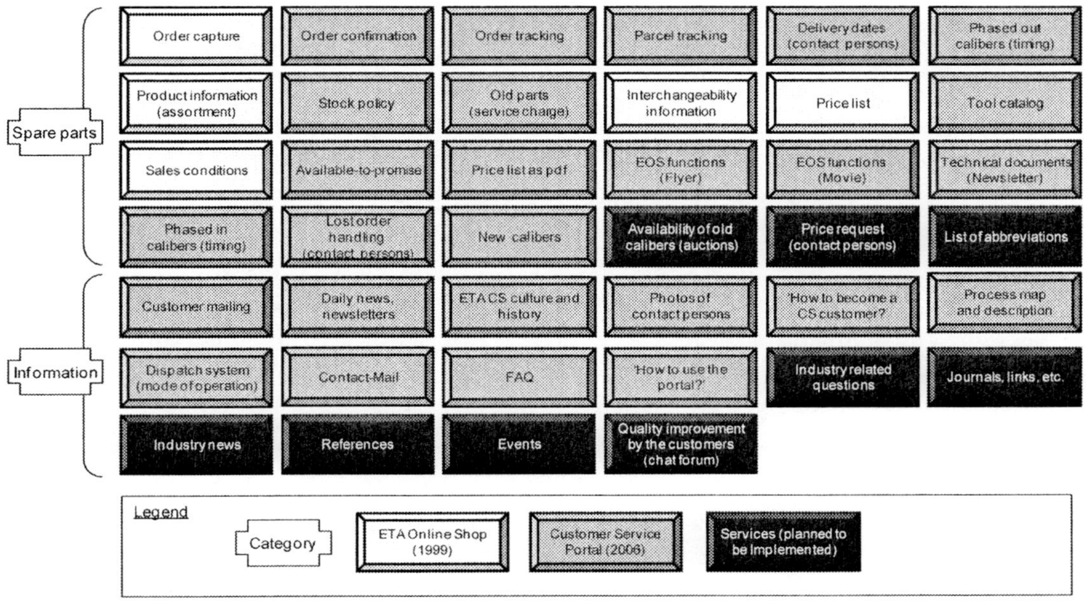

part data to technical product documentation, product availability and personalized prices for ETA's global customer base. For this purpose, ETA's Customer Service Portal mainly uses mechanisms of data integration (cf. Linthicum, 2000; Ruh, Maginnis, & Brown, 2001). The import of product master data into the ETA Online Shop requires multiple steps: Since ETA's product lifecycle management (PLM) system represents the leading system for product master data, text files with spare parts master data are periodically imported into the enterprise resource planning (ERP) system. The ERP-system stores additional sales- and logistics-oriented product data, such as prices, sales conditions and replenishment lead times. The e-shop obtains the major part of its product data from an import database that acts as a staging area and pre-processes the product data obtained from the ERP and additional spare parts interchangeability information from a separate database. Orders placed in the EOS are copied to the ERP-system involving a "human firewall" for manual verification of orders.

Limitations of Portal-Based Integration

Considerations regarding the further development of ETA's e-service portfolio were initiated by customer feedback: Several larger customers, especially larger brands, complained about capturing spare parts orders twice, in their own ERP-system and in the Customer Service Portal. Consequently, they requested machine-machine integration. ETA's customer service reacted by initiating a pilot study with Watch Co., one of its larger customers, in order to re-examine the current order management process and discuss different integration options. Figure 4 illustrates the actual order management process flow between ETA-CS and Watch Co. using the activity diagram notation defined in (Österle, 1995). It shows the various manual or IT-supported activities (nodes), their temporal dependencies (edges) and processors in form of organizational units (vertical boxes or swim lanes).

The analysis revealed several pain points relating to three different integration areas:

- *Manual exchange of product master data.* Since large customers usually maintain spare parts data in their own ERP-system, the purchasing department of Watch Co. has to manually copy product data from ETA's online catalog and regularly screen the Customer Service Portal for updates such as price changes. Given that customers use their own numbering systems for spare parts identification, product numbers are assigned manually on both sides.
- *Multiple order capture and manual order confirmation.* Based on defined demand signals, the purchaser at Watch Co. generates orders in the ERP-system, and, after the internal approval, re-enters these order into the EOS. ETA-CS confirms the order via fax and the purchaser copies the relevant information into the ERP-system.
- *Manual dispatch information and invoice transfer.* Following the order confirmation, the purchaser at Watch Co. regularly checks ETA's order tracking module on the Customer Service Portal for adjusted order information such as changing delivery dates. This data is manually entered into Watch Co.'s ERP-system in order to allow timely inventory management and demand planning. The final dispatch advice is printed at ETA-CS and shipped together with the invoice and the goods. After receiving the goods, Watch Co.'s employees manually transfer the relevant information of those documents into the ERP-system.

The actual pain points in the case of ETA-CS underline some general limitations of process portals which are typical to B2B collaboration. Since process portals support activity sequences, they can be categorized as multi-step, human-machine interaction systems based on the typology introduced earlier in this article. Given that process portals do only require interorganizational agreements on the transport protocol and that infrastructure requirements are low, i.e. an Internet connection and browser, they can easily be deployed to external partners. While they are usually directly integrated with company-internal backend systems, human-machine interaction on the business partner's side generates manual activities and thereby impact external business processes. They usually satisfy the requirements of smaller business partners with a low transaction volume and relatively basic internal IS support, but impose some severe limitations in high-volume B2B relationships with larger customers or suppliers:

- Portal users have to adapt to the business processes and semantics defined by the portal provider. The complexity of serving the predefined process rises as companies are forced to utilize multiple process portals with different business partners.
- Large portal users which deploy an ERP-system internally, have to re-enter information and complain that this is not only a time-consuming but also an error-prone activity.
- Since portals rely on a human interface for transferring data, the real-time capability of portal-based integration is limited.

Process Vision and Related Benefits

Based on this pilot study, ETA developed a process vision for closer integrating large, ERP-equipped customers and providing them a direct machine-machine communication channel. Objectives were to periodically transmit electronic product master data from ETA to customers, to receive orders directly out of the customers' ERP-systems and to transmit a machine readable order confirmation, dispatch advice and invoice. In addition, ETA wanted to ensure the reusability of the implemented EOS based order tracking services for every customer, and to maintain all electronic services in parallel on the Customer Service Portal

Figure 4. Order management process between ETA-CS and Watch Co.

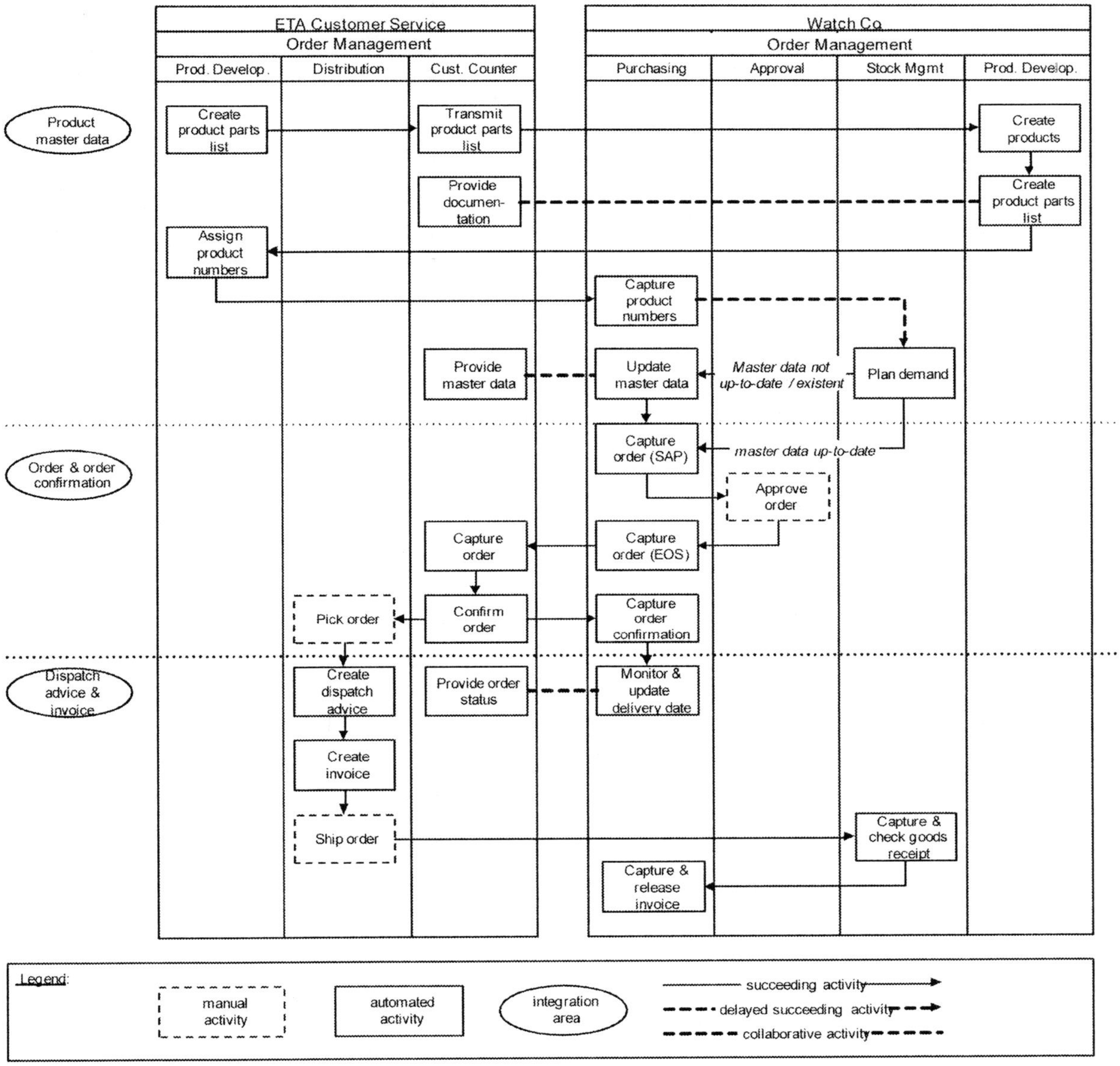

which is considered as the central "one face to the customer" communication platform.

The analysis of costs and benefits associated with the future order management process demonstrates that most benefits were realized on the customer side (Table 1). However, the crucial factor for justifying the project is that ETA-CS anticipates several strategic benefits by establishing a sustainable machine-machine integration platform. These benefits include better process transparency and improvements in process quality and efficiency due to up-to-date and accurate master data.

Table 1. Benefit estimation for future order management process

	ETA-CS		Watch Co.	
	Benefits	**Quantification**	**Benefits**	**Quantification**
Product master data	fewer free text orders	15 min. expense for decoding one free text order position	fewer free text orders	15 min. expense for decoding one free text order position
	fewer wrong deliveries per invoices	15 wrong invoices per year	no manual update of master data	5 min. expense per product and year
			fewer wrong deliveries per invoices	1% wrong deliveries per invoice and per year
Order & order confirmation			no double order capture	3 min expense per order + 2 min expense per position
			no manual capture of delivery date	1 min expense per position
			no monitoring and update of delivery dates	2 hours expense per year
Dispatch advice & invoice	Elimination of paper-based invoice	10 min expense per order	no manual capture of delivery note	1 min expense per position
			no manual capture of invoice	1 min expense per position

DESIGN OF THE FUTURE B2B ARCHITECTURE

Multi-Channel Architecture

Responding to the results of the pilot study with Watch Co., ETA decided to supplement its existing Customer Service Portal with a selected set of directly accessible machine-machine services. These services are targeted at distinct customer groups, mainly intra-group brands and local distribution companies. Since ETA has invested in establishing the Customer Service Portal as electronic collaboration platform with customers over the past eight years, it was decided that ETA should maintain it as central communication platform and continue to provide the entire set of services via the CSP. Only selected services were to be offered initially via FTP/XML as well and might be offered as Web Services in the future. In order to allow for coordinated customer interactions over different electronic and physical channels, ETA decided to design a multi-channel architecture:

- Existing communication channels with customers, namely phone, fax, e-mail and the portal, would in future be complemented by an additional electronic channel for structured XML message exchange via either FTP or Web services.
- The organizational channels remained the same since customer interactions are already coordinated internally by the so-called 'Customer Counter' at ETA-CS.

E-Service Portfolio

With regard to the future e-service portfolio, the key design decisions concerned interaction management, i.e. determining which services to offer additionally via the machine-machine interaction channel. ETA systematically reconciled the services currently offered on the Customer Service Portal with the additional requirements gathered in the pilot study. Accordingly, master data services such as product information, pricing and availability as well as functions for order entry and confirmation were assigned to the new channel. In addition to offering existing portal services via the direct channel, ETA is extending its service portfolio to electronic delivery notes and invoices that will be accessible via all electronic channels.

Integration Architecture

The transition to a multi-channel approach required ETA SA to rethink their integration architecture, notably the integration of the internal ERP and PLM applications with the electronic communication channels. By introducing multiple electronic communication channels, the overall complexity of ETA's integration architecture is multiplied as illustrated by the following examples: New interfaces between the existing applications have to be implemented and managed, e.g. in order to transmit invoices and delivery notes from the ERP-system to the Customer Service Portal. Additionally, the lacking separation of business and interaction logic in today's software architecture inevitably leads to a duplication of application logic in different systems when assigning services to multiple channels (cf. Lippert, Wolf, & Züllighoven, 2001). As an example, the presentation and personalization of product information in the CSP cannot be directly reused for the direct channel. Instead, the logic of product data selection has to be re-implemented and subsequently maintained in the import database. Channel synchronization represents an additional challenge: Integration relationships between internal applications at ETA have to be synchronized in order to guarantee a consistent view of the order management process covering all communication channels.

The future integration architecture developed at ETA-CS strengthens the role of its ERP-system as the leading system for order management related functions by allowing direct order entry into the ERP-system without the deviation over the ETA Online Shop. Order confirmations, delivery notes as well as invoices will be directly transferred from ETA's ERP-system to the ERP-system of its customers. The same data is transferred in parallel to the Customer Service Portal in order to maintain a central information platform providing full visibility of the order status. Master data needed by the customer such as product, condition and availability information will be periodically extracted from the import database and transferred to the customers' ERP-systems.

With regard to the integration mechanism, data items to be exchanged between ETA-CS and its customers will be defined and encoded in XML and transferred via FTP or Web Services. The whole integration process will be managed by the enterprise application integration (EAI) broker.

Vision: Towards a Service-Oriented Multi-Channel Architecture

Since complexity rises with any additional channel, ETA SA started to think about increasing the sustainability and flexibility of their multi-channel architecture. Based on the concept of an SOA, this included the distinction between four separate architectural layers (cf. Newcomer & Lomow, 2004; Zimmermann, Doubrovski, Grundler, & Hogg, 2005): Whereas the application layer contains the existing application resources, services encapsulate reusable business logic that needs to be accessible by different electronic communication channels. The interaction layer comprises so-called interaction services, i.e. functional elements that handle the communication channel-specific interaction between the business services and corresponding front-ends on the presentation layer (cf. Lippert, et al., 2001; Zimmermann, et al., 2005).

Figure 5 applies the concept of a service-oriented multi-channel architecture to the case of ETA-CS. The application layer comprises the ERP-system, the import database for product master data and the e-shop application. The service layer contains services encapsulating the key business logic relevant to ETA-CS customers, such as product data, order capture and order

Figure 5. Concept of a service-oriented multi-channel architecture

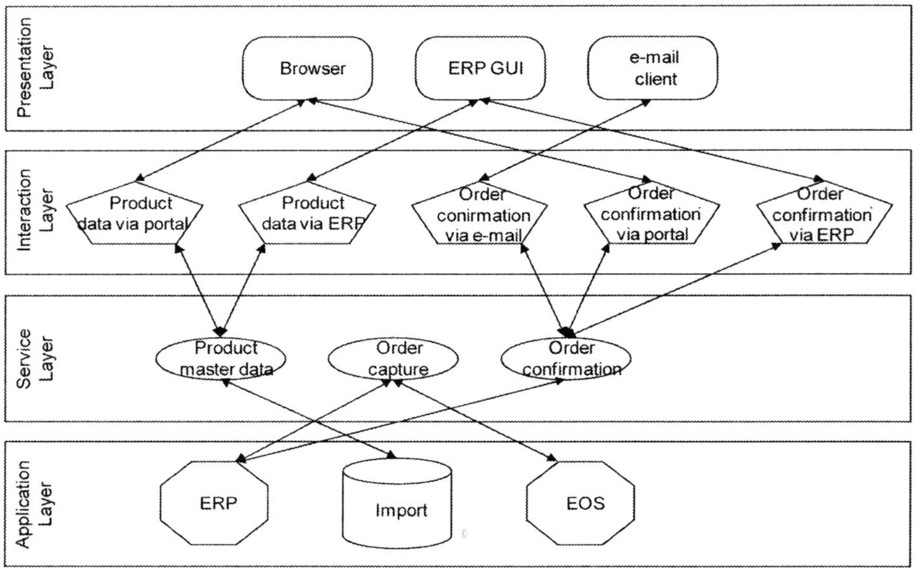

confirmation. These services do not incorporate any channel or presentation-specific logic. Interaction services implement the requirements of the specific communication channel and evoke the business services as needed. As an example, the generic order confirmation service is transformed by means of interaction services to be presented in the portal, sent by email or directly transferred to a customer ERP-system. The outlined separation of interaction and business logic increases maintainability and reduces redundancies in the existing system architecture. Another advantage of the service-oriented architecture approach is that existing business services can easily be offered via additional communication channels, providing improved process flexibility for ETA. The major obstacle to the outlined architecture is that the distinct separation of the different layers is not incorporated in today's packaged applications and thus a major redesign of the existing application landscape will be required.

SUMMARY AND DISCUSSION

The objective of our research was two-fold: First, we wanted to understand how technology development and changing customer requirements impact the type and nature of business-to-business e-services over time. Second, we were interested in the implications for interorganizational process and system design.

Based on the case of ETA-CS, our research identifies three distinct stages in the evolution of B2B e-services (cf. Table 2). The initial stage represents a first generation e-commerce solution providing transaction-related services, i.e. electronic order entry functionality, to customers via a Web front-end. The second stage comprises the realization of a comprehensive customer process portal as single electronic channel and is associated with a significant extension of the e-service portfolio (service innovation). The customer process portal complements transaction-related services with pre- and post-purchase services in order to cover the entire customer process. While the portal-based solution has limited real-time integration capabilities the third stage is character-

Table 2. Three stages in the evolution of B2B e-services

	Stage 1: E-commerce / e-Shop	**Stage 2: Customer process portal**	**Stage 3: Multi-channel architecture**
Focus	Basic portfolio of transaction-related services	Service innovation: pre- and post-purchase related services, personalization	Channel innovation: Introduction of direct (machine-machine) channels
Electronic services	Transaction-related services, e.g. • Product search / catalog • Electronic order entry	Transaction-related services, e.g. • Product search / catalog • Electronic order entry Pre- and post-purchase related services, e.g. • Order tracking • Repair tracking • Technical documentation • Personalized notification	C.f. stage 2 + transaction-related services via additional channels, e.g. • Electronic exchange of master data exchange / product catalog • Electronic exchange of orders, order confirmations and delivery notes
Coverage of customer process	Procurement	Procurement Maintenance & Repair Knowledge Transfer	Procurement Maintenance & Repair Knowledge Transfer
Electronic channel(s)	e-Shop here: ETA Online Shop (EOS) for ordering watch spare parts	Process portal providing comprehensive e-service for the entire customer process here: ETA Customer Process Portal (CSP)	Two electronic channels reflecting the needs of customer segments: • Human-machine (Portal) • Machine-machine (XML messages, Web Services)
Electronic interaction	Human-machine	Human-machine	Human-machine and machine-machine
Integration architecture	Data integration between ERP-system and e-Shop	Data and / or functional integration between portal, e-Shop, ERP and other systems (here: PLM system and document management system)	Service-oriented architecture (services encapsulate business logic from the operational systems that need to be accessible by different electronic communication channels)

ized by the introduction of an additional electronic machine-machine channel (channel innovation). In this stage, electronic services are offered via both, the portal-based and the machine-machine channel, in order to allow for seamlessly integrated business processes with customers. These different stages can be considered to occur chronologically as technology develops and firms learn about their customers' acceptance of e-services. The move from one stage implies major redesigns of e-services and the implementation of new functionalities or channels based on a revised integration architecture.

Most importantly, our study suggests that service innovation and channel innovation are the two main dimensions of B2B e-services evolution. With regard to service innovation, our findings are mostly consistent with research on the evolution of e-services in consumer markets, while some aspects are to be weighted differently in view of the specific nature of B2B relationships: The initial focus on basic services, in particular the order entry transaction, has been incrementally extended to cover the entire set of transaction-related and pre- and post-purchase services as outlined by (Otim & Grover, 2006) in their comprehensive framework. However, B2B customers differ from consumers with regard to their perception of interaction value (Piccoli, et al., 2004), i.e. the overall utility realized by customers using Web sites or other forms of e-services. Consequently, studies on B2C e-services put a strong emphasis on Web site design (e.g. site layout, personalization) and enjoyment of the virtual experience during the pre-purchase phase (e.g. interactive configuration, recommender systems). Business customers, on the other hand, are more goal-oriented and concerned about the efficiency gains and problem solutions e-services can provide in the everyday world of work. ETA has taken these specific B2B requirements into account when prioritizing and designing e-services, in particular the e-catalog, order entry and status tracking functionalities. Similar considerations are true with regard to the personalization and customization of e-services. While personalized services are tailored to individual user preferences or profiles in the B2C case, knowledge about the business relationship is used to personalize B2B e-services. Good examples are personalized views of the online shop displaying customer-specific product identifiers and sales conditions, or personalized notifications announcing the end of spare parts production for a previously purchased watch movement.

With regard to channel innovation and the evolving nature of electronic interactions, our study also reveals some specificities of the B2B field: Given today's worldwide diffusion of Internet protocols, Web sites or portals have become the most popular form of e-service delivery and usually satisfy the requirements of many consumers and business partners. The case of ETA-CS demonstrates the limitations of such human-machine integration in B2B relationships with higher transaction volumes, most notably the manual data re-entry and portal monitoring efforts. The study suggests that with greater company-internal IS integration and a growing number of portals, companies will increasingly prefer machine-machine interaction. Hence, companies like ETA will move from a mono-channel provider strategy (Payne & Frow, 2004) toward a customer segment strategy, which offers combinations of human-machine and machine-machine channels at particular customer groups.

With the extension of the e-service portfolio and the need to serve additional machine-machine channels, companies face a two-fold challenge: First, higher levels of electronic interaction require a comparably larger number of additional interorganizational agreements than portal-based interactions. In the absence of widely accepted industry standards, companies have to align business processes, message formats and the underlying master data with every individual business partner. This adds significant complexity to the conceptualization and realization of machine-machine e-services as opposed to portal-based e-services which are manageable within the organizational boundaries. Consequently, B2B e-services will increasingly require a collaborative effort for designing interorganizational process and system linkages. Second, it is imperative for ETA CS to integrate the activities in the different channels to produce the most positive customer experience possible and to manage the internal customer-facing processes (Payne & Frow, 2004). Our study demonstrates that service and channel innovation require a well designed application and integration architecture. Whereas service innovation requires significant initial efforts to prepare internal data and functionality to be accessed by external customers, the introduction of an electronic channel multiplies custom interfaces to existing applications and duplicates implementation efforts. We conclude that the technical convergence of internal, ERP-based and external, Internet-based integration based on open Web services standards is an important enabler for realizing multi-channel integration. Thus, a service-oriented architecture increases not only maintainability, but also reduces costs and effort for realizing multiple electronic communication channels. Hence, our study highlights the reasons why Web services are seen as the foundation for future IOS (Daniel & White, 2005). It also confirms the view that a well-designed B2B architecture is required to cope with the growing complexity of electronic channels (Al-Naeem, Rabhi, Benatallah, & Ray, 2004).

This work contributes to the existing body of research on e-services and interorganizational systems by exploring the evolution of B2B e-services. As in any case study research, the results from the present research cannot be interpreted without taking the study's limitations into account. This applies in particular to the risk of focusing on unique conditions and events, rather than on general concepts and trends. With this regard, we have been able to relate our findings to the recent studies by Otim & Grover (2006) and Piccoli et al. (2004). Interestingly, our conclusions with regard to e-channel innovation, and in particular the necessity to offer a combination of machine-machine and human-machine channels, are also in line with earlier experiences from the EDI era where WebEDI was introduced to satisfy the requirements of smaller customers (Reimers, 2001). In any event, we encourage researchers to conduct more extensive qualitative and quantitative studies in order to validate and refine the different stages of B2B e-services we have identified (c.f. Table 2). These studies should try to describe and measure the state of e-services adoption according to the two dimensions of service innovation and channel innovation. Since existing studies focus on one specific channel for delivering electronic services, the most interesting questions relate to the combination of several electronic channels for specific business partner segments. In view of the differences in the adoption of e-services in various industries, it would be particularly valuable to identify contingencies and analyze their impact on the level of service innovation and channel innovation.

In view of the converging technologies for internal and external integration, much more research is needed with regard to the design of future B2B architectures. Researchers need to investigate how service-oriented concepts and Web service interoperability can be leveraged for multi-channel integration: How are e-services systematically translated into their technical implementation as Web service? To what extent will service orientation reduce the efforts for implementing e-services and increase the flexibility of serving multiple electronic channels? Which service design criteria have to be applied in order to ensure that services are reusable with multiple channels and partners? Given that the interorganizational alignment is one of the main issues in setting up machine-machine interaction, a very active research community currently investigates how the service-oriented paradigm can be leveraged to enhance existing B2B standardization. This involves the development of semantic Web services (e.g. de Bruijn, Fensel, Keller, & Lara, 2005), e.g. using ontologies, and more flexible approaches for interorganizational process design using the abstraction concept of "public processes" (Legner & Vogel, 2008).

ACKNOWLEDGMENT

This research has been conducted in the context of the Competence Center Business Networking, notably CC BN2 and CC BN3, at University of St. Gallen, Switzerland. The author would like to thank all industry partners for funding this research and the ETA project team members, notably Dr. Rudolf Zurmühlen, Thomas Bühler, Dr. Marc Cäsar and Jan Schemm, for their valuable input.

REFERENCES

Al-Naeem, T., Rabhi, F. A., Benatallah, B., & Ray, P. K. (2004). Systematic Approaches for Designing B2B Applications. *International Journal of Electronic Commerce*, *9*(2), 41–70.

Alt, R. (2008). *Überbetriebliches Prozessmanagement - Gestaltungsmodelle und Technologien zur Realisierung integrierter Prozessportale*. Berlin: Logos.

Alt, R., & Fleisch, E. (2000). Business Networking Systems - Characteristics and Lessons Learned. *International Journal of Electronic Commerce*, *5*(2), 7–27.

Alt, R., Reichmayr, C., Cäsar, M. A., & Zurmühlen, R. (2002). Evolution of Electronic Catalogs to Customer Process Portals - A structured Approach at ETA S.A. In *Proceedings of the 15th Bled Electronic Commerce Conference, Bled.*

Alt, R., Reichmayr, C., & Zurmühlen, R. (2000). Developing eCommerce within Business Networks - The Case of ETA SA. In *Proceedings of the 13th Bled Electronic Commerce Conference, Bled.*

Archer, N., & Yufei, Y. (2000). Managing business-to-business relationships throughout the e-commerce procurement life cycle. *Internet Research: Electronic Networking Applications and Policy*, *10*(5), 385–395. doi:10.1108/10662240010349390

Bakos, J. Y. (1991). A strategic Analysis of Electronic Marketplaces. *MIS Quarterly*, *15*(3), 295–310. doi:10.2307/249641

Barnes, D., Hinton, M., & Mieczkowska, S. (2005). Enhancing Customer Service Operations in E-Business: The Emotional Dimension. *Journal of Electronic Commerce in Organizations*, *3*(2), 17–32.

Benbasat, I., Goldstein, D. K., & Mead, M. (1987). The Case Research Strategy in Studies of Information Systems. *MIS Quarterly*, *11*(3), 369–386. doi:10.2307/248684

Bhappu, A. D., & Schultze, U. (2006). The role of relational and operational performance in B2B customers′ adoption of self-service technology. *Journal of Service Research*, *8*(4), 372–385. doi:10.1177/1094670506286571

Bitner, M. J., Brown, S. W., & Meuter, M. L. (2000). Technology Infusion in Service Encounters. *Journal of the Academy of Marketing Science*, *28*(1), 138–149. doi:10.1177/0092070300281013

Bucklin, C. B., DeFalco, S. P., DeVincentis, J. R., & Levis, J. P. I. (1996). Are you tough enough to manage your channels? *McKinsey Quarterly*, *1*, 104–114.

Cenfetelli, R. T., Benbasat, I., & Al-Natour, S. (2008). Addressing the What and How of Online Services: Positioning Supporting-Services Functionality and Service Quality for Business-to-Consumer Success. *Information Systems Research*, *19*(2), 161–181. doi:10.1287/isre.1070.0163

Chen, Q., & Chen, H.-M. (2004). Exploring the success factors of eCRM strategies in practice. *Database Marketing & Customer Strategy Management*, *11*(4), 333–343. doi:10.1057/palgrave.dbm.3240232

Cristobal, E., Flavian, C., & Guinaliu, M. (2007). Perceived e-service quality (PeSQ): Measurement validation and effects on customer satisfaction and web site loyalty. *Managing Service Quality*, *17*(3), 317–140. doi:10.1108/09604520710744326

Cullen, A. J., & Webster, M. (2007). A model of B2B e-commerce, based on connectivity and purpose. *International Journal of Operations & Production Management*, *27*(2), 205–225. doi:10.1108/01443570710720621

Daniel, E. M., & White, A. (2005). The future of inter-organisational system linkages: findings of an international Delphi study. *European Journal of Information Systems*, *14*(2), 188–203. doi:10.1057/palgrave.ejis.3000529

Davenport, T. H., & Brooks, J. D. (2004). Enterprise systems and the supply chain. *Journal of Enterprise Information Management*, *17*(1), 8–19. doi:10.1108/09576050410510917

Day, G. S., & Hubbard, K. J. (2003). Customer Relationships Go Digital. *Business Strategy Review*, *14*(1), 17–26. doi:10.1111/1467-8616.00240 de Bruijn, J., Fensel, D., Keller, U., & Lara, R. (2005). Using the Web Service Modeling Ontology to Enable Semantic E-Business. *Communications of the ACM*, *48*(12), 43–47. doi:10.1145/1101779.1101807

Dias, C. (2001). Corporate Portals: a literature review of a new concept in information management. *International Journal of Information Management*, *21*, 269–287. doi:10.1016/S0268-4012(01)00021-4

Eisenhardt, K. M. (1989). Building Theories from Case Study Research. *Academy of Management Review*, *14*(4), 532–550. doi:10.2307/258557

Gregor, S. (2006). The Nature of Theory in Information Systems. *MIS Quarterly*, *30*(3), 611–642.

Hagel, J. III, & Brown, J. S. (2001). Your Next IT Strategy. *Harvard Business Review*, *79*(October), 105–113.

Hong, I. (2002). A new framework for interorganisational systems based on the linkage of participants' roles. *Information & Management*, *4*(2), 261–270. doi:10.1016/S0378-7206(01)00095-7

Johnston, H. R., & Vitale, M. R. (1988). Creating competitive advantage with Interorganizational Information systems. *MIS Quarterly*, *12*(2), 153–165. doi:10.2307/248839

Kaplan, S., & Sawhney, M. (2000). E-Hubs: The New B2B Marketplaces. *Harvard Business Review*, *78*(3), 97–103.

Kubicek, H. (1992). The Organization Gap in Large-Scale EDI Systems. In R. J. Streng, C. F. Ekering, E. van Heck & J. F. H. Schultz (Eds.), *Scientific Research on EDI "Bringing Worlds Together"* (pp. 11-41). Amsterdam: Samsom.

Legner, C., & Vogel, T. (2008). Leveraging Web Services for Implementing Vertical Industry Standards: A Model for Service-Based Interoperability. *Electronic Markets*, *18*(1), 39–52. doi:10.1080/10196780701797623

Linthicum, D. S. (2000). *Enterprise Application Integration*. Upper Saddle River, NJ: Addison-Wesley.

Lippert, M., Wolf, H., & Züllighoven, H. (2001). Domain Services for Multi-Channel Application Software. In *Proceedings of the 34th Annual Hawaii International Conference on System Sciences (HICSS-34), Maui.*

Löwer, U. M. (2005). *Interorganisational Standards: Managing Web Services Specifications for Flexible Supply Chains*. Heidelberg: Physica-Verlag.

Lu, J., & Zhang, G. (2003). Cost benefit factor analysis in e-services. *International Journal of Service Industry Management*, *14*(5), 570–595. doi:10.1108/09564230310500237

Madlberger, M. (2006). Exogenous and Endogenous Antecedents of Online Shopping in a Multichannel Environment: Evidence from a Catalog Retailer in the German-Speaking World. *Journal of Electronic Commerce in Organizations*, *4*(4), 29–51.

Mahadevan, B. (2003). Making sense of emerging markets structures in B2B e-commerce. *California Management Review*, *46*(1), 86–100.

McAfee, A. (2005). Will Web Services Really Transform Collaboration? *MIT Sloan Management Review*, *46*(2), 78–84.

Merrilees, B., & Fenech, T. (2007). From catalog to Web: B2B multi-channel marketing strategy. *Industrial Marketing Management*, *37*(1), 44–49. doi:10.1016/j.indmarman.2006.06.014

Miles, M. B., & Huberman, A. M. (1994). *Qualitative Data Analysis: An Expanded Sourcebook* (2. ed.). Thousand Oaks: Sage Publications.

Newcomer, E., & Lomow, G. (2004). *Understanding SOA with Web Services*. Maryland: Addison-Wesley.

Österle, H. (1995). *Business in the Information Age - Heading for new Processes*. Berlin: Springer.

Österle, H. (2001). Enterprise in the Information Age. In H. Österle, E. Fleisch & R. Alt (Eds.), *Business Networking: Shaping Collaboration Between Enterprises* (Vol. 2, rev. and extended ed., pp. 17-54). Berlin: Springer.

Otim, S., & Grover, V. (2006). An empirical study on Web-based services and customer loyalty. *European Journal of Information Systems, 15*, 527–541. doi:10.1057/palgrave.ejis.3000652

Palmer, J. W. (2000). Electronic commerce in retailing: Convenience, search costs, delivery and price across retail formats. *Information Technology and Management, 1*(1-2), 25–43. doi:10.1023/A:1019152403937

Parasuraman, A., Zeithaml, V. A., & Berry, L. L. (1988). SERVQUAL: A multiple item scale for measuring consumer perceptions of service quality. *Journal of Retailing, 64*(1), 12–40.

Payne, A., & Frow, P. (2004). The Role of Multichannel Integration in Customer Relationship Management. *Industrial Marketing Management, 33*(6), 527–538. doi:10.1016/j.indmarman.2004.02.002

Piccoli, G., Brohman, M. K., Watson, R. T., & Parasuraman, A. (2004). Net-Based Customer Service Systems: Evolution and Revolution in Website Functionalities. *Decision Sciences, 35*(3), 423–455. doi:10.1111/j.0011-7315.2004.02620.x

Puschmann, T., & Alt, R. (2005). Developing an Integration Architecture for Process Portals. *European Journal of Information Systems, 14*(2), 121–134. doi:10.1057/palgrave.ejis.3000527

Ranganathan, C., & Ganapathy, S. (2002). Key dimensions of business-to-user web sites. *Information & Management, 39*(6), 457–465. doi:10.1016/S0378-7206(01)00112-4

Reimers, K. (2001). Standardizing the new e-business platform: Learning from the EDI experience. *Electronic Markets, 11*(4), 231–237. doi:10.1080/101967801753405508

Ruh, W. A., Maginnis, F. X., & Brown, W. J. (2001). *Enterprise Application Integration*. New York: John Wiley & Sons.

Salomann, H., Kolbe, L., & Brenner, W. (2006). Self-Services in Customer Relationships: Balancing High-Tech and High-Touch Today and Tomorrow. *e-Service Journal, 4*(2), 65-84.

Schierholz, R., Ostrowski, A., Glissmann, S., Kolbe, L. M., & Brenner, W. (2007). Don't call us, we'll call you - Performance Management in Multi-Channel Environments. *Journal Information Science and Technology, 3*(2), 44–61.

Stone, M., Hobbs, M., & Khaleeli, M. (2002). Multichannel customer management: The benefits and challenges. *Journal of Database Marketing, 10*(1), 39–52. doi:10.1057/palgrave.jdm.3240093

Subramaniam, C., Shaw, M. J., & Gardner, D. M. (2000). Product Marketing and Channel Management in Electronic Commerce. *Information Systems Frontiers, 1*(4), 363–378. doi:10.1023/A:1010061924822

van Riel, A. C. R., Liljander, V., & Jurriëns, P. (2001). Exploring consumer evaluations of e-services: a portal site. *International Journal of Service Industry Management, 12*(4), 359–377. doi:10.1108/09564230110405280

Walker, R. H., Craig-Lees, M., Hecker, R., & Francis, H. (2002). Technology-enabled service delivery - An investigation of reasons affecting customer adoption and rejection. *International Journal of Service Industry Management*, *13*(1), 91–106. doi:10.1108/09564230210421173

Weston, F. C. (2003). ERP II: The extended enterprise system. *Business Horizons*, *46*(6), 49–55. doi:10.1016/S0007-6813(03)00088-0

Wilson, H., & Daniel, E. (2007). The multi-channel challenge: A dynamic capability approach. *Industrial Marketing Management*, *36*(1), 10–20. doi:10.1016/j.indmarman.2006.06.015

Yeung, W. L., & Lu, M.-t. (2004). Functional Characteristics of Commercial Web Sites: a Longitudinal Study in Hong Kong. *Information & Management*, *41*(4), 483–495. doi:10.1016/S0378-7206(03)00086-7

Yin, R. K. (2002). *Case Study Research. Design and Methods* (3rd ed., Vol. 5). London: Sage Publications.

Zimmermann, O., Doubrovski, V., Grundler, J., & Hogg, K. (2005). Service-Oriented Architecture and Business Process Choreography in an Order Management Scenario: Rationale, Concepts, Lessons Learned. In *Proceedings of Oopsla'05, San Diego, California*.

APPENDIX

LIST OF ABBREVIATIONS

B2B: Business-to-business
B2C: Business-to-consumer
CSP: Customer Service Portal (ETA's process portal for customers)
EAI: Enterprise application integration
EDI: Electronic data interchange
ERP: Enterprise resource planning
ETA-CS: ETA's customer service organization
ETA SA: ETA, one of the world's largest manufacturers of watch movements and components
ETA EOS: ETA Online Shop (ETA's first generation e-commerce solution)
IOS: Interorganizational system(s)
PLM: Product lifecycle management
SG: Swatch Group, a Swiss-based group in the watch industry
SOA: Service-oriented architecture

Chapter 13
Conceptualizing Competences in E-Services Adoption and Assimilation in SMEs

Ada Scupola
Roskilde University, Denmark

ABSTRACT

This article investigates the competences deemed necessary both at top managerial and individual levels for the successful adoption and assimilation of business-to-business e-services in small and medium size enterprises. To this end, an in-depth case study of a business-to-business e-service system, a Web-based travel reservation system, was conducted. The results show that three main competences, namely vision, value and control, are important at top management level for the primary adoption of e-services. For secondary adoption and assimilation, three categories of competences were identified as being important either to have or to develop at the individual level, namely technical, interpersonal and conceptual skills.

INTRODUCTION

The aim of this study is to explore the competencies that are needed by small and medium size corporations (SMEs) in order to successfully adopt and assimilate e-services. SMEs are defined here as firms with up to 250 employees (OECD, 2000). An extensive body of literature has argued and investigated the importance of e-commerce technologies and e-services for the competitive advantage of the small and medium size firms (e.g., Grandon & Pearson, 2003; OECD, 2000). In addition, many studies have investigated the barriers to the adoption and diffusion of the Internet, e-commerce and e-business in small and medium size firms, as well as factors affecting the adoption and diffusion of these technologies in this type of firm (e.g., Mirchandani & Motwani, 2001; Scupola, 2003). Some of these studies have focused on competencies as it is thought that they might positively influence the firm's strategic commitment to the adoption and assimilation of technologies such as e-commerce or e-services (Chaston & Mangles, 2002; Eikebrokk & Olsen, 2002; McGowan, Durkin, Allen, Dougan, & Nixon, 2001; Wainwright, Green, Mitchell, & Yarrow, 2005).

In this article, e-services are defined as services that are provided or consumed through the use of Internet-based systems. The consumption or the provision of a service requires the interaction between the service provider and the user. Traditionally, this has been based on personal interactions, most often face-to-face interactions. In e-services, the consumption and the provision take place through the intermediation of an Internet-based system and therefore are separated in time and space (Fuglsang & Sundbo, 2006). E-services can involve a number of different relations: business to consumer, business to business, government services offered on the Internet or e-government. E-services also include the selling of physical goods on the Internet such as an airline ticket that is purchased online, but delivered by surface mail. There are three main characteristics of e-services:

- The service is accessible via the Internet or other electronic networks,
- The service is consumed by a person via the Internet or other electronic networks, and
- There might be a fee that the consumer pays the provider for using the e-service, but that might not always be the case. For example, some e-services offered by the government are free (Scupola, 2008). The purpose of this article is to investigate the competences required for the successful adoption and assimilation of business-to-business e-services in small and medium size enterprises. The basic research question tackled here is: What are the competences required both at the top managerial and individual levels for the successful adoption and assimilation of business-to-business e-services in small and medium size corporations? To answer the research question, a theoretical model of e-service competencies is developed and a case study of an e-service system is conducted. The results show that three main competences, namely vision, value and control, are important at top management level, while a number of competences are important at the individual level.

The article is structured as follows. This section has presented the main purpose and motivation of the study. The next section discusses adoption and assimilation. This is followed by a literature review relating to the concepts of competence, IT competence, and the development of a model to investigate e-services competencies. It also discusses the methodology employed. The remaining part of the article presents the analysis, the discussion and the conclusions.

Adoption and Assimilation of E-Services

For the purpose of this study, adoption is defined as "the decision to make full use of an innovation as the best course of action available" (Rogers, 1995, p. 21). According to Zaltman, Duncan, and Holbeck (1973) innovation adoption within organizations often occurs in two stages. The first is a firm-level decision to adopt the innovation also called primary adoption; the second is the actual implementation or individual adoption by users also called secondary adoption. At the first level, managers identify objectives to change some aspect of their business and look for innovations that fit their objectives. Then they make the primary adoption decision (Gallivan, 2001). Once the primary adoption decision has been made, the implementation and use of the innovation at the individual level takes place. According to Gallivan (2001), management may proceed by taking three fundamentally different paths to ensure secondary adoption: (1) they can mandate that the innovation be adopted throughout the organization at once; (2) they can provide the necessary infrastructure and support for users to adopt the innovation, while allowing it to diffuse voluntarily; or (3) they may target specific pilot projects within the firm, observe the processes and outcomes that unfold, and decide whether to introduce the innovation more widely later on. This two-stage adoption model has also been defined as a contingent adoption decision, meaning that employees cannot adopt the innovation until primary adoption has occurred at a higher level of authority, often managerial or top management (Zaltman et al., 1973).

Assimilation is defined as the extent to which the use of a technology diffuses across organizational work processes and becomes routinized in the activities associated with those processes (Tornatzky & Klein, 1982). Moreover, it is important to look at assimilation because the adoption of a technology at a company level does not automatically lead to assimilation and use. Fichman and Kemerer (1999), for example, show that most information technologies exhibit an assimilation gap that is their rates of organizational assimilation and use lag behind their rates of organizational adoption.

The Concept of Technology Competence

The concept of competence has been much discussed in different types of research such as management, human resources and information systems and there is much confusion regarding the definition of competence. Furthermore, the concept of competence has been used at different analytical levels: for example, task-specific competences, firm specific competences or industry specific competences (Nordhaug, 1998). Competence has often also been identified with performance. But if performance has been used as a proxy for competence due to the difficulty of measuring competence, the mixing of competence with performance often implies mixing the outcome with the process. However, a large amount of literature distinguishes between competence and performance, especially in relation to competence as a specific skill. Marcolin, Compeau, Munro, and Huff (2000) define user competence as "the user's potential to apply technology to its fullest possible extent so as to maximize the users' performance on specific job tasks." Competence is also seen as a personality trait, and might include generic knowledge, motivation, social role, or skill of a person linked to superior performance on the job (Haynes, 1979). Competence has also been associated with knowledge, and in this view competence is not only task specific, but embodies the ability to transfer knowledge across tasks, thus becoming interactive and dynamic (Brown, 1994). In human capital theory, the concept of competence has been related to specific firm technologies and to the execution of tasks that are related to the technology and the routines required to use that technology. Furthermore, there are many different typologies of competences found in the literature. For example, Yukl (1989) develops a typology that consists of technical, conceptual and interpersonal competences.

IT COMPETENCE MODELS AND IT COMPETENCE

A number of models have been developed to investigate IT competencies in small firms. Wainwright et al. (2005) develop a competency based model to be used for comparing practice and performance with respect to ICT within small firms. They propose that an ICT competence and capability approach can be a viable research avenue for investigating IT performance in small firms. Chaston and Mangles (2002) propose that the resource-based view of competitive advantage may provide the basis for assessing the ability of an organization to exploit the Internet to enhance market performance. They develop a competency model to define the competencies and capabilities that may influence the execution of an Internet marketing strategy. Their results show that the proposed competencies such as positioning, innovation and so forth, are critical for small firms to achieve market performance.

Finally, as with the concept of technology competence, the concept of competence related to IT has been defined in different ways and has focused on different levels of analysis (Table 1 provides a summary of studies of IT competence).

The stream of literature focusing on the organizational level (e.g., Van der Heijden, 2000) is influenced by parallel studies in the management literature and especially by the resource based view of strategic advantage (e.g., Barney, 1991; Prahalad & Hamel, 1990). The other stream of research focuses on the individual level, for example, Basselier, Reich, and Banbasat (2001) study IT competences in business managers, and identifies a concept of IT knowledge that is decomposed into explicit IT knowledge and tacit IT knowledge. Explicit IT knowledge is knowledge that can be read, taught or explained to others. Tacit IT knowledge is more difficult to explain and to do so Basselier et al. use the constructs of experience and cognition. Experience relates to the managers' know how, while cognition involves more then doing things, and refers to working models of the world that an individual forms, including acumen, beliefs and viewpoints. However, this kind of literature has not focused on other factors related to competences such as communication, social or leadership skills.

Table 1. A summary of studies of IT competences

Author	Definition of Competence	Individual/Organizational Level
Basselier et al. (2001)	IT related explicit and Tacit Knowledge	Individual level (Business Manger)
Lee & Trauth (1995)	Critical Knowledge and Skills	Individual level (IS professionals)
Bharadwaj, Sambamurthy & Zmud (2000)	Critical IT Capabilities	Organizational Level
Van der Heji-den (2000)	IT core capabilities; organization specific routines, skills, resources and processes	Organizational Level and IT management)
Sambamurthy & Zmud (1994)	Capabilities, skills and tacit know-how	Organizational Level and IT management
Chaston & Mangles (2002)	Internet related capabilities; organization specific routines, resources and processes	Organizational Level
Wainwright et al. (2005)	ICT related competence; Organizational ICT performance and practice	Organizational Level

In this study, being interested in the small firm's competencies deemed important for the successful adoption and assimilation of e-services both at primary and secondary levels, the focus is placed on top management and individual level competencies. Top management, through their beliefs and visions can offer guidelines to managers and employees about the opportunities and risks in assimilating technological innovations (Gallivan, 2001; Chatterie, Grewal, & Sambamurthy, 2002). For example, in firms where top managers believe that e-services offer a strategic opportunity, their beliefs serve as powerful signals to the rest of the firm's employees about the importance placed on e-services. This makes the employees and the managers use their time and energy in making sense of e-services or exploring ways in which the technology's functionality could be leveraged to improve the company's efficiency, routines or business value.

In this study, therefore, managerial competences become the basis for the firms' vision, norms, beliefs and strategies in adopting and assimilating e-services (Durkin & McGowan, 2001; Middleton & Long, 1990). Individual competences are relevant in relation to the specific task of adopting and using an e-service system (Brown, 1994; Haynes, 1979).

A MODEL OF COMPETENCY FOR E-SERVICE ADOPTION AND ASSIMILATION

Theoretical Background of the Model

Durkin and McGowan (2001) develop an Internet competency model to investigate which competencies the manager of a small firm must have in order to decide the extent to which the Internet is adopted in marketing activities. They argue that the extent to which the Internet is adopted and offers a fundamental advantage for the marketing activity is contingent upon the development at the managerial level of four main competencies: vision, value, technical ability and control. These competencies are not independent of each other, but are interrelated. For example, at management level there is sequentiality in the sense that first the company has to develop a vision, then develop the value competency, the technical ability and finally control (Durkin & McGowan, 2001).

Chatterie et al. (2002) describe how the structuration theory of technology assimilation states that firms act as institutions in shaping the behaviours and cognitions of the individuals in the corporation in facilitating or preventing them from assimilating a technology. This can be done in three ways:

1. Structures of signification, where prevailing institutional structures yield meaning and understanding. Individuals apply these structures as guides to understanding how they should behave/act with respect to the assimilation of new technology.
2. Structures of legitimization, where prevailing institutional structures validate specific behaviours as being appropriate in the organization and consistent with the goals and values of the organization. Individuals draw upon these structures as normative templates to reassure themselves about the organizational legitimacy of their assimilation actions.
3. Structures of domination, where institutional structures regulate individual actions and behaviours. Individuals draw upon these structures to ensure that their acts of assimilation do not violate institutional rules and to avoid being the target of organizational sanctions (Chatterie et al., 2002, p. 68).

Finally, Yukl (1989) states that three sets of competences or skills are relevant at the individual level in explicating a specific task or activity: technical, interpersonal and conceptual skills. Technical skills or competences represent knowledge about methods, techniques and processes required to conduct a specific activity and the ability to use the tools and equipment necessary to explicate that activity. Interpersonal skills include knowledge about social skills, the ability to communicate, and the ability to cooperate, as well as empathy. Finally, conceptual skills include creativity, efficiency in problem solving, analytical capability and capacity to understand opportunities and problems.

The Research Model

By drawing on the literature review above and on the work by Durkin and McGowan (2001), Chatterie et al. (2002) and Yukl (1989), a model of competencies for e-service adoption and assimilation is proposed. In accordance with McGowan et al. (2001) it is argued that the faster primary adoption of e-services in the small and medium size firm is contingent upon the development of management level competencies such as vision, value, technical ability, and control. By drawing on Yukl (1989), it is argued that faster secondary e-services adoption and assimilation, by contrast, is contingent upon the development of individual level competencies related to the specific task that has to be performed (in this case the use of e-services) such as technical, interpersonal and conceptual skills (Figure 1). Finally, after the firm has made the decision for primary adoption, it can enforce secondary adoption by developing structures of signification, legitimization and domination as is asserted by the structuration theory of technology assimilation (Chatterie et al., 2002). This is shown in Figure 1 by the arrow between the two boxes representing the development of competencies. Like Durkin and McGowan (2001) vision is defined as the managerial capability to see what contribution e-services can make to the general business level. The value competency requires that top management clearly understand what the business value of e-services is. Value competency requires more proactivity and implies that the manager goes a step further and contextualizes the initial awareness and general vision of his own company by evaluating appropriateness and receptivity (McGowan et al.,

Figure 1. A model of competencies for e-service adoption and assimilation

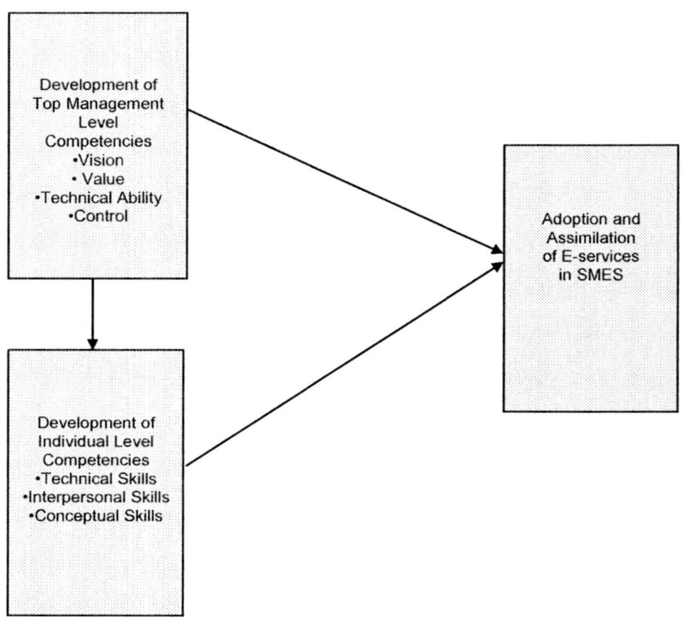

2001). Technical ability implies that the manager is comfortable with new technology and its operation, even though he/she is not a real expert in the technology. It is possible that he/she could attend courses to learn how to use the technology. By developing the "control" competency, the manager enforces rules and policies to ensure that e-services are adopted at the individual level. In fact, as the structuration theory of technology assimilation states, firms act as institutions in shaping the behaviours and cognitions of the individuals in the corporation in facilitating or preventing them from assimilating a technology (Chatterie et al., 2002, p. 68).

By drawing on Yukl (1989), three sets of competence or skills are defined to be relevant at individual level in e-service adoption and assimilation: technical, interpersonal and conceptual skills. Technical skills are mainly identified here with knowledge competency, including the capability to transfer knowledge from one situation to another (Brown, 1994). Following Basselier et al. (2001), e-service knowledge is divided into two components: tacit knowledge and explicit knowledge. Explicit knowledge is formal knowledge that can be acquired through formal training, and can be read or explained. Tacit knowledge by contrast is gained through experience and experiential training. Furthermore, experience increases the memory of how to undertake an activity, which in turn increases competency levels in relation to that experience. The combination and variety of experiences influence the level of tacit knowledge. In this study interpersonal skills include knowledge about social skills, the ability to communicate, the ability to cooperate as well as empathy and judgement. In common with McGowan et al. (2001), it can be argued that judgement is a function of both tacit and explicit knowledge. Finally, conceptual skills include creativity, efficiency in problem solving, analytical capability and the capacity to understand opportunities and problems.

Methodology

To investigate the competences required for the adoption and assimilation of e-services in small and medium size companies, a case study was conducted where the object of analysis was the e-service (Yin, 1989). The e-service was/is a Web-based travel reservation system developed by the travel agency TQ3 which sells business-to-business travel solutions. In the last few years, they have developed a Web-based system, called Webbuster, which can be used by the client companies to make the reservations and buy the travel tickets by themselves, thus bypassing the employees at TQ3. To understand the e-service in question and the competences required for its successful adoption and assimilation in the SMEs, interviews were conducted both with the e-service provider and with the customer companies or adopters of the e-service. Face-to-face qualitative semistructured interviews were used to collect the data. These interviews were complemented by the critical incident technique (see below). The sampling was purposeful. Interviewees were chosen in small firms that had been successful in assimilating e-services. Retrospectively, the study tried to understand the competencies deemed important in both primary and secondary adoption and assimilation. The respondents had to be involved in the adoption process at managerial level or had to be users of the e-services at individual level. Managers and travel bookers at TQ3 were interviewed because they could provide useful information about their customers' competences in the adoption and assimilation of the e-service. A total of 14 interviews were conducted in a total of six companies. The interviews lasted between 60 and 90 minutes, and were tape recorded and fully transcribed by a research assistant. Table 2 provides a brief overview of the companies interviewed.

The semistructured interviews were based on a guide aimed at collecting information both about the managerial level and the individual level

Table 2. Companies interviewed

Company/ Information	Type of Business	Number of Employees
A (TQ3)	Provider of Web-based Travel Solutions	12,000 Worldwide
B	Paint and Varnish Business	No Data
C	Research and Development of Human and Animal Medical Products	140 in Denmark
D	Producer of Cleaning Equipment	No Data
E	Engineering consulting	100
F	Production and service of plastic card products	150 in Denmark

competences. The second part of the interview specifically applied the critical incident technique to elicit critical incidents aimed at highlighting the competences required to successfully adopt and assimilate the e-service in the companies. By relying on Yin (2003), the data were analyzed by following the "general strategy of relying on theoretical orientation" of the case study. Following Miles and Huberman (1994, p. 58), a provisional "start list" of codes based on the conceptual framework was created prior to the field work to guide the analysis. The coding was manual. Specific analytic techniques included making matrixes of categories and placing evidence within such categories (Yin, 2003; Miles & Huberman, 1994) and finding relevant critical incidents (Flanagan, 1954; Fuglsang, 2007).

The Critical Incident Technique (CIT)

The critical incident technique is a research method used to collect data specifically related to competences. This technique was first developed by the U.S. Air Force to select competent pilots (Flanagan, 1954) and has been used in many studies investigating competencies (e.g., Fuglsang & Sundbo, 2006). The critical incident technique gives a detailed description of what happened, why it happened and what specific actions were taken to solve the problem. Here CIT has been used to collect, organize and structure data about e-services related competences. According to Flanagan (1954):

> *By critical incident is meant any observable human activity that is sufficiently complete in itself to permit inferences and predictions to be made about the person performing the act. To be critical, an incident must occur in a situation where the purpose or intent of the act seems fairly clear to the observer and where its consequences are sufficiently definite to leave little doubt concerning its effects.* (Flanagan, 1954, p. 327)

Following Flanagan, in this study concrete incidents with a positive or negative impact with respect to e-services adoption and assimilation (the aim of the situation) have been identified. The incidents that were found were then analyzed and grouped according to the competency model developed above and reported in the analysis where appropriate to illustrate the point.

Analysis and Results

Top Management Level

The findings mainly support the model of competencies developed in Figure 1. As suggested by the model, vision was the first competency deemed necessary in order for the company to take into consideration the possibility of adopting e-services and making a proactive effort to understand their value (McGowan et al., 2001). In all the companies interviewed, vision involved an understanding of how a Web based travel booking system could add value to the company and contribute to the company's business and strategy. A critical incident illustrating the importance of vision is as follows (TQ3):

Interviewer. What do you think it is important at your customer?
Interviewee. There I believe vision. I am happy when people (top management) say "It is that way we have to go" and then you have to be realistic. .... You need to know if you are geared, but vision, that I like. When people are visionary and can see possibilities, it is nice. I also visit clients that simply work like we did 100 years ago, and they never go forward...

Value competency mainly consisted of finding out what value the Web-based reservation system could bring to the company. This value consisted mainly of decreases in operational expenses, including travelling expenses. In fact, the main idea of TQ3 Web-based reservation system is that it should create value both for the producer and the consumer as it should reduce costs for both. For TQ3, the Web-based system should decrease personnel costs and allow the company to concentrate on the customers with special needs. For the e-service users it should reduce the company's total travel costs as the company can get a discount when they book their travel directly through Webbuster, bypassing TQ3. This was clearly pointed out by the respondents at TQ3, as well as by employees interviewed in the customer companies. A critical incident reported by an employee at TQ3 was as follows:

Interviewee. So many of our things are developed with the purpose of saving money, where we can save money at our company is one of the things, do something new, and where we can save money for our customers is the other thing.

Once the competencies of vision and value have been developed and the decision to adopt an e-service has been made, senior management started developing a new competency to encourage and enforce assimilation at individual level: the "control competency" (Durkin & McGowan, 2001). This competency manifested itself in the development of a company policy to enforce the use of the e-service at the individual level. This included both the person that is in charge of making the travel reservation and the employees that had to travel. The following critical incident illustrates this point (Company D):

Interviewee: But there is a little back and forth sometimes that they prefer one solution, and then you say "Yes, that is possible, but it is so expensive, but ..... we have a travel policy in the company....

In addition, all the respondents that were also users of the e-service in the small companies said that they had to use the Web-based reservation system due to the wishes of top management to reduce travel costs and the company policy of cutting costs. This was also justified by the fact that employees in the companies interviewed were travelling a lot. A critical incident illustrating this point is as follows (company C):

Interviewer. We talked about policy, why has the company introduced that policy?
Interviewee. Simply to save money. Because at TQ3, there you get 5% saving circa per ticket you reserve on the Internet. And if you have a budget of millions crowns, then it is a lot of money that you can save in one year. And then you have a total online solution, so it saves time both for us and TQ3.

The analysis shows that the top management competency "technical ability" was not important for primary adoption. However the study shows that the technical ability and capacity of the whole corporation was important in order to be able to initiate primary adoption. This is shown by the following critical incident (TQ3):

Interviewee. Yes, it is clear, technical they also have to be, and have the right IT equipment and be updated on the IT-side, that they almost must have to be to be able to consider Webbuster.

After the decision to adopt e-services at top management level had been made and the competencies were developed at management level to ensure adoption and assimilation at the individual level, a set of competences at the individual level were deemed necessary to ensure secondary adoption and to make the e-service become an integral part of the company's routines and culture (assimilation). These are presented in the following section and are based on the model of Figure 1.

Individual Level

Technical Skills. The main competence found relevant within this category was knowledge competency both tacit and explicit. The users of the TQ3 e-service system interviewed in this study were all first time users of such an e-service system, and therefore the acquisition of both explicit and tacit knowledge in learning how to use the e-service system was extremely important. Explicit knowledge was acquired through a formal one day training seminar organized by TQ3 after the system was installed at the customer site. Tacit knowledge was acquired through learning by doing, once the user came back to the company and started to use the e-service system. This is shown by the following critical incident (company D):

Interviewee: You do by trying it, when you for ten times have been sitting and writing your user name, then you become tired of it. Then you ask yourself whether it is possible to make it easier. (Secretary, Company D)

As suggested by Brown (1994), knowledge competency also includes a component relating to the transfer of the skills learned and knowledge acquired in a given context to a new context. For example, in this study this was shown by a transfer of general Internet skills to the specific e-service system in question, the transfer of previous expertise in the travel sector and IT to online travel booking. A critical incident in company B illustrates this:

Interviewer: Are you more used (then your colleagues not willing to use Webbuster) to use this type of technology?
Interviewee: Yes. And it is also that, I believe, that make the difference. I do not believe it is a problem to sit and reserve a flight on the Internet, where others believe that it is difficult....I use also net bank at home and everything, whatever I have to book as ferry tickets or what we do...concert tickets, that I do often on the Internet. All those kind of things...and then I use it also here in all possible contexts (Secretary, Company B).

Interpersonal Skills

The main competencies in this category were communication and empathy. Communication competency manifested itself as the capability to communicate with others. Examples include the ability to understand what other people say and make oneself understood, to negotiate with the travel agency as to what to do with the ticket if the system breaks down during the reservation process, or calling the help desk or the local travel agency for help. A critical incident illustrating this is the following (company D):

Interviewer... I think that they were very good at speaking so that we could understand that.
Interviewee. So they could understand what your problem was and you could understand, what they said?
Resp. Yes, exactly.

Another example of how it might not be easy to communicate is provided by the following critical incident (company F):

Interviewer. ...That means that you have to communicate with them, that have to travel, when you go into the system, and say, what possibilities there are.
Interviewee... the best for me is to make them to stay behind the computer monitor, so they can see what possibilities there are....so I have the communication immediately.

Other examples can be calling the help desk or the local travel agency for help as shown by the following critical incident (Company E):

Interviewee Then I call this number (help desk) and say who I am ..in the process of doing something on Webbuster and I cannot go further..... And I can also contact our own travel agency if I have a problem ...

In our study empathy manifested itself as the capability to understand and take into consideration other colleagues' needs and wants with regard to travel and economic restrictions. In fact, the travel booker found herself in the dilemma of accommodating the travel wishes of the colleagues, for example travelling at specific times and dates, have specific seats, minimize travel time, as well as taking into consideration the company's policy with regard to reducing travel expenses. Sometimes this can be difficult. The following critical incident illustrates the capability to take into consideration colleagues' preferences (company F):

Interviewee. It is very different what people believe it is good. I try to find it out what they prefer. That makes it also easier for me...

Conceptual Skills

In this category two main competencies were found relevant: creativity and judgement.

Creativity manifested itself as the capability to search for new information, to use and understand other sources, the capability to navigate and relate to the IT system, to understand when the information and the knowledge at hand is not enough, and the capability to handle multiple sources at the same time. Clearly this requires a certain degree of reflection, whether conscious or unconscious, about what is known and how to go to get further information or knowledge. An example of looking for more information in the system manual to solve a certain task is given by the following critical incident (company C):

Interviewer: Had you tried, before calling them (help desk), to look at the manual or other sources?
Interviewee. Yes, I had been in the system and looked at what I have printed here (the manual).

The following critical incident (company B) illustrates the capability to look for different solutions:

Interviewee. ...the system could simply not find that out, it comes back and says, that there are some problems with the number, so I try some different ways.

A critical incident illustrating the capability to handle multiple sources simultaneously is as follows (Company C):

Interviewee. So it has also something to do with being able to keep many bolds in the air, certainly....Because you have so many tickets running at the same time. And you have so many arrangements.

Judgement competency manifested itself as the capability to judge different kinds of situations and make a decision. Examples of this is include knowing when to stop trying to solve a problem by oneself and ask others instead. To judge when it is good to use the online system to book a trip and when it is better to do it manually. To judge when it is the system that has made a mistake or when it is the user making a mistake. To judge the validity of the information the user gets from the system. An example of the capability to question the validity of the information the user gets from the system is provided by the following critical incident (company F):

Interviewee. ....Yes, I believe that they make it look nicer then it is (talking about the hotel information that she can find in the system), some hotels do, for sure..they exaggerate.

DISCUSSION

The analysis has shown that for a successful adoption and assimilation of e-services in small and medium size companies it is important to develop appropriate competencies both at top management and individual levels. In this study three competences were deemed important at top management level: vision, value and control. Through vision and value top management can understand and highlight the strategic importance of e-services and take a proactive approach to creating the condition for its primary adoption. Through control, top management can create the conditions for employees to adopt the e-service (secondary adoption), thus reducing the assimilation gap. In fact according to Chatterije et al. (2002) top management can either encourage or discourage individual adoption by mandating rules and policies about the assimilation of a technology. In all the companies interviewed in this study this was achieved by formulating a company policy to reduce travel expenses. This policy consisted of the following: first the Web-based system had to be used for all the trips as this allowed for a discount in respect to the price charged by the travel agency; second the employees had to travel as cheaply as possible; third all the travel reservations had to be undertaken by one person. The company travel policy functions as structures of domination by regulating the individual actions and behaviours. The travel booker draws upon this policy to ensure that his/her assimilation actions in using the Web-based travel system do not violate the institutional rules. Contrary to the model described in Figure 1, this study found that the "technical ability" competence was thought unimportant for the adoption of e-services at managerial level. In the companies interviewed top managers were not necessarily acquainted with e-service technology, neither were they interested in or took a proactive approach to learning it. They were mostly interested in the value that it could create for the company and then they delegated the technical tasks to the individuals that had to use it, while developing rules to enforce such use at individual level. However, the analysis shows that the IT capability of the firm is important for primary adoption (Bharadwaj et al., 2000).

In order to use and assimilate the e-service, a number of individual level competences have been identified in this study. These competences can be technical as, for example, knowledge regarding how to use the e-service system, interpersonal such as the capability to communicate, or conceptual such as judgement and creativity competences.

CONCLUSION AND LIMITATIONS

This study is important because it has shown that both top management and individual level competences are important in e-services adoption and assimilation in small and medium size enterprise. This has been done by developing a competency model and by applying it to a specific e-service system: a Web-based travel reservation system. The empirical findings mainly confirm the validity of the model. Furthermore, this study has identified a number of competences at top management and individual levels that are necessary for the successful adoption of e-services. Competences at top management level are important in order to make the initial decision to adopt the technology and enforce or facilitate its use in the company. Competences at the individual level are important for the successful assimilation of e-services in the company. Some can be acquired through formal training, others have to be acquired by experiential learning.

Finally, this study presents a number of limitations. First of all, the number of interviews is too small to be able to make any generalization regarding competencies in e-services adoption and assimilation in SMEs. Further research could extend this study to a larger sample of companies for the same e-service system or could test the model in other types of e-services. The companies participating in the study were all successful adopters of the Web-based travel system. Further research could focus, for example, on companies that did not succeed in adopting the e-service, to understand major reasons for failure. Finally, there might be other approaches to understanding and conceptualizing competences than the one used in this study. Nevertheless, this study gives some good insights into the competences required at top management and individual levels to adopt and successfully assimilate e-services in small and medium size companies. These results might be of interest to other researchers as well as corporations' managers interested in adopting e-services as a strategic or value adding technology.

REFERENCES

Basselier, G., Reich, B.H., & Benbasat, I. (2001). Information technology competence of business managers: A definition and research model. *Journal of Management Information Systems, 17*(4), 159-182.

Bharadwaj, A.S., Sambamurthy,V., & Zmud, R.W. (2000). IT capabilities: Theoretical perspectives and empirical operalization. In *Proceedings of the 21st International Conference on Information Systems*, Brisbane, Australia, (pp. 378-385).

Brown, R.B. (1994). Reframing the competency debate. *Management Learning, 25*(2), 289-299.

Chaston, I., & Mangles, T. (2002). E-commerce in small UK manufacturing firms: A pilot study on internal competencies. Journal of Marketing Management, *18*(3-4), 341-360.

Chatterjee, D., Grewal, R., & Sambamurthy, V. (2002). Shaping up for e-commerce: Institutional enablers of the organizational assimilation of Web technologies. *MIS Quaterly, 26*(2), 65-90.

Durkin, M.G., & McGowan, P. (2001). "Net Effect"-views from the periphery: Exploring the role and importance of the Internet in marketing activity in entrepreneurial firms. *Irish Marketing Review, 14*, 15-25.

Eikebrokk, T.R., & Olsen, D.H. (2002). Understanding e-business competencies in SMEs. *Seeking success in e-business—multidisciplinary approach* (pp. 575-602). Kluwer Academic Publishers.

Flanagan, J. (1954). The critical incident technique. *Psychological Bulletin, 51*(4), 327-358.

Fuglsang, L. (2007). Critical incident Teknikken i Teknikker i Samfundsvidenskaberne, In L. Fuglsang, P. Hagedorn-Rasmussen, P. Bitch Olsen (Eds.), *Roskilde universitetsforlag (*pp. 260-277).

Fuglsang, L., & Sundbo, J. (2006). Flow and consumers in e-services self-services: New provider-consumer relations. *The Service Industries Journal, 26*(4), 361-379.

Gallivan, M, J. (2001). Organizational adoption and assimilation of complex technological innovations: Development and application of a new framework. *Database for Advances of Information Systems, 32*(3), 51-86.

Grandon, E., & Pearson, J.M. (2003). Strategic value and adoption of electronic commerce: An empirical study of Chilean small and medium businesses. *Journal of Global Information Technology Management, 6*(3).

Haynes, J.L. (1979). A new look at managerial performance: The AMA model of worthy performance. *Management Review, 2-3.*

Lee, D.M.S., & Trauth, E.M. (1995). Critical skills and knowledge requirements of IS professionals: A joint academic/industry investigation. MIS Quarterly, *19*(3), 313-340.

Marcolin, B., Compeau, D.R., Munro, M.C., & Huff, S.L. (2000). Assessing user competence: Conceptualization and measurement. *Information Systems Research, 11*(1), 37-60.

McGowan, P., Durkin, M.G., Allen, L., Dougan, C., & Nixon, S. (2001). Developing competencies in the entrepreneurial small firm for use of the Internet in the management of customer relationships. *Journal of Industrial Training, 25*(2-3-4), 126-136.

Middleton, B., & Long, G. (1990). Marketing skills: Critical issues in marketing education and training. *Journal of Marketing Management, 5*(3), 325-43.

Miles, M.B., & Huberman, A.M. (1994). *Qualitative data analysis* (2nd ed.). Sage Publications.

Mirchandani & Motwani. (2001). Understanding small business electronic commerce adoption: An empirical analysis. *Journal of Computer Information Systems, Spring*, 70-73.

Nordhaug, O. (1998). Competence specificity in organizations. *International Studies of Management and Organizations, 28*(1), 8-29.

OECD. (2000). *OECD small and medium size enterprise outlook.* Retrieved November 15, 2007, from http://www.oecd.org

Sambamurthy,V., & Zmud, R.W. (1994). *IT management competency assessment: A tool for creating business value through IT.* Morristown, NJ: Financial Executives Research Foundation.

Scupola, A. (2003). The adoption of Internet commerce by SMEs in the south of Italy: An environmental, technological and organizational perspective. *Journal of Global Information Technology Management, 6*(1), 51-71.

Scupola, A. (Ed.). (2008). *Cases on managing e-service.* Hershey, PA: IGI Publishing (Forthcoming).

Van der Hejiden, H. (2000). Measuring IT core capabilities for electronic commerce: Results from a confirmatory analysis. In *Proceedings of the 21st International Conference on Information Systems*, Brisbane, Australia, (pp. 142-163).

Wainwright, D., Green G., Mitchell, E., & Yarrow, D. (2005). Towards a framework for benchmarking ICT practice, competence and performance in small firms. *Performance Measurement and Metrics, 6*(1), 39-52.

Yin, R.K. (1989). *Case study design.* Newbury, CA.

Yukl, G.A. (1989). *Leadership in organizations* (2nd ed.). Englewood Cliffs, NJ: Prentice Hall.

Zaltman,G., Duncan, R., & Holbeck, J. (1973). *Innovations and organizations.* New York: John Wiley and Sons.

This work was previously published in Journal of Electronic Commerce in Organizations, Vol. 6, Issue 2, edited byM. Khosrow-Pour, pp. 78-91, copyright 2008 by IGI Publishing (an imprint of IGI Global).

Chapter 14
Adding Value to SMEs in the Courier Industry by Adopting a Web-Based Service Delivery Model

Paul Darbyshire
Victoria University, Australia

ABSTRACT

The aim of this research is to design a framework for a Web system that is intended for linking small and medium transport companies with their customers. The unique aspects of the framework are two-fold. The framework utilizes Web services, which means that it can be applied to existing software and hardware environments. This reduces the need for specialized integration and development, the cost of which becomes a further barrier to SMEs in adding value to customers through existing systems. The framework is additionally designed to link both communities of SMEs and customers in a fledgling digital ecosystem arrangement. Such arrangements offer inherent added value to both types of participants.

INTRODUCTION

One of the great promises of the Internet is that it would help level the playing field somewhat for small to medium businesses by giving them access to market share and allowing them to compete more effectively with larger corporations. We know that technology does make small businesses more competitive and helps in reducing the exclusivity of markets once thought to be beyond their reach. However, the playing field will never be really level and "small businesses are seldom on the cutting edge when it comes to a new technology" (Campbell, 2004). These companies simply cannot afford to provide the value-added services offered by their larger counterparts.

The courier industry is a very good example of this. Small and medium size courier companies generally offer a broad range of services and can provide valuable local service and knowledge to their customers. Utilization of the Web for e-commerce by these companies gives a number of advantages for the logistics trade, such as new market penetration, targeted advertisement of new and existing services, increased quality of services, acceleration of information processing, and customers' feedback improvement. However, these companies are often overlooked in favor of the larger providers that can offer a more expensive but value-added service. Services such as tracking of shipment deliveries, immediate delivery notification, and other information including GPS information is expensive to provide but gives customers value-added services they are willing to pay for. Another barrier to adoption is the high demand placed on resources for a Web system's ability to integrate with existing systems. This in turn increases the cost of qualitative logistics for the Web-system development.

While small and medium businesses may not be on the cutting edge of technology, generally their smaller size gives them agility to more rapidly adopt technology solutions. It is possible for small and medium courier companies to overcome many of these barriers by integrating two relatively new paradigms in ICT (information and communication technology) to develop a new service delivery model in which to do business. The evolution of e-commerce-type systems has been relatively rapid given the short time since the development of the Web. One of the emerging trends evident with Web 2.0 tools is that of communities. In particular, communities of consumers and businesses (Lemphers, 2007) coming together in a digital environment dedicated to a particular service. Additionally, by basing the new model on Web-services technologies, we can decrease the burden on existing systems by allowing transport companies to seamlessly integrate Web services' functionality into their enterprise software applications. By bringing together many smaller to medium transport companies in this way, more value-added services can be implemented through the technologies with the cost spread between more players.

The suggested new service delivery model provides integration between transport companies through Internet technologies such as XML, SOAP, UDDI, WSIL, and Web services. In this way the transport companies can receive more opportunities to attract customers and improve the quality of services and therefore increase their profits. The architecture for the selected business model of the Web-system design is based on thorough requirements analysis and research of contemporary Internet technologies. The core of the system uses J2EE that is cross-platform, multiuser, object-oriented, and scalable; thus, the framework may be applied to various software and hardware environments. The XML Web-services mechanism allows the framework to integrate with external Web systems. It uses UDDI for delivery agents to reach their customers and partners with information about their services. Service description information is stored in WSIL format, so it can be distributed to any location using an XML document. SOAP is used for the XML data exchange mechanism between delivery agents, customers, and external Web systems. It provides users with remote calls to the Web system's functionality.

The aim of this research is to develop a framework for a new service delivery model that assists small and medium transport companies at providing their services through the Internet. The proposed model is for an advanced logistics B2B solution that integrates multiple transport companies and provides them with a unified Web interface to extend their business over the Internet. In the remainder of this article, a brief literature review is given followed by the proposed developed framework for the target model. A costing for the development of the model is then given, followed by the conclusions and opportunities for further research.

BACKGROUND LITERATURE

The transport and courier industry has grown rapidly over the last few years. In 1977 in the United States, 3.2% of the value of all goods shipped were sent by parcel carriers or courier companies. In 1997 this percentage had climbed to 12.3%, with revenue to the courier industry being over US $38 Billion (Morlok, Nitzberg, & Balasubramaniam, 2000). Most of this trade (approximately 90% to 95%), is dominated by the four biggest carriers, with their growth rate during this period approximately 13.21%. This industry represents an important component of the US economy at over 10% of GDP. Figures for other developed countries are not so clear cut, but the World Trade Organization indicates an important industry worldwide with large revenues still dominated by four large players (WTO, 1998).

Recent trends which include the implementation of services such as parcel tracking, delivery notification, and customer tracking via Web systems have become possible through the development e-commerce systems, satellite tracking with Global Positioning Systems (GPS), and advances in customer support and information systems (Roy, 2001). While there are many smaller courier companies that make acceptable revenues it is particularly difficult for them to compete with the larger players in terms of these value-added services which the larger companies have introduced over the past decade. However, recent trends in business models and technologies do suggest a possible solution for the smaller players in this industry. As a result of the Web 2.0 technologies enabling further developments in e-commerce, many companies are taking advantage of network and relationship building software. This comes at a time when emerging social trends of networking seem to be dominating the Web among younger members of the Web community with sites such as Facebook, Friendster, Habbo, MySpace, and so on.

According to Chang and Uden (2007), SMEs will need to embrace the concept of digital ecosystems in order to remain competitive. "A digital ecosystem is a self-organising digital infrastructure aimed at creating a digital environment for networked organisations that supports the cooperation, the knowledge sharing, the development of open and adaptive technologies and evolutionary business models" (European Commission, 2004). Embracing such technologies becomes important for SMEs, particularly when competing with much larger players in a fixed market as is evident in the courier industry. These systems provide further opportunities and economic benefits through networking with other SME's, creating virtual enterprises and crowd sourcing. Indeed integrating with other SME's in such ecosystems will be necessary in the new digital business landscape where community is key (Lemphers, 2007).

In a study involving three on-demand service delivery models, Warkentin and Bajaj (2001) indicated that a key to success in this area would be in developing strong networks of cross-industry partners. They also propose a Web-based business model incorporating e-commerce systems, community networking, and content. There are many advantages to be gained in developing these cross-industry relationships and aligning goals across physical and virtual channels can create synergies and virtual organizations more effectively able to compete with larger market players (Steinfield, Adelaar, & Lai, 2002). Indeed, Chang, West, and Hadzic (2006) describe a digital ecosystem for a collaborative group of logistic enterprises to compete in a global environment. The digital ecosystem described provides many value-added benefits for the various logistics enterprises through the creation of partnerships and sharing of resources. However, the description of the ecosystem does not include the customer. By including the customer into such an ecosystem, we can create a truly horizontal e-marketplace and provide more value-added benefits to the customers.

Overview of Existing Systems

Transportation software is defined at the Web site of Chozam Inc. as software that gives business the ability to effectively manage trucking, barge traffic, and multimodel transport operations in an organized, logical fashion. Utilizing transportation software allows streamlining business while maintaining a customizable system that includes dispatch, sales, rating, payroll, accounting, and total logistics. Transportation software also gives the ability to connect with shippers, vendors and logistics organizations as well as manage vehicle maintenance (Chozam Inc., 2005).

In the Uniteds State and abroad there are a number of various software solutions for logistics business, which primarily can be broken down into software that supports various types of transport operations (order creation, miles calculation), and software that allows for better management of assets as they interact within the supply chain (supply chain management).

TMW Systems is the largest provider of enterprise management software for the trucking industry (Canada IT, 2006) and developer of trucking software for dispatch and transportation solutions (TMW Systems, 2005). TMW Systems offers two operations and dispatch software products that currently manage the transportation operations of over 500 common carriers and private fleets. The Framework for the TMW Systems suite is shown below in Figure 1. TMW Systems' customers gain efficiency by being able to better manage their assets as they interact within the supply chain. While TMW Systems seamlessly integrates and interfaces both software products with complementary software partners that are the best in the industry in their areas of specialization, such integration is not flexible enough due to the software demands placed on customers. The utilization of Web services can help provide a more flexible integration mechanism.

Manhattan Associates' Transportation Management solutions are able to automate, standardize, and optimize critical processes—streamlining, planning, and execution—and substantially reduce procurement time and associated transportation costs. The Transportation Management solutions can be leveraged on demand or as licensed solutions and offer comprehensive functionality covering the entire spectrum of processes related to a global intraenterprise transportation network (Manhattan Associates 2005). Modularity of Manhattan Associates' Transportation Management is

Figure 1. Three layers of the TMWSuite's framework (TMWSuite, 2005)

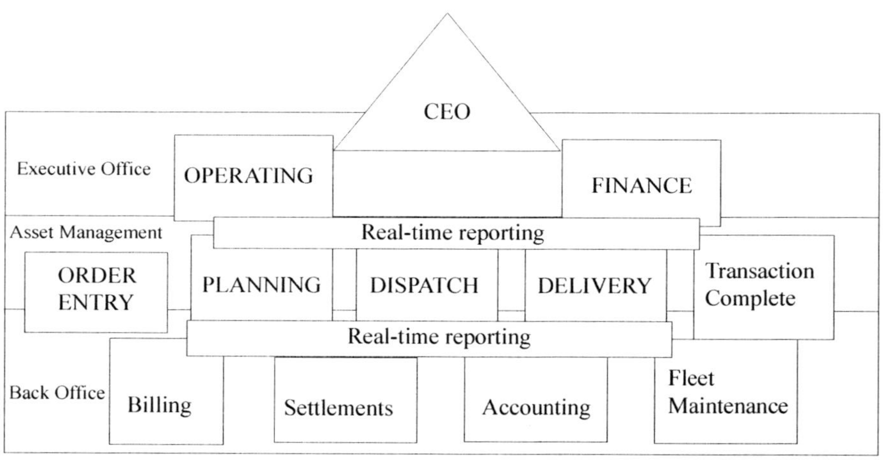

very good architectural decision, but the system is also intended for one big transport company. It cannot be used by a number of small to medium transport companies; also, it does not provide any external XML interfaces.

GT Nexus' unified software applications combine advanced supply chain modeling and event management capabilities with global transportation and trade document management to enable full control over the flow of inventory from order point to final delivery (GT Nexus, 2005). GT Nexus' strategy is to build a comprehensive logistics software solution, invest in the global infrastructure and service provider connections to support the solution, then deliver the whole thing as a service, over the Web, for a fraction of the cost, and none of the risk, of traditional supply chain software. Such strategy makes GT Nexus to be a service provider that is a comprehensive business scheme. The greatest advantage of this scheme is its ability to attract customers without big finance and with less risk for a customer.

While there are many other similar solutions in the marketplace, currently there is no transport software in the market that provides an electronic marketplace for small and medium transport companies. Some existing software systems provide very useful features, but not one of them offers budget e-logistics solution for orders taking and execution. Additionally, none of these provide user-friendly interfaces for customers' shipment tracking, pick-up, and drop-off as in DHL's Small Business Center. The value-added benefits available in these larger solution packages are available as a direct result of the cost of these existing systems. As a result, these systems are not a viable option to smaller SMEs.

Web-Services-Based Logistics

At present, one can choose appropriate tools and technologies from a wide range that is well presented in the printed literature and on the Internet. For a successful design and development of a Web-based communications framework for the transport industry a lot depends on the right choice of proper technology.

Web services provide a layer of abstraction above existing software systems, such as application servers, CORBA, .NET servers, and messaging and packaged applications. Web services work at a level of abstraction similar to the Internet and are capable of bridging any operating system, hardware platform, or programming language, just as the Web is. Unlike existing distributed systems, Web services are adapted to the Web. The default network protocol is HTTP. Most existing distributed computing technologies include the communication protocol as part of their scope. With Web services, the communication protocol is already there, in the far-flung, worldwide Web.

Cerami (2002) defined Web service as "any service that is available over the Internet, uses a standardized XML messaging system, and is not tied to any one operating system or programming language." Cerami names several alternatives for XML messaging, XML Remote Procedure Call (XML-RPC), SOAP, and HTTP GET/POST for pass of XML documents. He also advises that a Web service may have two additional properties: a Web service should be self-describing; a Web service should be discoverable (Cerami, 2002).

New applications become possible when everything is Web-service enabled. Web services are a trend which will facilitate Enterprise Reconstruction through Industry Deconstruction (Bieberstein et al., 2005). Once the world becomes Web-service enabled, all kinds of new business paradigms, discussion groups, interactive forums, and publishing models will emerge to take advantage of this new capability (Newcomer, 2002). However, this will entail a drastic rethink on Web based components for business applications. The Web-service architecture provides an interesting alternative for drastically decoupling presentation form content. For example, a site could consist of nothing but container pages that pass parameters to the real logic via SOAP or XML-RPC. This makes it easy to change presentation and also lets humans and computers "share" a single Web service (Cerami, 2002). However, while remote interfaces make the components distributable, it must be remembered that they have an overhead cost of remote communication, serialization, and network latency.

As Zhang and Chang (2002) wrote, in the last couple of years various on-line shipping tools have been developed for e-commerce application developers. They bring the example of the transportation industry; United Parcel Service of America (UPS) that provides several on-line XML Tools and HTML Tools (UPS On-line E-Commerce Tools, http://www.ec.ups.com/), and Federal Express (FedEx) that provides in-house Web tools (FedEx API, http://www.fedex.com/) for developers in order to facilitate the development of online shipping tools. But they have not seen a common service interface to allow users to easily hook up with existing tools. The developers of client applications usually are forced to construct by hand multiple requests for different back-end servers requiring a great deal of time and effort, although integration software vendors have been addressing this issue for years. Different shipping carriers might require distinctive implementations and could have proprietary platforms and their own constraints.

Sun *ONE Web Services Developer's Guide* (Sun, 2003) describes a Logistics sample application. Since the sample application follows a *session facade* design pattern to handle all business logic with *remote session beans*, these can be exposed as Web services. This adds another dimension of extending the architecture, so that different portions of the application can be accessed in multiple ways, with minimal coding. A well-layered architecture ensures code reuse.

MODEL FOR WEB-SERVICES ARCHITECTURE

The goal of the suggested service delivery model is to provide an Internet workplace for multiple transport companies that allow them to receive and fulfill orders, track shipments, notify customer about shipment delivery, send and receive payments, and obtain technical support in case of problems. The novelty of this model lies in the opportunities for transport companies to work together simultaneously and use the services of each other, through a common XML interface for instant messages between them. Also transport companies can use the services of external delivery agents, such as DHL, FedEx, and so on. The Framework also provides a customer interface for examination of transport companies and the services they provide.

The main user categories of the Web system are

- Delivery Agent: a transport company that uses the Web system's services to extend its business
- External Agent: a large transport company or insurance company that provides an interface for interaction with the Web system and is able to use its services
- Customer: a person or a company that requests delivery service(s)
- System Administrator: a main person responsible for proper functioning of the Web system; he or she must have access to every data and every interface in the system
- Helpdesk Engineer: a person responsible for proper functioning of the Web system; he or she has access to order data and database transactions and is able to solve common problems or give users help

The framework is a B2B model that provides transport companies, or delivery agents (*D*s), and their customers (*C*s) up-to-date information

Figure 2. High-level Web system overview

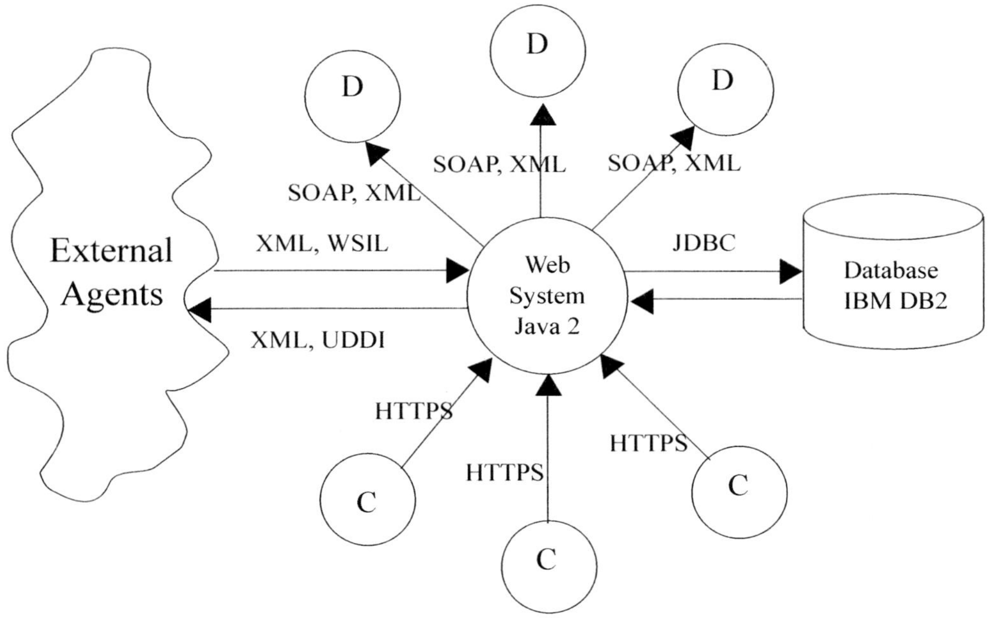

facilities via prevalent Internet technologies. The high-level Web system overview is shown in Figure 2.

Special performance requirements are suggested for the Web system; however, it obviously must be realized that any such system based on Internet delivery has no control over Internet performance. Notwithstanding, desirable qualities are that it must maintain a load of simultaneous work of multiple users and demonstrate high-speed response. The following performance indicators are suggested:

- Number of simultaneously working users—1000
- Account creation time—at most 0.5 second
- Order creation time—at most 1 second
- Sending off payment time—at most 0.5 second
- Sending off message time—at most 0.5 second
- Log-in time—at most, 0.1 second
- Show list of orders time—at most, 1 second
- Report creation time—at most, 3 seconds

The communications framework uses Web services for implementation of distributed computing. Targeted Web services offer productivity gains and cost reductions to logistics businesses that are integrated into the Web system. Web services can provide them with a common tool to interact with customers, contractors, and subcontractors in a seamless way.

Web services are an essential part of the new service delivery model's technical architecture, because they allow spreading the functionality of Web system's business model over enterprise applications for transport companies. The Web system's context involves the five types of users outlined previously, two of which (delivery agents and external agents) access Web system's functionality via Web services. This approach allows seamless integration of their enterprise applications with the projected Web system at the service level of the software application. The Web services mechanism includes the framework for a Web services data flow mechanism, business entities implementation and common alliance interface to allow users and businesses to easily hook up with existing tools.

Business Functionality

The Business model of the proposed framework for integration between multiple carrier services includes a description of the main actors of the system (omitted here for brevity), business logic, and use of technologies that help to implement the solution. At a high level, system functionality is represented by a set of modules and interfaces that provide the following features:

From the transport company (delivery agent) perspective:

- Enable delivery agents to place and edit their Web profile describing what services they offer, what locations, what price, and other specific conditions they have.
- Give delivery agents complete information about their orders, at what stage they are, in what location, and estimated time of delivery.
- Notify delivery agents about shipments delivery.
- Provide delivery agents with a variety of daily, weekly and monthly reports about their business transactions carried out via the Web system.
- Provide delivery agents with an interface to the services of other agents for the purpose of subcontracting work.
- Provide delivery agents with an interface to insurance companies for the purpose of shipment insurance.
- Provide delivery agents with an interface for solving technical problems.

From the customer perspective:

- Give customers complete information about the transport companies, services they offer, and any specific conditions related to their services.
- Help customers to make their choice of most efficient transport vendor, guided by given conditions.
- Allow customers to make order(s) using selected transport companies.
- Accept various kinds of payment.
- Give customers complete information about the state of their shipments.
- Notify customers about delivery of their shipments.
- Provide customers with an interface for solving technical problems.

From the Web-service perspective:

- Provide full information about a customer's order to external delivery agents.
- Provide information exchange between delivery agents.
- Secure interaction between delivery agents.
- Provide necessary information about subject of insurance to insurance company.
- Provide interaction between insurance company, delivery agents, and customers.
- Sends notification to indicated system administrator.

Application Architectural Design

The framework is developed around a layered architecture design implementing the model-view-controller (MVC) design pattern. The MVC model design pattern maps the traditional (input, processing, output) roles to controller, model, view components which exist in a relationship designed to decouple data access, business logic, and data presentation and user interaction. Thus, the design is based upon abstracting the application into distinct tiers:

Table 1. Layered architecture

Layer	Responsibility	Tool/Technology
Browser	Client tool to access Web-service delivery system.	Internet Explorer, Netscape Navigator Mozilla, Opera, and Firefox
Presentation Layer	Allows users to send their requests to the Web system and presents their data in the appropriate format. Provides standard navigation elements.	HTML, JSP
Processing Layer	Separates presentation logic from presentation view, security, user-input validation, Charest filter, and redirection.	IBM Websphere, Application Server, Servlets
Service Layer	Provides communication with the external system & sends and receives XML messages.	SOAP, UDDI, WSIL, Web services
Business Layer	Responsible for implementing business logic.	EJB
Data Access Layer	Provides access to the database, implements connection pool, data serialization, constructs queries to retrieve data from the database.	JDBC, JNDI
Database Layer	Provides mechanism for storing all data used by the Web system in the form of tables, indexes, and views. Provides mechanism for data protection from unauthorized access, backup/recovery, and database integrity.	IBM DB2 Universal Database

- The presentation layer
- The Web layer
- The service layer

Presentation logic is separated from the business logic in such a way that Servlets are used for control (controller), JSP pages for presentation (view), and EJBs are used to manage business components (model). A Servlet is a small, Java-based object that receives a request and generates a response based on that request. The layered architecture is presented in Table 1 and shows the application's logical layers. The name of each layer is shown in the first column, the second column describes what it is responsible for, and the third column lists tool(s) or technology that can be utilized at this level.

Figure 3 shows a pictorial representation of the model's layered architectural elements as described in Table 1. It is the Web services in the service layer that act as the interface between the various elements from the delivery and external agents.

The core of the framework is the service layer that provides the Web-services stack, which includes data representing computing commands, application programming interfaces (APIs), security levels, and data to be processed and transported over standard Internet transport mechanisms (HTTP, FTP). In other words, the standardized instruction sets of Web-services XML files hold the information necessary to invoke the application commands. Web services also represents a global standard for APIs accessible over the Internet. An API is an interface for letting a program communicate with another program. The following base protocols formed the initial specification for Web services:

- Simple Object Access Protocol (SOAP)—defines the runtime message that contains the service request and response. SOAP is independent of any particular transport and implementation technology.

Figure 3. Layered architecture depiction

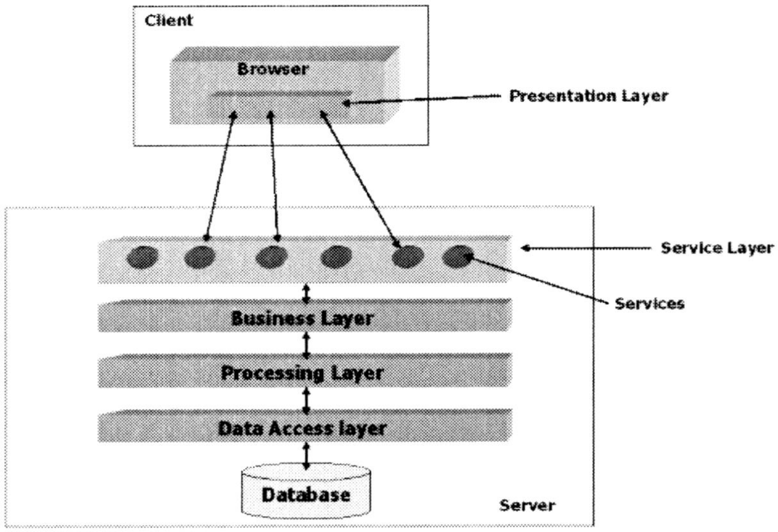

Figure 4. Web-services base protocols application

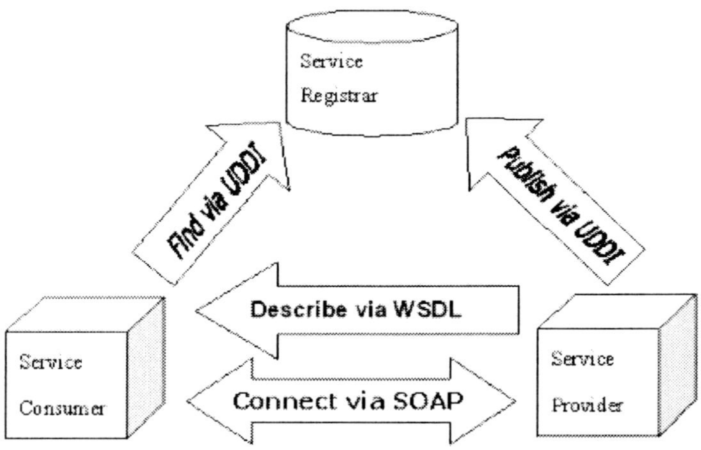

- Web Services Description Language (WSDL)—describes a Web Service and the SOAP Message. It provides a programmatic way to describe what a service does, paving the way for automation.
- Universal Discovery, Description, Integration (UDDI)—UDDI is a cross industry initiative to create a standard for service discovery together with a registry facility that facilitates the publishing and discovery processes.

The Application of these protocols for Web services is shown in Figure 4.

Data-Flow Mechanism

The data-flow mechanism is used by a basic architecture of Enterprise Java Bean (EJB)-based client/server Web system's application. The client side of an EJB architecture contains the EJB interfaces needed for invoking business-specific methods on an EJB as well as for managing handles to server-side objects, created by in-house developers for the transport companies. The server side of EJB architecture contains the instances of the actual EJB component implementation, as well as the container code that maps calls to and from clients and EJBs after appropriate service-management infrastructure logic has been executed. The data-flow mechanism of the framework is shown in the Figure 5.

The main components of the data-flow mechanism interact as follows:

- An EJB client application utilizes the Java Naming and Directory Interface (JNDI) to look up references to home interfaces and call home and remote Web-service interfaces to utilize all EJB-based functionality from the EJB implementation.
- The Web-services home interfaces provide operations for clients to create, remove, and find handles to Web-services remote interface objects. Underlying stubs marshal home interface requests and unmarshal home interface responses for the client.
- The Web-services remote interfaces provide the business-specific client interface methods defined for a particular EJB. Underlying stubs marshal remote interface requests and unmarshal remote interface responses for the client.

The EJB implementations are the actual EJB application components implemented by

Figure 5. The data-flow mechanism of the Web system

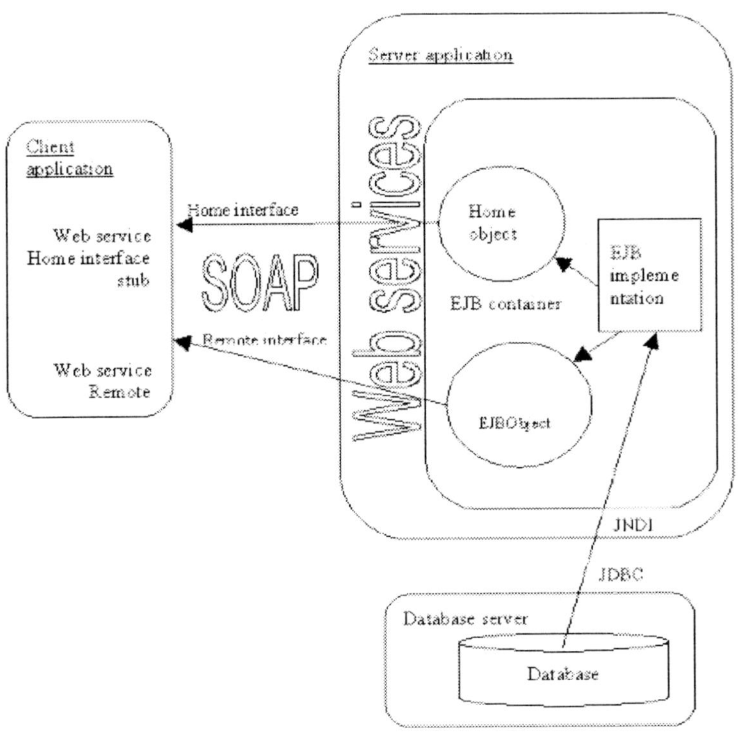

developers to provide any application-specific business method invocation, creation, removal, finding, activation, deactivation, database storage, and database loading logic. It uses the Java Naming and Directory Interface (JNDI) to get Java Database Connectivity (JDBC) connection from the connection pool and utilizes it for data retrieval. The EJB container manages the distributed communication skeletons used to marshal and unmarshal data sent to and from the client. Containers also may store EJB implementation instances in a pool and delegates to perform any service-management operations related to a particular EJB, before calls are delegated to the EJB implementation instance.

The database model, while not directly relevant to the article, is included in Appendix 1 for completeness.

WORKFLOW

To access system functionality through the Web services, a delivery agent (DA) must have implementation of the Web-services interface. So a DA's software should have a module that provides interaction between the DA's logistics software and projected Web-system's framework. The transport companies' software developers may easily create this module, because all interface methods must be described in the documentation for Web services.

The implementation module performs authorization and authentication for the DA, then the DA can select one of the following options: view own, new and operating orders, update order status, message exchange, work with payments, work with reports, use services of external agents, or update price list. The work-flow diagram for DA's activities through the Web service(s) is

Figure 6. Delivery agents complete work-flow model

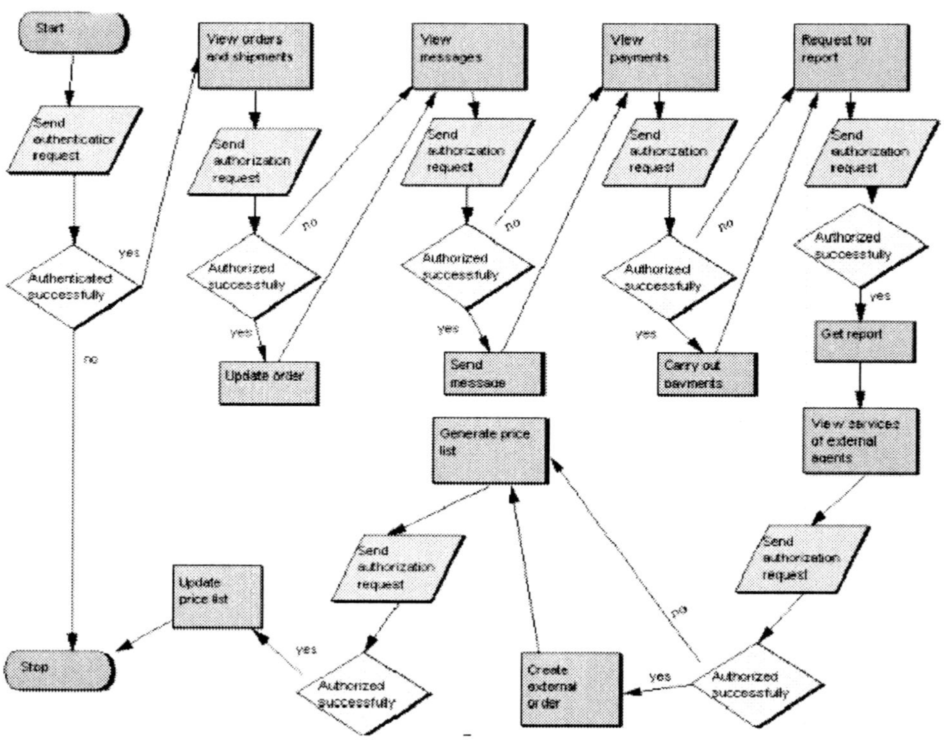

Figure 7. Work-flow diagram for delivery agent's activities through HTTPS protocol

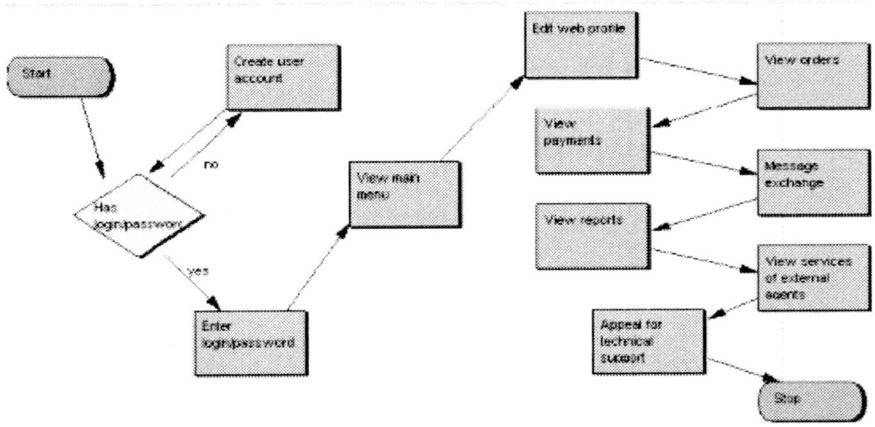

Figure 8. Work-flow diagram for customer's activities

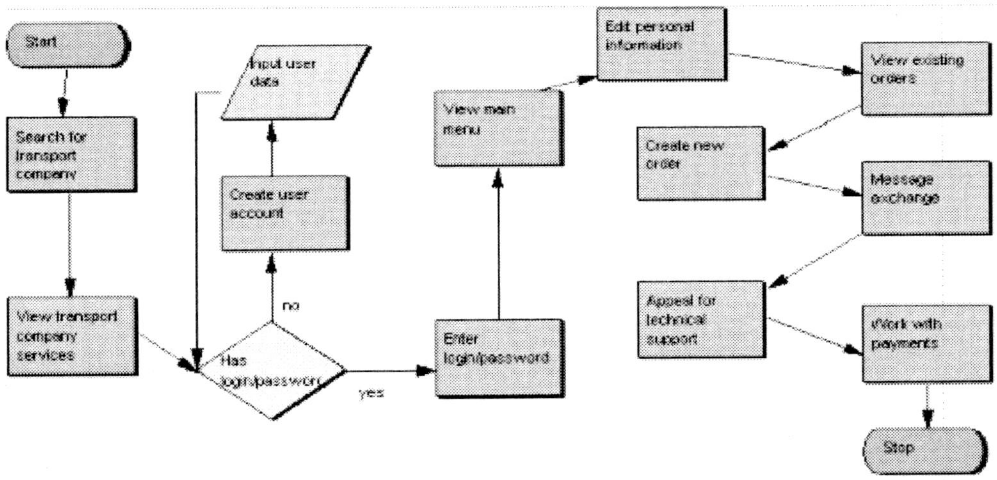

Figure 9. Work-flow diagram for external agent's activities

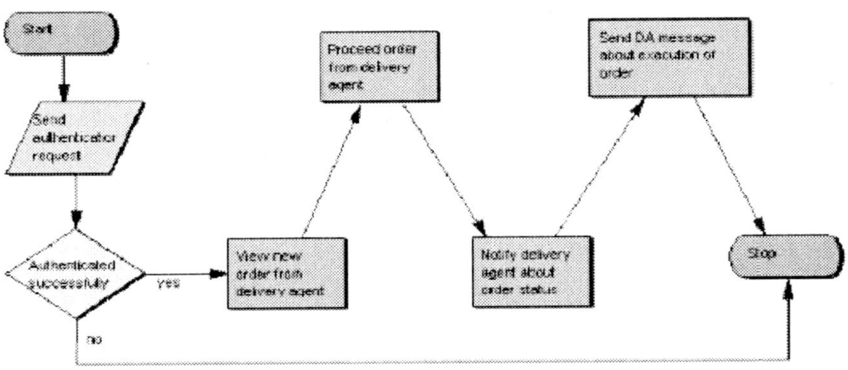

shown in Figure 6. The Web system's interface through the HTTPS protocol is used mainly for DA's registration and account configuration. However, a DA can also view some data of the Web system to facilitate fault finding. To start working via the HTTPS interface, a DA must be logged in to the system. The DA then gets to the main menu, where they select one of the following options: edit Web profile, view orders, view payments, message exchange, and view reports, view services of external agents, or appeal for technical support. The work-flow diagram for the DA's HTTPS activities is shown in Figure 7.

A customer can search for a delivery company and can view courier services without logging into the Web system. For other activities, a customer must register him- or herself in the Web system. To access personal functionality, a customer must be logged in to the system, to gain access to the main menu, where he or she can select one of the following options: edit personal information, create new orders, view existing orders, work with payments, message exchange, or appeal for technical support. The work-flow diagram for customer's activities is shown in Figure 8.

To gain access to the Web-system's functionality, an external agent (EA) must first be authenticated. The EA performs all activities from its own interface, so it uses only the functionality of the Web system, not the user interface. An EA can use one of the following options: get new order from delivery agent, precede order from delivery agent, notify delivery agent about order status, or send delivery agent message about execution of order. The work-flow diagram for the EA's activities is shown in Figure 9

FRAMEWORK DEVELOPMENT COSTING

Costs of the Web-system framework development project consist of

- Human resource costs
- Software costs
- Hardware costs
- Other costs

Human Resource Costs

Working out costs, especially human resources costs, can have very differing results from one location to the next. In the following tables, all values are in US dollars, based on a generic review of the monster.com job site. The results are shown in Table 2.

Software Costs

The software needed to be purchased for the Web-system framework development was considered. The staff will work on PCs using Windows XP Home Edition and Microsoft Office. Also, each project participant needs special software:

- Project manager and system architect use QSEE-SuperLite for modeling and analysis of system's architecture. QSEE-SuperLite is the superset of traditional CASE tools that supports modeling of computer systems, and it is freeware. Also, the project manager requires Microsoft Project for project planning.
- Software developers should use Eclipse IDE for code development. Eclipse is an open-source community whose projects are focused on providing an extensible development platform and application frameworks for building software. Eclipse provides extensible tools and frameworks that span the software development life cycle, including support for modeling, language development environments for Java, C/C++ and others, testing and performance, business intelligence, rich client applications, and embedded development. It is freeware.

Table 2. Human resources costs

Title	Skill Set	Salary (Month)	Period	Qty	Cost
Project Manager	Understands needs, identifies root causes of problems, and develops and implements creative and pragmatic solutions. Ability to create detailed functional and technical design specifications for software development efforts. Experience developing software tests for functional, stress, and regression testing that include manual, ad hoc, and automated testing elements.	$8,000	18 weeks	1	$32,000
System Architect	Software architecture leadership experience, including designing complex, multicomponent software systems. Experience establishing an enterprise framework for all phases of software development: requirements, design, development, testing, and delivery.	$8,000	18 weeks	1	$32,000
Software Developer	Familiarity with all aspects of the development life cycle. Minimum of 3 years software development experience. Working experience with SQL, XML, COM, MFC, Wi-Fi, USB, Windows XP. Strong communication and interpersonal skills. High-level problem-solving skills. Strong mathematical and mechanical background preferred. Ability to handle multiple tasks.	$5,000	9 weeks	4	$40,000
Software Consultant	Must be able to work independently and with a team. Experience in performing mass updates to mapping systems and maintaining high Quality Control standards. Able to anticipate users' needs and develop into documented solutions. Must have good interpersonal skills to work with customer.	$5,000	1 month	2	$10,000
Hardware Consultant	Cost estimations of hardware systems. Analysis of cost research data. Formulation of Cost Research. Modeling and simulation.	$4,000	18 weeks	1	$16,000
Test Engineer	Possesses general knowledge of QA systems and methodologies, use of debugging tools, knowledge of manual tests tools. Ability to write and implement test plans. Creates thorough technical/functional test plans/scripts and performs QA testing of end-user application development.	$4,500	9 weeks	2	$18,000
Support Engineer	Strong communication skills. Strong computer hardware and software skills. Ability to find bottlenecks where problems can occur. Ability to fix common problems and to find a way how to fix complex problems. Strong customer service skills.	$4,000	18 weeks	1	$16,000
Total				12	$164,000

Table 3. Software usage

Project participant	Qty	Software
Project Manager	1	Windows XP Home Edition, QSEE-SuperLite, Microsoft Office, Microsoft Project
System Architect	1	Windows XP Home Edition, Microsoft Office, QSEE-SuperLite, Eclipse IDE, JUnit
Software Developer	4	Windows XP Home Edition, Microsoft Office, WebSphere Application Server for developers, Eclipse IDE, JUnit
Software Consultant	2	Windows XP Home Edition, Microsoft Office
Hardware Consultant	1	Windows XP Home Edition, Microsoft Office
Test Engineer	2	Windows XP Home Edition, Microsoft Office, JUnit
Support Engineer	1	Windows XP Home Edition, Microsoft Office

- Test engineers and software developers use JUnit for testing. JUnit is a regression-testing framework written by Erich Gamma and Kent Beck. JUnit is open-source software, released under the Common Public License Version 1.0 and hosted on SourceForge.

The estimates of software used by each project participant are given in Table 3.

The Web platform and database software to be used will be IBM DB2 Universal Database and WebSphere Application Server. DB2 Universal Database is the first multimedia, Web-ready relational database management system that is strong enough to meet the demands of large corporations and flexible enough to serve small- and medium-sized businesses. It costs $475. WebSphere Application Server is the core Web-services J2EE 1.3-certified application server, optimized for ease of administration in a single-server deployment environment. It exists in two modifications; the server version costs $4,000 and the developers' version costs $750. The security software to be used will be Zone Alarm Pro 5.5; it costs $55 and has a free trial.

Therefore, calculations of software costs are shown in Table 4.

Hardware Costs

Most staff will work at desktop PCs, but the project manager and software consultants need laptops. Also, hardware for the Web server is needed: server, network hub, and firewall. Thus, calculations of hardware costs are given in Table 5.

Other Costs

Other costs of Web system development are Internet traffic; tea, coffee and meals for staff; and various unforeseen costs. It is estimated that other costs don't exceed $1,000. So, all costs of Web system development are presented in Table 6.

ADDING VALUE

The previous sections have concentrated on the technical aspects of the framework, including the database design, the work flow, and costing. In

Table 4. Calculation of software costs

Title	Price	Quantity	Cost
Windows XP Home Edition	$50	12	$450
Microsoft Office	$350	12	$4,200
QSEE-SuperLite	freeware	2	$0
Microsoft Project	$510	1	$510
WebSphere Application Server for Developers	$750	4	$3,000
Eclipse IDE	freeware	4	$0
JUnit	freeware	7	$0
DB2 Universal Database	$475	1	$475
WebSphere Application Server	$4,000	1	$4,000
Zone Alarm Pro 5.5	$50	1	$50
Total			$12,685

Table 5. Calculation of hardware costs

Title	Price	Quantity	Cost
Desktop PC	$850	9	$7,650
Laptop	$1,500	3	$4,500
Database server	$5,000	1	$5,000
Web server	$5,000	1	$5,000
Total			$22,350

Table 6. All costs of Web system development

Title	Sum
Human resources costs	$164,000
Software costs	$12,635
Hardware costs	$22,350
Other costs	$1,000
Total	$199,985

this section we explore the added value benefits which result as a direct consequence from the architecture itself, and those which are made possible (or more feasible) from the architectural design. Of course, it must be recalled that the latter benefits, (those made more feasible by the architecture) are achievable without the architecture developed in this research. This can be seen in value-added benefits within existing systems discussed previously. However, such systems are beyond the reach of SMEs and the raison d'etre of the architecture developed in this research is to facilitate added-value benefits for this group of entities.

Added Value From Digital Ecosystem

The structure of the architecture provides a basic digital ecosystem within the transport industry domain. Many of the added-value benefits will flow from this architecture. Previous research has identified a *Virtuous Circle,* expected to arise from the use of digital ecosystems (Giorgetti & Louarn, 2004). This is highlighted in Figure 10, where the digital ecosystem makes viable new organizational and business models, which in turn encourage and require cooperation, which improves competitiveness and efficiencies, which leads to development that feeds back into the digital ecosystem.... This becomes the *virtuous circle.*

The architecture developed within this research is somewhat different as it additionally incorporates customers into the digital ecosystem

Figure 10. Virtuous circle (Giorgetti & Louarn, 2004)

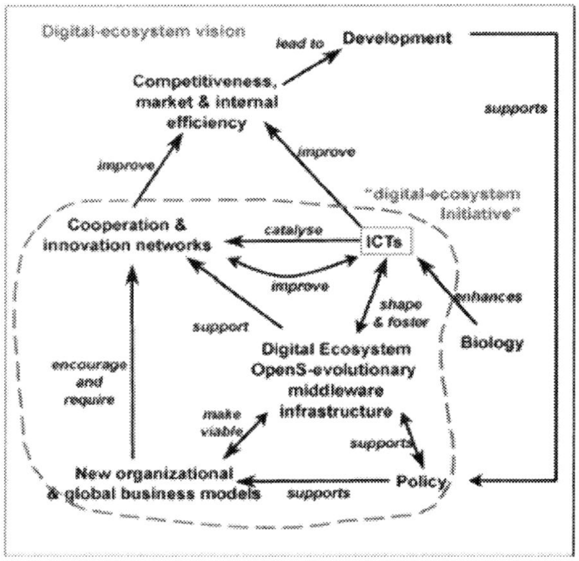

as well as the SMEs. When we include the customer into the ecosystem, we are creating a more horizontal e-marketplace which should help in providing the customer with even further value-added benefits. Similar to the benefits described by Giorgetti and Louarn (2004), the architecture is designed to foster cooperation between the SMEs within the ecosystem, giving rise to synergies. The result of the synergies then flow to the customers as added value within the ecosystem. The flow of added value within the digital ecosystem is depicted in Figure 11.

Before discussing specific added value for the different entities, it is worth noting that one of the main aims of the architecture is to group many SMEs within the digital ecosystem. This allows the cost of development to be amortized over many SMEs. When we can amortize the expense amongst many players, it then becomes viable (from a group basis) to develop further technologies which in turn can add value to both SMEs and customer. Some of these technologies such as parcel tracking, delivery notification and GPS information were initially discussed, but only as technologies viable for the large companies to implement. Alone, such technologies are beyond the reach of SMEs, but when participating in a digital ecosystem with many other others, their introduction now becomes viable. It was not in the intent of this research to build these technologies, but rather to develop and build the framework from which their development becomes viable for the stakeholders.

We can view the value-added benefits from two basic perspectives: that of the SME; that of the customer.

Added Value for SMEs

From Figure 11, we can see that the added value expected from the digital ecosystem arises from the cooperation between the SME and the synergies that then develop. Apart from the amortization of costs offsetting the cost of development for further added-value components discussed above, the cooperation should result in useful partnerships benefiting both SME and customers. SMEs can form alliances and utilize each others logistic chains and operation areas to provide a more holistic service to customers. The added

Figure 11. Conceptual view of added-value flow within digital ecosystem

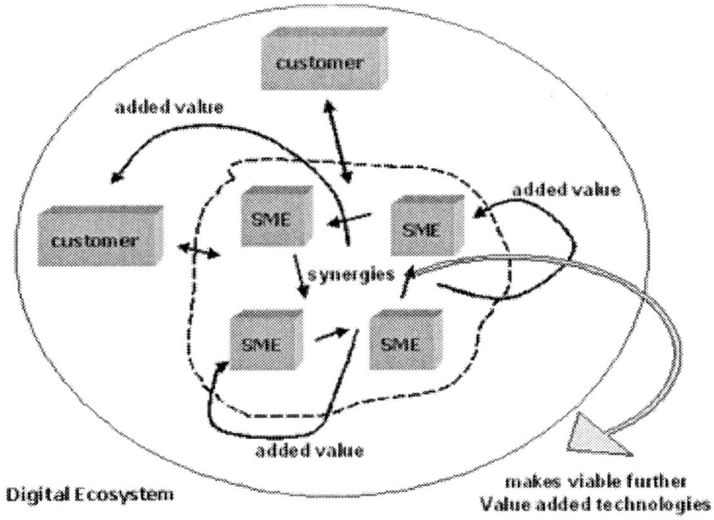

value to SMEs is the ability to retain customers by being able to supply services even though those services might not be supplied by the SME directly. Each of the SMEs will have the ability to *farm out* services to participating partner SMEs when customers require services not directly supplied by the SME or when the SME is unable to supply the service due other reasons. The customers still interface with the initial point of contact, giving them the possibility of still retaining the customer for future transactions.

The partnerships formed will additionally support new organization and business models, helping the SMEs to compete more effectively in an increasingly global system. This can be accomplished due to architecture of the ecosystem allowing the SMEs to interface to external larger companies. These larger companies can act as a global extension to the smaller SME, again allowing the SME to remain the virtual carrier to the customer.

Added Value for Customers

Customers benefit from added-value services in a number of ways. The natural competition resulting from the participating cooperating SMEs in the digital ecosystem will encourage transparent competitive pricing for the customer. Additionally, the mirrored benefits to the customer from those of the SMEs will result in a wider range and availability of services. This should flow on from the SMEs utilizing the logistic chains and services of each other. This will also provide a one-stop point of access for many customers, far more desirable then sourcing different services from multiple vendors. The ability of SMEs to link to more global carriers also provides access to global based services through their SMEs.

Further value-added services such as parcel tracking, delivery notification etc, could possibly be made available to customers through the prospect of these services now being viable to implement. However, such services would not directly be available through the adoption of the initial basic ecosystem by the SMEs, but rather as a consequence of further development due to viability. Nevertheless, such services would be added-value services as a consequence of the architecture, not normally available through SMEs operating alone.

CONCLUSION

The infrastructure described within this project consists of many stakeholders (delivery agents, drivers, companies, and consumers), customers, information technology, corporate security, and application owners to resolve complex technical and process issues. Problems needed to be resolved by the new architecture include those of scalability, extensibility, security, management, interoperability, provisioning, and administration. This project demonstrates an infrastructure initiative that can be a catalyst to rethink the development process and create a more efficient organization domain to meet the demands of agile business strategy, delivering a secure, scalable, reliable, and extensible Web-services environment.

The project addresses organizational and project challenges by understanding the technology, issues, and dynamics of evaluating, selecting, and deploying new Web-service infrastructure components; overcoming the challenges of deploying the new technologies, products, and best practices used to integrate business processes across the disparate application platforms and organizational boundaries; and exploring the evaluation criteria used to select Web-services infrastructure components for the chosen enterprise.

This project provides a framework to simplify the integration process between multiple-carrier services and allows improved transport services for the smaller and medium-sized companies using Web services to automate dispatching and real-time customer services. The designed architecture can be adopted as a new service-delivery model that incorporates implementation of user and functionality requirements, business processes, design patterns, and making use of existing technologies to define an accomplished B2B e-logistics solution.

The decomposition of the business domain depicts the design strategy for service-oriented architecture and method, defining the roles for Transport Company's (delivery agent) clients, Web services, administrators, external agents, and helpdesk engineers. These are processes of the services-oriented modeling architecture that consists of the three general steps: identification, specification and realization of services, and components and flows

This Web-services model allows any user access through the use of XML tags for information exchange using the other three standards: SOAP, UDDI, and WSDL. The technical architecture is implemented using XML as a way of designing and documenting the architecture of services, showing their relationships, dependencies, and so on, and is also a structure by which the right delivery services, at the necessary application levels, is enforced.

It allows both consumer and provider not to be tied in to a particular service set and truly minimizes the impact of change to switch service providers. Thus, the projected Web system allows transport companies to seamlessly integrate Web-services' functionality into their enterprise software applications, such as supply chain management, enterprise resource planning, and others.

The total projected costs of project implementation can be questioned as to the affordability for a small- to medium-sized courier company. However, as the project is designed to implement a small digital ecosystem facilitating the development of cross-industry relationships, these initial costs can be amortized across many companies, thus reducing the overall cost per company and making affordable and available value-added services usually associated with larger companies.

Future Research

Web services have an overwhelming need for increased privacy and protection in the industry and research community. The consumer and the relevant transport companies require privacy and protection when using Web-services. The W3C published a document, *Web Services Architecture,* which describes privacy requirements for future research along with other relevant research areas into XML, Platform for Privacy Preferences Project, for future research of Web-services based applications.

Selective components of the architecture will most likely be replaced or enhanced in differing ways. Research into different techniques to lower long-term costs of management and a higher long-term value for all players in the enterprise is required. To be consistent with open standards for creating, describing, and accessing data, a commitment to new and improved standards will be adhered to, especially as the Web-services technology architecture is still young. Many technological advances by major players have still not created a clear or mature standard.

The main conception of the architecture is to enable the transport industry to interchange specialist services within their domain, offering an improved level of services to clients and sub-contracted companies. As well as technological advances, a consideration of social and economical changes has to be acknowledged when deciding on future project directions.

REFERENCES

Bieberstein, N., et al. (2005). *Service-oriented architecture compass: Business value, planning, and enterprise roadmap.* Englewood Cliffs, NJ: Prentice Hall.

Campbell, A. (2004). *Technology levels the playing field.* Retrieved January 2, 2008, from http://trendtracker.blogspot.com/2004/04/trend-technology-levels-playing-field.html

Canada IT. (2006). *TMW systems acquires Maddocks Systems Inc.* Retrieved September 2006 from http://www.canadait.com/cfm/index.cfm?It=106&Id=24394&Se=2&Lo=2

Cerami, E. (2002). *Web services essentials.* New York: O'Reilly.

Chang, E., West, M., & Hadzic, M., (2006, August). A digital ecosystem for extended logistics enterprises. In *Proceedings of the 11th International Workshop on Telework.* Fredericton, Canada:

Chang, V., & Uden, L. (2007, November). The three e's: E-learning, e-knowledge and e-ecosystems for SMEs. In Proceedings of the Eighth International We-B (Working for E-Business) Conference. Melbourne Australia:

Chozam Inc. (2005). *Technology information center.* Retrieved August 30, 2005, from http://www.chozamtech.com/software/industry/transportation/transportation_software.html

European Commission. (2004). *Digital business ecosystems.* Retrieved April 12, 2008, from http://www.digital-ecosystems.org/

Giorgetti, M., & Louarn, M. L. (2004). *Technologies for digital ecosystems—Supporting regional growth and innovation.* Retrieved February, 2007, from http://www.ve-forum.org/apps/pub.asp?Q=1270

GT Nexus. (2005). GT Nexus' Web site. Retrieved August 30, 2005, from http://www.gtnexus.com/en/delivery_model/

Lemphers, D. (2007, November). *Communitize–Doing business with ourselves!* Keynote address at the Eighth International We-B (Working for E-Business) Conference, Melbourne Australia.

Morlok, E., Nitzberg, B., & Balasubramaniam, K. (2000). *The parcel service industry in the US: Its Size and role in commerce* (Tech. Rep., University of Pennsylvania , School of Engineering and Applied Science, Systems Engineering Department). Available online from http://www.pti.psu.edu/mautc/docs/MA-III-0007.pdf

Newcomer, E. (2002). *Understanding Web services: XML, WSDL, SOAP, and UDDI.* Boston: Addison-Wesley Professional.

Nickerson, J., Hamilton, B., & Wada, T. (2001). Market position, resource profile, and governance: Linking Porter and Williamson in the context of international courier and small package services in Japan. *Strategic Management Journal, 22*(3), 251-273. Retrieved April 12, 2008, from http://cniss.wustl.edu/summerschool/nickersonpaper.pdf

Roy, J. (2001). Recent trends in logistics and the need for real-time decision tools in the trucking industry. In *Proceedings of the 34th Annual Hawaii International Conference* (p. 9). Retrieved April 12, 2008, from http://www.hicss.hawaii.edu/HICSS_34/PDFs/DTIST02.pdf

Steinfield, C., Adelaar, T., & Lai, Y. (2002). Integrating brick and mortar locations with e-commerce: Understanding synergy opportunities. In *Proceedings of the 35th Annual Hawaii International Conference on System Sciences (HICSS'02)* (Vol. 8, p. 216.2).

Sun. (2003). *Logistics sample application: Design.* Retrieved July 28, 2005, from http://docs.sun.com/source/817-1243-10/dvlpdsgn.html

TMWSuite Systems Inc. (2005). TMW Systems, Inc. Retrieved August 30, 2005, from http://www.tmwsystems.com/fileUploads/contentManager-Documents/TMWSuiteBrochureFINALNOPics%5B0%5D.pdf

Warkentin, M.,& Bajaj, A. (2001). The On-demand delivery services model for e-commerce. In A. Gangopadhay (Ed.), *Managing business with electronic commerce: Issues and trends.* Hershey, PA: Idea Group.

WTO. (1998, June 12). Postal *and courier services* (S/C/W/39). Retrieved April 12, 2008, from http://www.wto.org/english/tratop_e/serv_e/w39.doc

Zhang, L-J., & Chang, H. (2002*). E-logistics processes integration framework.* Retrieved July 28, 2005, from http://www.Webservicesarchitect.com/content/articles/zhang01print.asp

APPENDIX 1: DATABASE MODEL

The database model can be used as a design plan by the database developer(s) to implement the data model in a specified database management system. The Web system uses the following main tables:

- Delivery Agent
- Courier Service
- Location
- Customer
- Order
- Shipment
- Payment
- Document
- Price List
- History

These tables are organized as follows:

- Delivery Agent (DA) has a number of Courier Services, which it provides. One DA may be linked to many Courier Services. Also, Delivery Agent has a number of Locations, in which it provides its services. One DA may be linked to many Locations. Delivery Agent has a number of Orders, and Customers are linked to that Orders. Customers are not linked to the DA directly. One DA may be linked to many Orders. Delivery Agent is not linked to the Shipment and Payment directly. There may be a number of Documents that are linked to the DA. So, one DA may be linked to many Documents.
- Courier Service may be linked to Delivery Agent, but Courier Service may exist even if it is not linked to any Delivery Agent. Courier Service may be linked to a Location, but it is unnecessary. If a Courier Service is linked to a Location, it may also be linked to another Location, where it is also provided. Courier Service is linked to an Order, one Courier Service may be linked to many Orders. Courier Service is not linked to Customer, Shipment and Payment directly; they are linked through an Order. Courier Service absolutely is not linked to Document. Courier Service may be linked to Price List (Price List's structure will be discussed later).
- There is a reference of Locations in the Web system. Orders and Delivery Agents can refer to Locations.
- Customer has a number of Orders that he or she has made. One Customer may be linked to many Orders, but one Order must be linked to one and only one Customer. Customer is linked to Delivery Agents through Orders. Customer is linked to Payment directly. One Customer may be linked to many Payments, but one Payment must be linked to one and only one Customer. There may be a number of Documents that are linked to the Customer. So, one Customer may be linked to many Documents.
- Order is the most important table in the system. Order refers to one Delivery Agent, one Courier Service, two Locations, one Customer, one Shipment, and many Payments. Delivery Agent, Locations, and Customer may exist without Order. ER diagram for Order is shown in Figure 12. Delivery Agent and Payments are optional for Order.

- Shipment is a subject of Order. For reasons of database normalization, Shipment is separated from Order; one Order may include some Shipments. Every Shipment must be linked to a Customer.
- Payment is linked to an Order. For reasons of database normalization, Shipment is separated from Order; one Order may include some Shipments. Every Shipment must be linked to a Customer. Customer has a number of Payments that he or she has made. So one Customer may be linked to many Payments, but one Payment must be linked to one and only one Customer.
- Document is linked to Delivery Agent, Customer, Order, or Shipment. One Document may be linked to only one of the following: Delivery Agent, Customer, Order or Shipment, but Delivery Agents, Customers, Orders and Shipments may be linked to many Documents.
- Price List is linked to Delivery Agent. Every Delivery Agent has one Price List. Every Price List has a number of Courier Services. One Courier Service may be linked to many Price Lists.
- History is linked to Delivery Agents, Courier Services, Locations, Customers, Shipments, and Payments. Every History record refers to one Delivery Agent and one Customer necessarily. History can have links to Courier Service, Location, Order, Shipment, and Payment but they are optional.
- Authorization is a specific table. Authorization is linked to Delivery Agents and Customers. Logins and Passwords are stored in Authorization table.

The complete Database Model is shown below in Figure 12.

Figure 12.Complete database model for proposed architecture

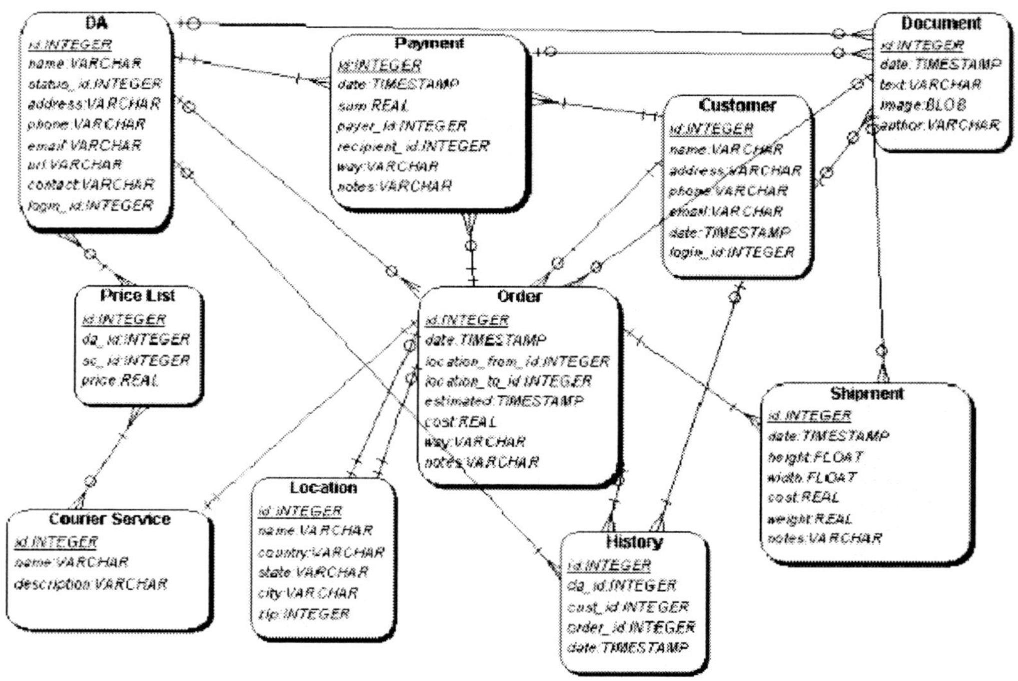

IBM DB2 Universal Database is the database management system recommended in this project. The database is accessed through the Java Database Connectivity (JDBC) driver. A connection pool is used, which contains a group of JDBC connections that are created when the connection pool is registered, usually when starting up the Web Server. The Web system borrows a connection from the connection pool, uses it, and then returns it to the connection pool by closing it. This DBMS is designed to work well is a Web-services environment. Data flow between the Web System Database and corresponding eLogistics applications uses a Web service in connection with the EJB-based architecture.

Security

The Delivery Agent and the Customer enter login and password in the corresponding electronic form at the main page to log into the Web system. The Web system then performs authentication; if it is successful, the user gets access to Web system's functionality. The System Administrator and the Helpdesk Engineer have their own service interface that is a thin client to the Web system's database. Using their corresponding Login ID and password at the main logon page of the Web system. The Web system performs authentication; and if successful, the secure user gets access to Web system's service functionality. The authentication is performed before every request from a secure user to the Web system. Interaction between authenticated users and the Web system is performed via Hyper-Text Transfer Protocol Secure (HTTPS). HTTPS is the secure version of HTTP, the communications protocol of the World Wide Web. HTTPS encrypts the session data using a version of the SSL (Secure Socket Layer) protocol. The level of protection depends on the correctness of the implementation by the Web browser and the server software and the actual cryptographic algorithms supported.

This work was previously published in Journal of Electronic Commerce in Organizations, Vol. 6, Issue 4, edited by M. Khosrow-Pour, pp. 47-76, copyright 2008 by IGI Publishing (an imprint of IGI Global).

Chapter 15
Factors Relating to the Adoption of Internet Technology by the Omani Banking Industry

Salim Al-Hajri
Higher College of Technology, Oman

Arthur Tatnall
Victoria University, Australia

ABSTRACT

The banking industry in Oman is of major importance to Oman's economy, yet Omani banks continue to conduct most of their transactions using traditional methods. A strong banking industry significantly supports economic development through efficient financial services, and their role in trying to achieve the objectives outlined by the Sultan of Oman will depend heavily on the industry's capabilities. Omni banks will need to introduce change at both procedural and informational levels that includes moving from traditional distribution channel banking to electronic channel banking. This chapter addresses the question: What are the enablers and the inhibitors of Internet technology adoption in the Omani banking industry compared with those in the Australian banking industry? The chapter does not attempt a direct comparison of the banking industries in these two very different countries, but rather presents a discussion of Internet technology adoption in Oman, informed by the more mature Australian experience.

INTRODUCTION

In developed countries such as Australia Internet technologies have been embraced by the banking industry and, for several years, banks have pursued strategies to encourage their clients to engage in Internet banking. In an article relating to on-line banking portals, Ayadi (2007) notes the diversity of competitive, technical and strategic challenges faced by banks on the Internet. In developing countries, financial institutions have been less inclined to adopt Internet technology and thereby capitalize on the benefits of simplicity, convenience and usefulness claimed to be associated with its implementation (Kurihara, Takaya, Harui, & Kamae, 2008). Oman is an example of a developing country committed to economic growth but with a banking industry that is yet to embrace Internet technology.

Traditionally, Oman has been heavily reliant on oil as its main income source. Fluctuations in

DOI: 10.4018/978-1-60566-964-9.ch015

world oil prices, however, can leave the country exposed to commodity price risk and during the 1990s Oman began to diversify its income sources. His Majesty, the Sultan of Oman, summarized his vision for Oman's economy as follows:

"The government has made major efforts in recent years to achieve comprehensive development in all fields. Now that it has succeeded, with the help of God, in moving the country from a traditional economy to a modern developed one, our future plan will be based on the balanced management of income and expenditure and the preservation of that balance so that Oman's present high prestige in the economic sphere, will be preserved. Therefore, the duty of the private sector in playing an active role in the development of economic process and in the achievement of national goals by taking appropriate initiatives, as is the case in all developed countries, is of great importance" (Qaboos bin Said's speech, June 2, 1995, reported in Oman Ministry of Development (1996)).

Since the development of the Omani banking industry in 1948 all banks have relied heavily on traditional distribution channels for banking services. However, in 1997 the Oman telecommunication company, Omantel, introduced Internet services to home and business users. According to the statistics provided by Omantel (2002) in their annual report, the number of Internet technology users had reached 12,348 in 1998 one year after of its introduction. By 2002 Omantel reported that the number of Internet users had reached 48,000, making up around 2% of the general population. Indeed, Internet technology infrastructure has grown steadily over the years but has only had a moderate effect on the Omani economy. The Omani banking industry has been very slow and cautious in moving from traditional distribution channel banking services to electronic distribution channel banking services (Al-Hajri, 2005).

Paradoxically, the growing number of users of Internet technology in the Omani economy has not yet motivated the Omani banking industry to adopt Internet technology the way the Australian banking industry has done. The Market Intelligence Strategy Centre (MISC, 2002) reported that in 2002 there were six million Internet banking users in the Australian banking industry reinforcing the observation that Internet technology is far more developed in the Australia than in Oman. Other analysts (Wright & Ralston, 2002) expect Internet technology to grow substantially in the Australian banking industry which means that the Omani banks will be left lagging further behind global competitive markets.

In every country the banking industry is important, as a strong banking industry can have a significant affect in supporting economic development through efficient financial services. In Oman the role of the banking industry in trying to achieve the objectives outlined by the Sultan depends heavily on the industry's capabilities and banks need to introduce change, both at the procedural level and at the informational level. This change includes moving from traditional distribution channel banking to electronic distribution channel banking. Given the prevalence of Internet technology adoption by the banking industry in developed countries, reasons for the lack of such innovation in developing countries such as Oman are of considerable importance.

There are, of course, huge cultural differences between Oman and Australia and a direct comparison of the two is not the intent of this paper. The backgrounds, economics, societies and indeed the very way of life in the two countries are all very different. Australian banks have already largely adopted Internet technologies while Omani banks are just beginning to do so. The goal of this paper is to discuss adoption of Internet technology in the banking industry in Oman, with reference to the Australian experience in order to identify possible ways forward. This paper will thus address the question of determining inhibitors and enablers to the adoption of Internet technology in the banking industry in Oman, and inform the

Oman experience by the more mature Australian experience.

TECHNOLOGY ADOPTION AND DEVELOPING COUNTRIES

Internet technology has played a major role in economic development in developed countries (Ghosh, 1998; Kalakota & Whinston, 1997; Kurihara et al., 2008; Raisinghani, 2000), but less so in developing countries and economies. A good deal has been written about adoption of new technologies and techniques in *agriculture* in developing countries (e.g. (Lee, 2005; Michelsen & Hartwich, 2004; Paarlberg, 2003), but not so much has been recently written on adoption of other technologies apart from a classic article published some fifteen years ago by Besley and Case (1993).

Montealegre (1999) describes a study of Internet adoption in four Latin American countries in which he notes a gradual but progressive course of institutional adoption actions including knowledge building, subsidy, knowledge deployment, innovation directive, and standard setting. Basant, Commander, Harrison and Menezes-Filho (2006) report on the determinants of Information and Communication Technology (ICT) adoption and its impact on performance on manufacturing firms in Brazil and India. They suggest that while Brazilian firms on average use ICT more intensively than their Indian counterparts, in both countries ICT intensity is strongly related to size, ownership structure, share of administrative workers and education. Rajapakse and Seddon (2007) explore the adoption of ERP software in Asia using Hofstede's dimensions of national cultures and conclude that contrasting dimensions of the cultural practices embedded in ERP software are a barrier to adoption. Molla and Licker (2005) report that their studies of eCommerce in developing countries emphasize the influence of contextual impediments including economic, technological, legal, and financial infrastructure as major determinants of adoption. Talukdar, Sudhir and Ainslie (2006) investigate new product diffusion across products and countries, and note that diffusion is much slower in developing than developed countries.

THEORETICAL FRAMEWORK

It has been suggested that "... explaining human behavior in all its complexity is a difficult task" (Ajzen, 1991:179). Further to this we will argue that the main complexity in understanding Internet technology adoption behavior, or that lack of it, within the context of the banking industry in Oman (a non-adopter) and Australia (an adopter) is that this involves people such as bank managers and customers, and people behave in different ways. In deriving a framework for this study, four existing research frameworks were considered: The Theory of Reasoned Action (TRA), The Theory of Planned Behavior (TPB), The Technology Acceptance Model (TAM) and Diffusion of Innovations.

The Theory of Reasoned Action (TRA)

Fishbein and Ajzen (1975) originally developed the Theory of Reasoned Action (TRA) in 1975, before later comprehensively refining it with empirical evidence to support its validity and reliability (Ajzen & Fishbein, 1980). In developing their theory they integrated various studies on attitude from social psychology with the aim of developing an integrated conceptual framework to predict and explain an individual's behavior in a general situational setting. In sum, their study focused on three major determinants of an individual's behavior: **Behavioral Intention**, **Attitude** and **Subjective Norm**. They illustrated their model using these three major variables in a hierarchical sequence to facilitate understanding. In summary they postulated that:

1. An individual's behavioral intention is the immediate determinant of behavior
2. His/her attitude and subjective norm are mediated through behavioral intention
3. His/her behavioral and normative beliefs are mediated through attitude and subjective norm respectively.

In predicting and explaining behavior, TRA has been applied in various fields including consumer behavior (Engel, Blackwell, & Miniard, 1995; Taylor & Todd, 1997; Thomson, Haziris, & Alekos, 1994), moral behavior (Vallerand, Pelletier, Deshaies, Cuerrier, & Mongeau, 1992), ethical/unethical behavior (Chang, 1998; Gibson & Frakes, 1997; Loch & Conger, 1996), environmental behavior (Goldenhar & Connell, 1993), coupon usage behavior (Bagozzi, Davis, & Warshaw, 1992; Shimp & Kavas, 1984), fast food consumption behavior (Bagozzi, Wong, Abe, & Bergami, 2000), adoption of strategic information systems (IS) behavior (Mykytyn & Harison, 1993), system investment decision behavior (Clark & Soliman, 1999), information technology (IT) adoption behavior (Karahanna, Straub, & Cherveny, 1999) and Internet use behavior (Bobbitt & Dabholkar, 2001).

Although TRA has gained wide acceptance in the behavioral sciences and the IS literature due to its well integrated paradigm, the likelihood that a person will actually perform the specific behavior described has been questioned by many researchers (e.g. Warshaw and Davis (1985) and Davis et al. (1989)). For example, in a situation where there is a gap between behavioral intention and actual behavior, some studies have found a low correlation between behavioral these (Bonfield, 1974; Harrell & Bennett, 1974).

The Theory of Planned Behavior (TPB)

After identifying problems with TRA Ajzen (1991) came up with a modification: the Theory of Planned Behavior (TPB). His main criticisms of TRA were that it was designed to predict and explain behavior, or actions, based on the assumption that the behavior was under a person's volitional control. He argues, however, that some behavior that is not under a person's volitional control might be problematic due to the differences in individuals' abilities and in external forces. This may have an influence on actual performance of the behavior and, therefore, consideration should be given to the concept of behavioral control to overcome these volitional control problems. To achieve this Ajzen extended the Theory of Reasoned Action by adding another construct called *Perceived Behavioral Control*, which refers to an individual's perception of the "... presence or absence of requisite resources and opportunities" (Ajzen & Madden, 1986:457) required to perform the specific behavior.

Like the theory of Reasoned Action, the Theory of Planned Behavior assumes that Behavioral Intention is a function of two determinants: the individual's attitude towards the behavior, and the subjective norm within the social environment. Ajzen (1991) argues, however, that the inclusion of Perceived Behavioral Control is important, and provided two rationales for this: it is expected that Perceived Behavioral Control will increase the likelihood of success in performing the behavior, and it is expected that it will provide a measure of actual control. Its validity as a measure to predict the likelihood of achieving the target behavior will depend on the individual's capability (e.g. having the requisite opportunities and resources) to exercise control over the specific behavior. This means that a person who believes that they have the resources and opportunities expects fewer problems in performing the specific behavior and hence perceives that they have complete control over the specific behavior.

Ajzen (1991) reviewed several studies employing TPB and found support for it, more evidence also being found in various situational settings. These studies include: task performance (Locke,

Fredrick, Lee, & Bobko, 1984), attending class and weight loss (Shifter & Ajzen, 1985), obtaining grades (Ajzen & Madden, 1986), investment decisions (East, 1993), consumer behavior (Sparks & Shepherd, 1992), dishonest behavior (Beck & Ajzen, 1991), ethical behavior (Kurland, 1995; Randall & Gibson, 1991), leisure activities (Ajzen & Driver, 1992), executives' adoption of IT in small business (Harrison, Mykytyn, & Riemenschneider, 1997), adoption of virtual banking (Liao, Shao, Wang, & Chen, 1999), learning to use the Internet (Klobas & Clyde, 2000) and Internet purchases (George, 2002). However, Liao et al. (1999) did not find support for the Theory of Planned Behavior and research studies by Randall and Gibson (1991), Beck et al. (1991) and Kurland (1995) found only weak support. These studies thus show rather mixed and confusing results.

Technology Acceptance Model (TAM)

The Technology Acceptance Model (TAM) is a theoretical model that evaluates "... the effect of system characteristics on user acceptance of computer-based information systems" (F. D. Davis, 1986:7). In a similar fashion to the Theory of Reasoned Action, TAM assumes that a computer user is generally quite rational and uses information in a systematic manner to decide whether to use, or not to use this technology in the workplace. The main goal of TAM was:

"... to provide an explanation of the determinants of computer acceptance that is general, and capable of explaining user behavior across a broad range of end-user computing technologies and user populations, while at the same time being both parsimonious and theoretically justified" (Fred D. Davis et al., 1989:985).

In attempting to fulfill this aim, Davis' (1986) conceptual framework proposed that a user's motivational factors are related to actual technology usage and hence, act as a bridge between technology design (e.g. system features and capabilities) and actual technology usage. This means that information obtained from the prediction of actual usage at the early stage will guide designers and implementers to enhance the chance of implementing technology successfully or even avoid the risk of failure. In the conceptual framework Davis (1986) assumes that stimulus variables (e.g. system features and capabilities) trigger organism factors (e.g. user motivation to use the technology) and in turn users respond by actually using the technology. Davis identified three major determinants of technology acceptance (or adoption) suggested by previous research studies that relate to cognition and effectiveness. He began with the Theory of Reasoned Action (Ajzen & Fishbein, 1980; Fishbein & Ajzen, 1975) and adapted this as a basis for causal links between **Perceived Usefulness**, **Perceived Ease of Use**, **Attitude Toward Using Technology** and **Behavioral Intention** to explain technology adoption.

Benefits claimed for the use of TAM are twofold: (1) it provides IS designers with information about how and where to modify design to enhance IS acceptance, and (2) it provides IS implementers with information about how and where to manage IS implementation.

Many studies (e.g. (F. D. Davis, 1989; Segars & Grover, 1993; Taylor & Todd, 1995b) have examined TAM across various technologies and have found TAM to be a reliable and valid model for predicting and explaining technology adoption or acceptance behavior. Various studies utilized TAM to assess users' acceptance or rejection of various computer technologies such as microcomputers (Igbaria, Guimaraes, & Davis, 1995), DOS and Windows (Speier, Morris, & Bridggs, 1995), word processing software (Adams, Nelson, & Todd, 1992; Fred D. Davis et al., 1989; F. D. Davis & Venkatesh, 1996; Hendrickson & Collins, 1996; Venkatesh & Davis, 1996), spreadsheets (Mathieson, 1991), groupware (Taylor & Todd, 1995a), database management systems (Szajna, 1994), e-mail (Adams et al., 1992; F. D. Davis,

1989; Szajna, 1996) and the World Wide Web (Lederer, Maupin, Sena, & Zhuang, 2000). Other studies extended TAM by including other variables such as perceived credibility (Wang, Wang, Lin, & Tang, 2003), gender difference (Gefen & Straub, 1997), and enjoyment (Al-Gahtani & King, 1999). In summary, although several authors (Fred D. Davis et al., 1989; Venkatesh & Davis, 2000) assert that TAM is a useful model for predicting user technology adoption behavior, the above studies showed mixed results. Moreover, most of these studies concentrated on the fundamental prescriptions of TAM with the exception of the study by Wang et al. (2003) which included Perceived Credibility.

Diffusion of Innovations

The theory of Diffusion of Innovations as described by Rogers (1995) is well known. Rogers describes diffusion of innovations as: "... the process by which an innovation is communicated through certain channels over time among the members of social systems. It is a special type of communication, in that the messages are concerned with new ideas" (Rogers, 1995:5).

In the diffusion process, Rogers stresses the importance of communication in an attempt to educate the customer about the innovation concerned. He identifies four main elements: an innovation, communication channels, time and the social system. Rogers suggests that consumers' reactions to the innovation can be measured based on how an innovation is different, how well it is communicated, how long it takes to persuade customers to adopt it, and whether the social system is structured towards achieving a common goal. According to Rogers the four elements of the diffusion of innovation provide an indication of how an innovation is passed to the consumer from first knowledge of this innovation to its final adoption or rejection. The adoption decision for an innovation such as Internet technology in the banking industry is defined as:

"... the process through which an individual (or other decision-making unit) passes from first knowledge of an innovation to forming an attitude toward the innovation, to a decision to adopt or reject, to implementation and use of the new idea, and to confirmation of this decision" (Rogers, 1995:21).

A decision not to adopt an innovation relates to the rejection of the available new idea. However, in order to explain the rate of adoption of innovations Rogers suggests measurement of the following **Perceived Characteristics of Innovations**: (1) relative advantage; (2) compatibility; (3) complexity; (4) trialability; and (5) observability. Rogers (1995) postulated that the adoption of innovations is influenced by these five characteristics, and that they can explain the rate of technology adoption.

Innovation Translation (from Actor-Network Theory)

Innovation Translation (Latour, 1986, 1996; Law & Callon, 1988), informed by Actor-Network Theory (ANT), makes use of a model of technological innovation which uses the concept of heterogeneity in a world that is full of hybrid entities (Latour, 1993) containing both human and non-human elements. It notes that innovations are often not adopted in their entirety but only after 'translation' into a form that is more appropriate for use by the potential adopter, and uses these notions in an explanation of the adoption, or non-adoption of technology (Tatnall & Davey, 2007).

Callon et al. (1983) have proposed that translation involves all the strategies that an actor goes through to identify other actors and to arrange them in relation to each other. Latour (1996) speaks of 'chains of translation' and suggests that with the translation model the initial idea hardly counts and the innovation is not endowed with autonomous power, but moves only if it interests one group of actors or another.

Table 1. Classification of participants according to country and level of management

Country	Number of banks	Interviews – Strategic	Interviews – Tactical	Interviews – Operational	Total Participants
Australia	4	4	4	4	12
Oman	5	5	5	5	15
Total	9	9	9	9	27

Developing a Framework for this Study

This study involved an exploratory investigation of Internet technology adoption in the Omani banking industry, informed by the situation in the Australia. The various technology adoption models described above suggest that a number of factors might affect adoption. Relative advantage and ease of use, suggested by Moore and Benbasat (1991), were firstly considered in the analysis. Rogers (1995) originally identified these two perceptions of technology adoption as relative advantage and complexity. Moore and Benbasat (1991) and Taylor and Todd (1995b) explained that these two constructs are similar to those of TAM (F. D. Davis, 1986). They highlighted the fact that relative advantage is similar to the perceived usefulness construct and perceived complexity is similar to perceived ease of use, due to the similarity in their definitions and in the operation of their measurements. Two additional perceptions, organizational performance and customer/ organizational relationship, not previously identified in the ICT literature, were also considered. From an analysis of this literature, and from some preliminary discussions with bank managers, a theoretical framework was devised that suggests that bank managers' perceptions of only four adoption factors:

- **Relative Advantage**
- **Organizational Performance**
- **Customer/Organizational Relationship**
- **Ease of Use**

might affect any decision to adopt Internet technology in the banking industry. The results of the study, specifically investigating the major enablers and inhibitors of Internet technology adoption in the banking industry in Oman, suggest that these four adoption factors do indeed provide a useful explanation of this adoption.

RESEARCH METHOD

The Australian banking industry was considered in this investigation because of the advanced development of its Internet technology adoption. By examining Internet technology adoption in the banking industry of a developing country in relation to a developed country, the study hoped to provide a richer understanding of the industry and to enrich the analysis of how to improve Internet technology adoption in the banking industry in Oman. Twenty-seven interviews were conducted with strategic, tactical and operational managers at each of nine major banks, five in Oman (Bank of Dhofar, Oman Arab Bank, Oman International Bank and National Bank of Oman) and four in Australia (National Australia Bank, WestPac, Commonwealth Bank and Australia New Zealand Bank).

Selection of interviewees was made from available and willing bank officers so that roughly equal number from each management group were considered. A significant difficulty was encountered in arranging interviews with Australian bankers, but eventually some bank officers of appropriate standing did agree to participate and gave willing of their time, however several hoped for interviews did not eventuate thus restricting the depth of the information gathered about some Australian banks. Given the exploratory nature of this investigation, data was gathered through semi-structured interviews with these managers, and available internal and public reports were used to facilitate understanding and to explore Internet technology adoption. Interview questions were devised to cover the issues identified from the literature and the research framework. Statements by participants are very important sources of evidence, and interviews were thus tape-recorded for later analysis. The average interview time was around 35 minutes.

A schedule of semi-structured interviews with these managers was set up to gather field evidence of perceptions of the four adoption factors: Relative Advantage, Organizational Performance, Customer/ Organizational Relationship and Ease of Use. Interviews were based on issues identified from the literature to explore in-depth the major enablers and inhibitors of Internet technology adoption in the banking industries of Oman and Australia. The four interview factors are further described below:

Perceived Relative Advantage

Discussion here concentrated on the extent to which a manager perceives that Internet technology would enable their organization to gain relative advantage in the industry. From NVIVO analysis of the interviews, issues related to perceived relative advantage were identified as: convenience of services; innovation of ideas, and management of services.

Perceived Organizational Performance

This explored the extent to which a manager perceives that Internet technology could improve organizational performance in the industry. Clearly, if an organization expects to improve its performance with Internet technology then the likelihood of their adopting it will be greater. From NVIVO analysis of the interviews, perceived organizational performance related issues were: profitability, market environment and employees' productivity.

Perceived Customer/ Organizational Relationship

Here the discussion emphasized the extent to which a manager perceives that Internet technology would improve their organization's relationship with customers. If an organization perceives that Internet technology can improve its relationship with its customers then the likelihood of their adopting it will be higher. From the literature it was possible to identify focused issues as: customer trust, customer commitment, and customer satisfaction.

Perceived Ease of Use

If a bank officer perceives that Internet technology is easy to use then they will most likely be motivated to adopt this technology. The issues identified here by NVIVO analysis of the interviews include: easy to navigate, easy to learn, and easy to manage.

Analysis of Data

The elaborated responses from the 27 managers were analyzed, and the findings reported according to suggestions for data reduction and display offered by various authors (Miles & Huberman, 1994; Strauss & Corbin, 1990). NVIVO (version 2) qualitative research software was utilized in analyzing the data. Specifically, NVIVO assisted the analysis in both reducing (organizing and sorting) and displaying the evidence (creation of thematic conceptual tables) in a way that made it easier to see. A case-by-case matrix display was prepared using NVIVO to compare participants' perceptions on various issues. A thematic conceptual matrix display was then produced to compare and identify similar or different patterns of concerns within both banking industries to facilitate understanding of Internet technology adoption. Another thematic conceptual matrix display was produced to compare and identify similar or different concerns within different levels of management (strategic, tactical and operational). From the analysis it was possible to:

- Identify major patterns to facilitate in-depth understanding of Internet technology adoption.
- Reduce the data through the process of generating categories and establishing links (e.g. open and axial coding) as suggested by Strauss and Corbin (1990) using a grounded theory approach where a code is attached to a segment of text and then links between the coded data are explored.
- Display thematic conceptual tables to compare issues within a case and across cases.
- Identify patterns that facilitate understanding and lead to useful conclusions.

MAJOR FINDINGS OF THE STUDY

The study found that Omani bank managers' perceptions of four issues, namely:

- Perceived Relative Advantage
- Perceived Ease of Use
- Perceived Organizational Performance
- Perceived Customer/Organizational Relationship

jointly provided a broader understanding of Internet technology adoption in the banking industry than that proposed by previous theories and models including the Theory of Reasoned Action, the Theory of Planned Behavior, the Technology Acceptance Model and the theory of Innovation Diffusion.

Table 2. Enablers relating to the factor of Perceived Relative Advantage

Major enablers in both Oman and Australia
Convenience of services (- convenience of available service and convenience of location)
Innovation of ideas (- rapid development of innovative ideas)
Management of services (- easy to follow up requests/complains)

The Omani and Australian banking industries were both found to perceive that Internet technology was important and that it could enable the industry to gain relative advantage. In relation to Perceived Relative Advantage, the enablers of Internet technology adoption found in both the Omani and Australian banking industry include:

A number of authors (e.g. (Dedrick, Gurbaxani, & Kraemer, 2003; King & Teo, 1996) have argued from a theoretical perspective that it is possible to increase the rate of Internet technology adoption if a manager perceives that Organizational Performance could be improved. The study showed that in the Omani banking industry, the majority of managers' responses indicated that this was indeed the case. Specifically it was found that perceived improvement in productivity of employees (e.g. improvement in business efficiency) could increase the rate of adoption. On the other hand, perceived problems in profitability, such as technology investment cost and the need for economies of scale for Internet technology use, could inhibit the banking industry from adopting Internet technology. The majority of the respondents in the Australian banking industry also suggested that perceived organizational performance was associated with Internet technology adoption. However, Australian managers expect to face a problem in the market environment in the difficulty of expanding their customer base. One reason for this is that the Australian banking industry is now facing intensive competition (Ramsey & Smith,

Table 3. Enablers and inhibitors relating to the factor of Perceived Organizational Performance

	Major enablers	*Major inhibitors*
Omani banks	Productivity of employees (- business efficiency)	Profitability (-technology investment cost and the need for economies of scale for Internet technology use)
Australian banks	Profitability (- reduction of communication cost	Market environment (- customer base expansion)
	Productivity of employees (- business efficiency)	

1999; Sathye, 1999). Moreover, major Australian banks have already adopted Internet technology as a strategic tool to expand their markets and offer the same services online, making it harder for others to differentiate their offerings. So the real challenge here is how banks will be able to expand their market on the Internet. Table 3 below shows the major enablers and inhibitors of Internet technology adoption in relation to Perceived Organizational Performance.

The role of perceived customer/organizational relationship in IT adoption has attracted considerable attention in the literature (Anderson & Srinivasan, 2003; Athanassopoulos, 2000; Sathye, 1999). In this study, participants were asked whether or not they perceived that Internet technology adoption could improve the relationship with their customers in relation to three focused issues identified in the literature: customer trust; customer commitment; and customer satisfaction. Results of the study suggests that the decision to adopt Internet technology is indeed based on what managers perceive about the customer/organizational relationship in the banking industry.

Omani managers expect to satisfy their customers, but this creates several challenges including: customer trust (e.g. Internet security) and customer commitment (e.g. customer loyalty). Most respondents did not think that Internet technology adoption could improve their customers' trust as they thought that their customers feared that their privacy might be invaded on the Internet. They also did not think that Internet technology could improve their customers' commitment as this related to a close personal relationship with bank branches. This finding has implications for the way banks conduct their business as they have the option of conducting banking business either on the basis of pure Internet banking or using branches and the Internet.

In the semi-structured interviews participants were asked whether or not they believed that Internet technology was easy to use in the context of the banking industry. The literature identified three major issues: easy to navigate, easy to learn and easy to manage. The findings of this study suggest that the decision to adopt Internet technology is consistent with the literature and that managers' perceptions about Ease of Use are very important. Australian managers perceived no difficulties in using Internet technology whereas their Omani counterparts saw some difficulty in navigating on the Internet. They highlighted part of this difficulty as lack of awareness and knowledge

Table 4. Enablers and inhibitors relating to Perceived Customer/Organizational Relationship

	Major enablers	**Major inhibitors**
Omani banks	Customer satisfaction (- reduce conflict)	Customer trust (- Internet security)
		Customer commitment (- customer loyalty)
Australian banks	Customer commitment (- customer loyalty)	Customer trust (- Internet security)
	Customer satisfaction (- reduce conflict)	

about Internet technology, and the accessibility of services. These findings suggest that the Omani banking industry faces a major challenge in using the Internet. Al-Wohaibi et al. (2002) claim that human resource deficiency is a major problem in Oman, however, Abdul-Ghadir and Kozar (1995) argued that computer knowledge, experience and educational level would contribute to the process of increasing the rate of Internet adoption in the Gulf countries. Other authors (F. D. Davis & Venkatesh, 1996; Venkatesh, 1999) have suggested that it would be more beneficial to focus on training rather than on system design.

DISCUSSION: IMPLICATIONS AND RECOMMENDATIONS FOR OMANI BANKS

The banks that stay behind will not be able to embrace Internet technology and the benefits it is claimed to bring and hence will not be able to compete in the industry or in international markets (Porter, 2001). This will also have consequences on the economy. For example, the Omani economy will not improve if banks stay behind, meaning that Internet technology is an important tool in the new economy and should not be ignored. The findings of this study have practical implications for Internet technology adoption in the Omani banking industry and below we offer some guidelines.

Perceived Organizational Performance

The major inhibiting concern here is profitability. This concern relates to high technology investment cost and the need for economies of scale for Internet technology use.

Implications

Internet technology adoption has major implications on banks' performance and hence the economy, and could affect the marketing mix in industry (Al-Hajri, 2005; Kettinger, Grover, Guha, & Segars, 1994). The privatization of the Oman Telecommunication Company (Omantel) is a good forward step allowing Omantel to play a role in the economy. This means that the banking industry can also expect changes such as improvement in telecommunication services, the expansion of telecommunication services, quality of delivery in telecommunication services and lower cost of delivering the service (Oman Telecommunication Company, 2002). Of course the change in telecommunication services will have positive implications on the growth of the Internet population as well as on the geographical distribution of Internet services. In addition it is expected that the number of Internet cafés will grow in addition to home and business Internet use. Moreover, the development of an Internet village is expected to offer awareness of Internet

Table 5. Enablers and inhibitors relating to the factor of Perceived Ease of Use

	Major enablers	Major inhibitors
Omani banks	Easy to learn (- increased automation of process)	Easy to navigate (- awareness/knowledge about Internet technology and accessibility of services)
Australian banks	Easy to navigate (- user friendly and accessibility of service)	
	Easy to learn (- awareness/knowledge about Internet technology)	
	Easy to manage (- customization of banking services and online tracking of banking/ financial services)	

technology in the region and e-government is also expected to play a major role in e-payment as it expands. Improvement in telecommunication infrastructure and services, Internet village and e-government will have positive implication on the economic scale of Internet use and the cost of offering Internet banking services. This suggests that banks should be motivated to invest more in Internet technology (Molla & Licker, 2005).

Recommendations

To reduce this inhibiting factor, King and Teo (1996) suggested a focus on economies of scale for Internet technology use. However, others (Lavender, 2004; Reichheld & Schefter, 2000) argued that most ICT decision-makers ignore intangible cost/benefits. Thus the Omani banking industry ought to consider intangible cost/benefits in their ICT decisions.

Perceived Customer/ Organizational Relationship

For this factor, the major inhibiting concerns are customers trust (e.g. Internet security) and customer commitment (e.g. customer loyalty).

Implications

Internet technology adoption has the potential to affect the relationship between customers and banks: it could improve or destroy this relationship.

Recommendations

To improve customer/bank relationships the banks ought to consider reducing factors inhibiting improvement in the relationship. For example, Whitman (2004) argued that people lack understanding of threats to information security and others (Gefen, 2002; Khalfan, 2004; Mukherjee & Neth, 2003) reported that Internet security was a major concern in trusting the Internet. Therefore the Omani banking industry needs to continually address Internet security concerns through learning, making backups, and implementing secure systems, databases and networks (Held & Bowers, 2001). On the other hand, some authors (Julian & Ramaseshan, 1994; Reichheld & Schefter, 2000; Vatanasombut, Stylianou, & Igbaria, 2004) mentioned that it is hard to win customer loyalty on the Internet so the Oman banking industry will need to focus on continually creating values as well as having a long-term plan.

Perceived Ease of Use

The major inhibiting concern here is ease of navigation on the Internet (e.g. lack of awareness/ knowledge about Internet technology and accessibility of service).

Implications

Successful adoption of ICT depends on how easy it is for people to use it (Fred D. Davis et al., 1989; Moore & Benbasat, 1991; Taylor & Todd, 1995b).

Recommendations

To reduce these inhibiting concerns Venkatesh and Davis (1996); Davis and Venkatesh (2004); and Abdul-Gader and Kozar, (1995) suggest that training could improve awareness/knowledge about Internet technology and hence increase favorable perceptions about ease of use. On the other hand Karahanna and Limayem (2000) suggested that users should have both physical and informational access to the ICT. To ensure this, customers of Omani banks need to address the issue of Internet navigation with telecommunication companies.

Al-Wohaibi et al. (2002) suggested that Omani government organizations need to focus on the development of ICT awareness in schools and universities if they want to increase the rate of successful ICT implementation. They also suggested that banking should consider IT training in its strategic plan as well as supporting it financially to make it work. Consistent with Al-Wohaibi et al. (2002) these suggestions have two important benefits. One is that the Omani banks would be able to develop Internet technology/bank services awareness smoothly when they introduce it to their customers in the industry through a series of seminars, training, online manuals and online support without having to invest a lot of money to train and educate their customers. The second benefit would be that Omani banks would also be able to educate their employees through a series of technical seminars and short-term courses to advance their employees' skills so that they could support their Internet technology infrastructure.

CONCLUSION

This study has shown that analysis of four perceptions of managers in the banking industry: Relative Advantage, Ease of Use, Organizational Performance and Customer/Organizational Relationship, can shed light on the reasons for adoption (or non-adoption) of Internet technology. As expected, the findings confirm that Australian managers perceived less inhibitors to the introduction and implementation of Internet technology than Omani managers, meaning that the Australian banking industry has less challenges to confront than that of Oman.

It has revealed that despite cultural differences there are many similarities in the attitude to technology adoption in the banking systems of the two countries. Bankers in both countries see the use of Internet technologies as offering advantages including: convenience of available service and location, being able to rapidly develop innovative ideas, and the ease with which requests and complaints could be followed up. Both saw ease of learning to use this technology, improvements in productivity of bank employees and customer satisfaction as important enabling factors. Omani and Australian banks each saw issues of customer trust relating to privacy and Internet security as inhibitors to use of Internet technologies.

There were also, however, some important differences. While profitability was regarded as an enabler by Australian banks, which had already installed this technology, it was regarded as an inhibitor by the Omani banks due to the high cost of setting this up. Customer commitment (loyalty) was seen as an enabler to use of the Internet by Australian banks but as an inhibitor in Oman where bank customers still appreciate the personal attention available in local bank branches. Finally, with their greater experience in using computers and the Internet Australian banks found the services easy to use, easy to navigate and quite accessible. In Oman, however, where such experience was lacking, issues of Internet navigation were seen as an inhibiting factor to use of the technology.

This study should provide useful information to the banking industry in Oman to assist it with decisions regarding the adoption of Internet technology. A report provided to Omani bank managers will make them more aware of the problems of technology adoption and of recommendations to take action to improve their competitive position in the global market.

REFERENCES

Abdul-Gader, K., & Kozar, K. A. (1995). The Impact of Computer Alienation on Information Technology Investment Decisions: An Exploratory Cross-National Analysis. *MIS Quarterly*, *19*(4), 535–559. doi:10.2307/249632

Adams, D., Nelson, R., & Todd, P. (1992). Perceived Usefulness, Ease of Use, and Usage of Information Technology: A Replication. *MIS Quarterly*, *16*(2), 227–247. doi:10.2307/249577

Ajzen, I. (1991). The Theory of Planned Behavior. *Organizational Behavior and Human Decision Processes, 50*(2), 179–211. doi:10.1016/0749-5978(91)90020-T

Ajzen, I., & Driver, B. (1992). Application of the Theory of Planned Behaviour to Leisure Choice. *Journal of Leisure Research, 24*(3), 207–224.

Ajzen, I., & Fishbein, M. (1980). *Understanding Attitudes and Predicting Social Behavior*. London: Prentice-Hall, Englewood Cliffs.

Ajzen, I., & Madden, T. (1986). Prediction of Goal-Directed Behavior: Attitudes, Intentions, and Perceived Behavioral Control. *Journal of Experimental Social Psychology, 22*, 453–474. doi:10.1016/0022-1031(86)90045-4

Al-Gahtani, S., & King, M. (1999). Attitudes, Satisfaction and Usage: Factors Contributing to Each in the Acceptance of Information Technology. *Behaviour & Information Technology, 18*(4), 277–297. doi:10.1080/014492999119020

Al-Hajri, S. (2005). *Internet Technology Adoption in the Banking Industry.* Victoria University, Melbourne.

Al-Wohaibi, M., Masoud, F., & Edwards, H. (2002). Fundamental Risk Factors in Deploying IT/IS Projects in Omani Government Organisations. *Journal of Global Information Management, 10*(4), 1–22.

Anderson, R., & Srinivasan, S. (2003). E-Satisfaction and E-Loyalty: A Contingency Framework. *Psychology and Marketing, 20*(2), 123–138. doi:10.1002/mar.10063

Athanassopoulos, A. (2000). Customer Satisfaction Cues to Support Market Segmentation and Explain Switching Behavior. *Journal of Business Research (Bangladesh), 47*, 191–207. doi:10.1016/S0148-2963(98)00060-5

Ayadi, A. (2007). Business Challenges of Online Banking Portals. In A. Tatnall (Ed.), *Encyclopaedia of Portal Technology and Applications* (Vol. 1, pp. 102-105). Hershey, PA: Information Science Reference.

Bagozzi, R., Davis, F., & Warshaw, P. (1992). Development and Test of a Theory of Technological Learning and Usage. *Human Relations, 45*(7), 659–670. doi:10.1177/001872679204500702

Bagozzi, R., Wong, N., Abe, S., & Bergami, M. (2000). Cultural and Situational Contingencies and the Theory of Reasoned Action: Application to Fast Food Restaurant Consumption. *Journal of Consumer Psychology, 9*(2), 97–106. doi:10.1207/S15327663JCP0902_4

Basant, R., Commander, S. J., Harrison, R., & Menezes-Filho, N. (2006). ICT Adoption and Productivity in Developing Countries: New Firm Level Evidence from Brazil and India. IZA Discussion Paper No. 2294

Beck, L., & Ajzen, I. (1991). Predicting Dishonest Actions Using the Theory of Planned Behavior. *Journal of Research in Personality, 25*(3), 285–301. doi:10.1016/0092-6566(91)90021-H

Besley, T., & Case, A. (1993). Modeling Technology Adoption in Developing Countries. *The American Economic Review, 83*(2), 396–402.

Bobbitt, L., & Dabholkar, P. (2001). Integrating Attitudinal Theories to Understand and Predict Use of Technology-Based Self-Service: The Internet as an Illustration. *International Journal of Service Industry Management, 12*(5), 423–450. doi:10.1108/EUM0000000006092

Bonfield, E. (1974). Attitude, Social Influence, Personal Norm and Intentions as Related to Brand Purchase Behavior. *JMR, Journal of Marketing Research, 11*(Nov), 379–389. doi:10.2307/3151284

Callon, M., Courtial, J. P., Turner, W. A., & Bauin, S. (1983). From Translations to Problematic Networks: An Introduction to Co-Word Analysis. *Social Sciences Information. Information Sur les Sciences Sociales*, *22*(2), 191–235. doi:10.1177/053901883022002003

Chang, M. (1998). Predicting Unethical Behavior: A Comparison of the theory of Reasoned Action and the Theory of Planned Behavior. *Journal of Business Ethics*, *17*, 1825–1834. doi:10.1023/A:1005721401993

Clark, J., & Soliman, F. (1999). A Graphical Method for Assessing Knowledge-Based Systems Investments. *Logistics Information Management*, *12*(1-2), 63–77. doi:10.1108/09576059910256277

Davis, F. D. (1986). *A Technology Acceptance Model for Empirically Testing New End-User Information Systems: Theory and Results*. MIT, Boston.

Davis, F. D. (1989). Perceived Usefulness, Perceived Ease of Use, and User Acceptance of Information Technology. *MIS Quarterly*, *13*(3), 318–340. doi:10.2307/249008

Davis, F. D., Bagozzi, R., & Warshaw, P. (1989). User Acceptance of Computer Technology: A Comparison of Two Theoretical Models. *Management Science*, *35*(8), 982–1003. doi:10.1287/mnsc.35.8.982

Davis, F. D., & Venkatesh, V. (1996). A Critical Assessment of Potential Measurement Biases in the Technology Acceptance Model: Three Experiments. *International Journal of Human-Computer Studies*, *45*(1), 19–45. doi:10.1006/ijhc.1996.0040

Davis, F. D., & Venkatesh, V. (2004). Toward Preprototype User Acceptance Testing of New Information Systems: Implications for Software Project Management. *IEEE Transactions on Engineering Management*, *51*(1), 31–46. doi:10.1109/TEM.2003.822468

Dedrick, J., Gurbaxani, V., & Kraemer, K. (2003). Information Technology and Economic Performance: A Critical Review of the Empirical Evidence. *ACM Computing Surveys*, *35*(1), 1–28. doi:10.1145/641865.641866

East, R. (1993). Investment Decisions and Theory of Planned Behaviour. *Journal of Economic Psychology*, *14*, 337–375. doi:10.1016/0167-4870(93)90006-7

Engel, J., Blackwell, R., & Miniard, P. (1995). *Consumer behaviour* (8th ed.). Fort Worth, Texas: Dryden Press.

Fishbein, M., & Ajzen, I. (1975). *Belief, Attitude, Intention, and Behavior: An Introduction to Theory and Research*. Reading: Addison-Wesley.

Gefen, D. (2002). Customer Loyalty in e-Commerce. *Journal of the Association for Information Systems*, *3*(2), 27–51.

Gefen, D., & Straub, D. (1997). Gender Differences in Perception and Adoption of E-Mail: An Extention to the Technology Acceptance Model. *MIS Quarterly*, *21*(4), 389–400. doi:10.2307/249720

George, J. (2002). Influences on the Intent to Make Internet Purchases. *Internet Research: Electronic Networking Applications and Policy*, *12*(2), 165–180. doi:10.1108/10662240210422521

Ghosh, S. (1998). Making Business Sense of the Internet. *Harvard Business Review*, (March-April): 126–133.

Gibson, A., & Frakes, A. (1997). Truth or Consequences: A Study of Critical Issues and Decision Making in Accounting. *Journal of Business Ethics*, *16*, 161–171. doi:10.1023/A:1017914713375

Goldenhar, L., & Connell, C. (1993). Understanding and Predicting Recycling Behavior: An Application of the Theory of Reasoned Action. *Journal of Environmental Systems*, *22*, 91–103.

Harrell, G., & Bennett, P. (1974). An Evaluation of the Expectancy Value Model of Attitude Measurement for Physician Prescribing Behavior. *JMR, Journal of Marketing Research, 11*(3), 269–278. doi:10.2307/3151142

Harrison, D., Mykytyn, P., & Riemenschneider, C. (1997). Executive Decisions About Adoption of Information Technology in Small Business: Theory and Empirical Tests. *Information Systems Research, 8*(2), 171–195. doi:10.1287/isre.8.2.171

Held, J., & Bowers, J. (2001). *Securing E-Business Applications and Communications*. Boca Raton, FL: Auerbach.

Hendrickson, A., & Collins, M. (1996). An Assessment of Structure and Causation of IS Usage. *The Data Base for Advances in Information Systems, 27*(2), 61–67.

Igbaria, M., Guimaraes, T., & Davis, G. (1995). Determinants of Microcomputer Usage Via a Structural Equation Model. *Journal of Management Information Systems, 11*(4), 87–114.

Julian, C., & Ramaseshan, B. (1994). The Role of Customer-Contact Personnel in the Marketing of a Retail Bank's Services. *International Journal of Retail and Distribution Management, 22*(5), 29–34. doi:10.1108/09590559410067316

Kalakota, R., & Whinston, A. (1997). *Electronic Commerce: A manager's Guide*. Reading, Mass: Addison-Wesley.

Karahanna, E., & Limayem, M. (2000). E-mail and V-mail Usage: Generalizing Across Technologies. *Journal of Organizational Computing and Electronic Commerce, 10*(1), 49–66. doi:10.1207/S15327744JOCE100103

Karahanna, E., Straub, D., & Cherveny, N. (1999). Information Technology Adoption Across Time: A Cross-Sectional Comparison of Pre-Adoption and Post-Adoption Believes. *MIS Quarterly, 23*(2), 192–213. doi:10.2307/249751

Kettinger, W., Grover, V., Guha, S., & Segars, A. (1994). Strategic Systems Revisited: A Study in Sustainability and Performance. *MIS Quarterly, 18*(1), 31–58. doi:10.2307/249609

Khalfan, A. (2004). Information Security Considerations in IS/IT Outsourcing Projects: A Descriptive Case Study of Two Sectors. *International Journal of Information Management, 24*(1), 29–42. doi:10.1016/j.ijinfomgt.2003.12.001

King, W., & Teo, T. (1996). Key Dimensions of Facilitators and Inhibitors for the Strategic Use of Information Technology. *Journal of Management Information Systems, 12*(4), 35–53.

Klobas, J., & Clyde, L. (2000). Adults Learning to Use the Internet: A Longitudinal Study of Attitudes and Other Factors Associated with Intended Internet Use. *Library & Information Science Research, 22*(1), 5–34. doi:10.1016/S0740-8188(99)00038-9

Kurihara, Y., Takaya, S., Harui, H., & Kamae, H. (Eds.). (2008). *Information Technology and Economic Development*. Hershey: Idea Group Reference.

Kurland, N. (1995). Ethical Intentions and the Theories of Reasoned Action and Planned Behavior. *Journal of Applied Social Psychology, 25*(4), 297–313. doi:10.1111/j.1559-1816.1995.tb02393.x

Latour, B. (1986). The Powers of Association. In J. Law (Ed.), *Power, Action and Belief. A New Sociology of Knowledge? Sociological Review monograph 32* (pp. 264-280). London: Routledge & Kegan Paul.

Latour, B. (1993). *We Have Never Been Modern* (C. Porter, Trans.). Hemel Hempstead: Harvester Wheatsheaf.

Latour, B. (1996). *Aramis or the Love of Technology*. Cambridge, Ma: Harvard University Press.

Lavender, M. (2004). Maximizing customer Relationships and Minimizing Business Risk. *International Journal of Bank Marketing, 22*(4), 291–296. doi:10.1108/02652320410542563

Law, J., & Callon, M. (1988). Engineering and Sociology in a Military Aircraft Project: A Network Analysis of Technological Change. *Social Problems, 35*(3), 284–297. doi:10.1525/sp.1988.35.3.03a00060

Lederer, A., Maupin, D., Sena, M., & Zhuang, Y. (2000). *The Role of Ease of Use, Usefulness and Attitude in the Prediction of World Wide Web Usage*. Paper presented at the ACM SIGCPR Conference on Computer Personnel Research, Boston, MA.

Lee, D. R. (2005). Agricultural Sustainability and Technology Adoption: Issues and Policies for Developing Countries . *American Journal of Agricultural Economics, 87*(5), 1325–1334. doi:10.1111/j.1467-8276.2005.00826.x

Liao, S., Shao, Y., Wang, H., & Chen, A. (1999). The Adoption of Virtual Banking: An Empirical Study. *International Journal of Information Management, 19*(1), 63–74. doi:10.1016/S0268-4012(98)00047-4

Loch, K., & Conger, S. (1996). Evaluating Ethical Decision Making and Computer Use. *Communications of the ACM, 39*(7), 77–83. doi:10.1145/233977.233999

Locke, E., Fredrick, E., Lee, C., & Bobko, P. (1984). Effect of Self-Efficacy, goals, and Task Strategies on Task Performance. *The Journal of Applied Psychology, 69*(2), 241–251. doi:10.1037/0021-9010.69.2.241

Mathieson, K. (1991). Predicting User Intentions: Comparing the Technology Acceptance Model with the Theory of Planned Behavior. *Information Systems Research, 2*(3), 173–191. doi:10.1287/isre.2.3.173

Michelsen, H., & Hartwich, F. (2004). *University-based Agricultural Research: A Comparative Study in Sub-Saharan Africa* Addis Ababa: International Service for National Agricultural Research

Miles, M., & Huberman, A. (1994). *Qualitative Data Analysis: An Expanded Sourcebook* (2nd ed.). Thousand Oaks, CA: Sage Publications.

MISC. (2002). 6 Million Internet User Registration Milestone Reached, After 1st Quarter, As Banking Usage (Transactions) Grow 15% in Latest Quarter. Retrieved 17 December 2004, from www.marketintelligence.com.au

Molla, A., & Licker, P. S. (2005). Ecommerce Adoption in Developing Countries: a Model and Instrument. *Information & Management, 42*(6), 877–899. doi:10.1016/j.im.2004.09.002

Montealegre, R. (1999). A Temporal Model of Institutional Interventions for Information Technology Adoption in Less-Developed Countries. *Journal of Management Information Systems, 16*(1), 207–232.

Moore, G., & Benbasat, I. (1991). Development of an Instrument to Measure the Perception of Adopting an Information Technology Innovation. *Information Systems Research, 2*(3), 192–222. doi:10.1287/isre.2.3.192

Mukherjee, A., & Neth, P. (2003). A Model of Trust in Online Relationship Banking. *International Journal of Bank Marketing, 21*(1), 5–15. doi:10.1108/02652320310457767

Mykytyn, P., & Harison, D. (1993). The Application of the Theory of Reasoned Action to Senior Management and Strategic Information Systems. *Information Resources Management Journal*, *6*(2), 15–25.

Oman Ministry of Development. (1996). *Vision for Oman Economy* (Qaboos bin Said speech, June 2 1995). Muscat: Ministry of Development.

Oman Telecommunication Company. (2002). *Annual Book 2002*. Muscat: Oman Telecommunication Company.

Paarlberg, R. L. (2003). Technology Adoption in Developing Countries: The Case of Genetically Modified Crops. Retrieved April 2007, from http://www.au.af.mil/au/awc/awcgate/cia/nic2020/technology_adoption_nov6.pdf

Porter, M. (2001). Strategy and the Internet. *Harvard Business Review*, (March): 63–78.

Raisinghani, M. (2000). *Electronic Commerce at the Down of the Third Millennium - Electronic Commerce: Opportunities and Challenges*. Thousand Oaks: Sage Publications.

Rajapakse, J., & Seddon, P. B. (2007). ERP Adoption in Developing Countries in Asia: A Cultural Misfit Unpublished Working paper. University of Melbourne.

Ramsey, J., & Smith, M. (1999). Managing Customer Channel Usage in the Australian Banking Sector. *Managerial Auditing Journal*, *14*(7), 329–338. doi:10.1108/02686909910289812

Randall, D., & Gibson, A. (1991). Ethical Decision Making in the Medical Profession: An Application of the Theory of Planned Behavior. *Journal of Business Ethics*, *10*(2), 111–122. doi:10.1007/BF00383614

Reichheld, F., & Schefter, P. (2000). E-Loyalty: Your Secret Weapon on the Web. *Harvard Business Review*, *78*(4), 105–113.

Rogers, E. M. (1995). *Diffusion of Innovations* (4th ed.). New York: The Free Press.

Sathye, M. (1999). Adoption of Internet Banking by Australian Consumers: An Empirical Investigation. *International Journal of Bank Marketing*, *17*(7), 324–334. doi:10.1108/02652329910305689

Segars, A., & Grover, V. (1993). Re-Examining Perceived Ease of Use and Usefulness: A Confirmatory Factor Analysis. *MIS Quarterly*, *17*(4), 517–525. doi:10.2307/249590

Shifter, D., & Ajzen, I. (1985). Intention, Perceived Control and Weight Loss: An Application of the Theory Planned Behaviour. *Journal of Personality and Social Psychology*, *49*, 843–851. doi:10.1037/0022-3514.49.3.843

Shimp, T., & Kavas, A. (1984). The Theory of Reasoned action Applied to Coupon Usage. *The Journal of Consumer Research*, *11*(Dec), 795–809. doi:10.1086/209015

Sparks, P., & Shepherd, R. (1992). Self-Identity and the Theory of Planned Behaviour: Assessing the Role of Identification with Green Consumerism. *Social Psychology Quarterly*, *55*(4), 388–399. doi:10.2307/2786955

Speier, C., Morris, M., & Bridggs, C. (1995). *Attitudes Toward Computers: The Impact of Performance*. Paper presented at the AIS, Pittsburgh, PA.

Strauss, A., & Corbin, J. (1990). *Basics of Qualitative Research: Grounded Theory Procedures and Techniques*. Newbury Park: Sage Publications.

Szajna, B. (1994). Software Evaluation and Choice: Predictive Validation of the Technology Acceptance Instrument. *MIS Quarterly*, *18*(3), 319–324. doi:10.2307/249621

Szajna, B. (1996). Empirical Evaluation of the Revised Technology Acceptance Model. *Management Science*, *42*(1), 85–94. doi:10.1287/mnsc.42.1.85

Talukdar, D., Sudhir, K., & Ainslie, A. S. (2006). Investigating New Product Diffusion across Products and Countries. Unpublished Working paper. Yale SOM Working Paper No. MK-06.

Tatnall, A., & Davey, B. (2007, 19-23 May 2007). *Researching the Portal.* Paper presented at the IRMA: Managing Worldwide Operations and Communications with Information Technology, Vancouver.

Taylor, S., & Todd, P. (1995a). Assessing IT Usage: The Role of Prior Experience. *MIS Quarterly, 19*(4), 561–570. doi:10.2307/249633

Taylor, S., & Todd, P. (1995b). Understanding Information Technology Usage: A Test of Competing Models. *Information Systems Research, 6*(2), 144–176. doi:10.1287/isre.6.2.144

Taylor, S., & Todd, P. (1997). Understanding the Determinants of Consumer Composting Behavior. *Journal of Applied Social Psychology, 27*(7), 602–628. doi:10.1111/j.1559-1816.1997.tb00651.x

Thomson, K., Haziris, N., & Alekos, P. (1994). Attitudes and Food Consumption. *British Food Journal, 96*(11), 9–14. doi:10.1108/00070709410074632

Vallerand, R., Pelletier, L., Deshaies, P., Cuerrier, J., & Mongeau, C. (1992). Ajzen and Fishbein's Theory of Reasoned Action as Applied to Moral Behavior: A Confirmatory Analysis. *Journal of Personality and Social Psychology, 62*(1), 98–109. doi:10.1037/0022-3514.62.1.98

Vatanasombut, B., Stylianou, A., & Igbaria, M. (2004). How to Retain Online Customers. *Communications of the ACM, 47*(6), 65–69. doi:10.1145/990680.990682

Venkatesh, V. (1999). Creation of Favorable User Perceptions: Exploring the Role of Intrinsic Motivation. *MIS Quarterly, 23*(2), 239–260. doi:10.2307/249753

Venkatesh, V., & Davis, F. (1996). A Model of the Antecedence of Perceived Ease of Use: Development and Test. *Decision Sciences, 27*(3), 451–481. doi:10.1111/j.1540-5915.1996.tb01822.x

Venkatesh, V., & Davis, F. (2000). A Theoretical Extension of the Technology Acceptance Model: Four Longitudinal Field Studies. *Management Science, 46*(2), 186–204. doi:10.1287/mnsc.46.2.186.11926

Wang, Y., Wang, Y., Lin, H., & Tang, T. (2003). Determinants of User Acceptance of Internet Banking: An Empirical Study. *International Journal of Service Industry Management, 14*(5), 501–519. doi:10.1108/09564230310500192

Warshaw, P., & Davis, F. (1985). Disentangling Behavior Intention and Behavioral Expectation. *Journal of Experimental Social Psychology, 21*, 213–218. doi:10.1016/0022-1031(85)90017-4

Whitman, M. (2004). In Defence of the Realm: Understanding the Threats to Information Security. *International Journal of Information Management, 24*, 43–57. doi:10.1016/j.ijinfomgt.2003.12.003

Wright, A., & Ralston, D. (2002). The Lagging Development of Small Business Internet Banking in Australia. *Journal of Small Business Management, 40*(1), 51–57. doi:10.1111/1540-627X.00038

Chapter 16
E-Government Service Delivery Performance:
An Adaptation of the Balanced Scorecard to the Public Sector

Assion Lawson-Body
University of North Dakota, USA

Jared Keengwe
University of North Dakota, USA

Laurence Mukankusi
University of North Dakota, USA

Abdou Illia
Eastern Illinois University, USA

Glenn Miller
University of North Dakota, USA

ABSTRACT

E-government service delivery performance has been discussed in literature as a way governments use information technologies (IT) to deliver valuable services to their citizens at a lower cost. This article examines the effectiveness of Web site-supported Balanced Scorecard's four dimensions (innovation and learning, internal process, veteran value proposition, and financial) in improving e-government service delivery performance. The study used content analysis to analyze the data obtained from a sample of 19 county veteran service officers (CVSOs) to test the hypotheses. CVSOs use Web sites to serve veterans on a Government-to-Citizen (G2C) basis.

INTRODUCTION

Financial accounting measurements are extensively used to evaluate government performance (Hsu, 2005; Wright, Smith, Jesser, & Stupeck, 1999), but they do not reveal adequately the benefits of investing in IT. Financial accounting measurements do inform managers of historical outcomes, but they do not indicate why those results were achieved or what managerial actions have to be performed to improve future results (Wright et al., 1999). Because the excessive reliance on financial accounting performance measurements is inadequate and can be misleading, Wright et al. (1999) suggested a Balanced Scorecard theory as a consistent performance measurement tool for the use of IT, Internet and electronic service (e-service). Despite the problems encountered by Internet users over the last years, there has been a continually increasing interest in the use of the internet for electronic service deliveries (Levenburg & Klein, 2006). This increasing interest has led governments to provide citizens with new electronic services (Kvasny & Keil, 2006; Levenberg & Klein, 2006).

In government, IT and information are public property, not a proprietary resource to be protected and exploited for competitive advantage (Carter & Belanger, 2005; Dufner, Halley, & Reed, 2002); therefore, performance measurement is important to assess e-government efforts. A government needs to track what is working and what is not and assure citizens that the government's time and funds are being well spent (Stowers, 2004). Some traditional performance measures may be appropriate for e-government, but e-government service is different in its delivery modes and expectations. E-government service delivery applications are relatively new and government service providers and researchers can only learn so much from traditional cost-benefit analyses. Therefore, new performance measures need to be developed to provide accountability for jurisdictions' e-government efforts. In addition, the significance of aligning the usability of Web sites with government service delivery strategies is widely recognized (Griffith, Sawyer, & Neale, 2003), but the lack of appropriate methodologies prevents government units and their constituents from integrating Web site projects with e-government service delivery performance.

The early experiences of companies using Balanced Scorecard have demonstrated that it meets several managerial needs (Kaplan & Norton, 1992; Pandey, 2005). The Balanced Scorecard serves as a management tool for the following purposes (Wright et al., 1999): 1) identifying the performance drivers that affect electronic service delivery and 2) establishing a set of cause-and-effect relationships among business performance factors (Hsu, 2005). Therefore, Balanced Scorecard is not only a tool for performance assessment, but also a key driver to a business organization's success (Hsu, 2005).

This article addresses the issue of delivering electronic services using the concept of an accepted management tool--the Balanced Scorecard. In other words, the Web site-supported Balanced Scorecard dimensions may be used to improve e-government service delivery performance. Like many other innovative technologies, government Web sites can generate sustainable electronic service delivery performance if the technology is used for improving the operational efficiency of government services.

Various ways to categorize e-government exist. Some authors classify e-government into six categories: Government to individuals as a part of a delivery system (G2C), Government to individuals as a part of the political process (G2IP), Government to business as a citizen (G2BC), Government to business in the marketplace (G2BMKT), Government to employees (G2E), and Government to government (G2G) (Carter & Belanger, 2005). This research deals with G2C e-government, focusing on government Web sites that county veterans service officers (CVSOs) use to deliver electronic services to veterans. Veterans are the nation's population who have been discharged or retired after serving active duty with the United States Armed Forces. CVSOs are county employees. Veterans interact with CVSOs when requesting government benefits or having questions about them.

This article is organized as follows. Section II contains the description of the Balanced Scorecard theory. In Section III, government service delivery performance is examined. In Section IV, the research model and hypotheses are presented. In Section V, the methodology is outlined. In Section VI, the results and findings are examined. In Section VII, the discussion of this article is presented. In Section VIII, the conclusion is presented.

BALANCED SCORECARD THEORY

By using the Balanced Scorecard, companies must create a system that simultaneously aligns and integrates four interrelated perspectives: 1) learning and innovation focus, 2) internal process focus, 3) customer focus, and 4) financial focus (Kaplan & Norton, 1992; Wright et al., 1999; Hsu, 2005). What happens in one perspective often has an impact on what is going on in another. Therefore, the performance indicators in the Balanced Scorecard system should be set on the basis of business success factors (Hsu, 2005). At the heart of the Balanced Scorecard concept is its theory of business: managers first decide where the market is and who their customers are and then initiate particular ways to achieve business goals and missions considering operational processes and creative learning and innovation (Hsu, 2005).

Prior to this study, many authors such as Wright et al. (1999) and Huang and Hu (2004) have examined the integration of IT or Web services with Balanced Scorecard. Building and expanding on previous efforts (Huang & Hu, 2004), we address the issue of e-government services delivery performance using the Balanced Scorecard approach.

E-GOVERNMENT SERVICE DELIVERY PERFORMANCE

Governments use IT to deliver valuable services to their citizens at a lower cost. E-government services refer to the emerging area of IS and IT services that are delivered electronically (Ramesh & Tiwana, 2001). This delivery enables and improves the efficiency with which government services are provided to citizens, employees, businesses, and agencies (Carter & Belanger, 2005; Moon & Norris, 2005). The way that government agencies design and deliver services and configure and deploy underlying information and communications technologies is central to the performance of e-government service delivery.

E-government service delivery can take many forms depending on the level of digitization of the services rendered to citizens. E-government service delivery can emphasize the electronic form that includes pure digital dimensions. That form of e-government service delivery can be considered as e-government which excludes all transactions realized through traditional fax, television and telephone channels, unless these transactions are initiated via the Internet. In other words, this form of e-government contains totally automated service provisions. For instance, applying for a tax refund online and being reimbursed by a deposit in a bank account, is pure e-government service delivery because the delivery and the payment are digital.

There is also e-government service delivery which can take an electronic form that includes a mix of digital and physical dimensions. That form of e-government service delivery encompasses partially automated service provisions. In fact, the e-government service delivery process may not be entirely electronic because online transactions require at some levels human intervention. This conception of e-government service delivery can be seen as the process in which government services, because of their intrinsic characteristics, cannot be contained in an electronic procedure.

Those government services are delivered via media like paper or fax. As a matter of fact, it is e-government that supports the traditional service delivery models in which the citizens analyze and decide about public actions. Only certain aspects of many decisions made by the citizens are electronically supported, for instance, delivery of social benefits, delivery of social security numbers in the United States, administration of the driving license examination, allowing a citizen to plead guilty or not to government law violation, delivery of health services, and so forth. Government services can need a separate shipping process from the electronic delivery process, for example, cars, consumer appliances, and food stamps or other items.

Using a case study, Park, Lee, and Yoo (2005) found that the importance of government performance indicators significantly depends on the characteristics of the services. Jain (2003) found that IT investments had a negative relationship with performance of the states in each of the performance dimensions, such as financial management, human resource management, information technology management, capital management, and result management. The same author found a positive relationship between IT investments and projected state budget deficits. Finally, Jain (2003) found that the more U.S. state governments invest in IT, the worse they perform. However, there are also indications that as time goes on, the relationship between IT investment and performance shows improvements in the public sector.

Many authors have examined how governments can invest in e-commerce to improve their service delivery performances (Dufner et al., 2002; Jain, 2004). After conducting six case studies and contacting over 50 government organizations, Cohen and Eimicke (2001) found that e-government service delivery could change human resource deployment patterns and improve organizational performance. Another source of e-government service delivery performance is savings generated from reduced costs. When costs were assessed in the cases presented in their study, Cohen and Eimicke (2001) found that typically services delivered over the Internet were less expensive to deliver than those delivered in person. The resulting cost savings in electronic service delivery can even be more significant because many valuable internal services may be made available to citizens, thus creating new revenue streams. Those cost reductions should allow and facilitate citizen access to crucial services using a Web portal.

Although some important work has been undertaken, the Internet-government service delivery performance relations are not adequately represented in current IS research. The current study is a first effort toward this end.

However, even though in this study we emphasize the economical dimension of e-government, it is believed that IT is introduced in governments not only for reducing cost. We realize that e-government is characterized by different dimensions. It is crucial to indicate that governments can use IT to increase other factors such as transparency, accountability, security, and privacy. These additional challenges and important principles can be followed by government agencies and e-government experts should be aware of those factors.

E-government includes major internal and external aspects such as the establishment of a secure government intranet and central database for more efficient, transparent, and cooperative interaction among governmental agencies (Moon, 2002). Some authors argue that e-government has multidimensional properties because it can be dedicated to improve government reputation and to build an image of government as a modern nation. An important aspect of e-government is having information accessible to citizens and government agencies open to input from their constituents. This can transform e-government to an attractive location for people and business. Goals for e-government that promote effective governance include: improved government decision making processes, increased citizen trust in government, increased government accountability and transparency, and ability to accommodate the citizen's will in the information-age (Griffith et al., 2003).

The advent of e-government is also the purpose of a growing number of studies. Public government agencies have used Web technology widely in recent years (Moon, 2002). But e-government adoption among public agencies seems to be done less for the organization's benefit and more for the benefit of the organization's clientele (Jain, 2004). In fact, public workers are concerned with the mission of their agency and feel a personal connection with what their agency does. Demchak, Friis, and LaPorte (2000) studied the use of e-government and reported that delivering government service on the Web will increase government interaction with citizens because Web sites are built with openness and effectiveness (Gant & Gant, 2002). West (2000) also suggested that while government agencies are adopting the Web as a tool for delivering services, government at all levels is not making full and effective use of commonly available information technology (Gant & Gant, 2002). Finally, Hart and Teeter (2000) reported that there is increasing public acceptance for state governments to develop online services (Gant & Gant, 2002).

Another challenge for e-government is the necessity to maintain the privacy, security, and confidentiality of citizens' and businesses' records. Some authors indicate that issues of security and privacy are quite high on the list of concerns for e-government users and government constituents (Jain, 2004). It is important for all online transactions to remain confidential, for all parties to be sure of the identity of all other parties, and for no one else to be able to interfere with those transactions. This means that some cryptographic system or vehicle like digital signatures must be used. Technically, digital signatures are the user's known data after the data, message, or document has been encrypted with a mathematical algorithm (Stowers, 2004).

The completion of online transactions requires e-government users and partners to sign or authenticate themselves because transactions often involve the collection or disbursement of financial resources or the collection and disclosure of personally identifiable data (Jain, 2004). Some public organizations are experimenting with or deploying public key infrastructures (PKIs) as a potential solution for securing government online transactions (Jain, 2004).

THE RESEARCH MODEL AND HYPOTHESES DEVELOPMENT

In the research model shown in Figure 1, the dependent variable will be drawn from e-government service delivery performance. The independent variables will be drawn from the Web site-supported four interrelated Balanced Scorecard business perspectives. Because these four interrelated business perspectives are supported by Web sites used by CVSOs on a G2C basis in order to serve veterans, they have been adapted as follows: Web site-supported innovation and learning perspective, Web site-supported internal process perspective, Web site-supported veteran value proposition perspective, and Web site-supported financial perspective. Each of them is discussed below.

Web Site-Supported Innovation and Learning Perspective

This perspective emphasizes innovation, creativity, skill, knowledge, and capability (Pandey, 2005). It is based upon performance indicators of how the organization invests in its business unit's people (Edgington & Chen, 2002; Griffith, et al., 2003). The foundation of any strategic activity is the learning and innovation perspective, which defines the core competencies and skills, the

Figure 1. Research Model (Adapted from the Balanced Scorecard Framework of Huang and Hu, 2004)

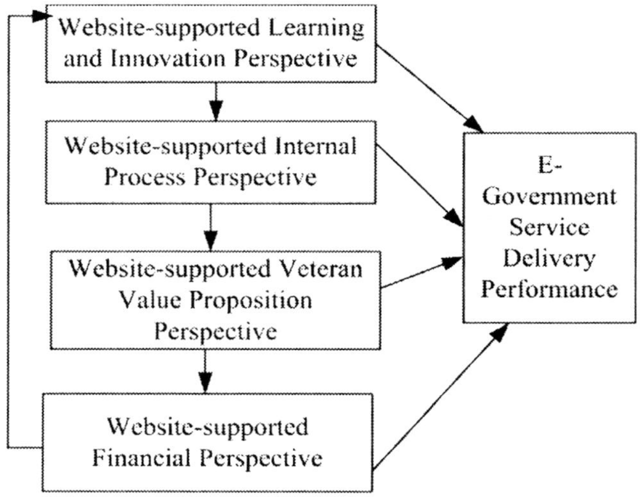

technologies, and the corporate culture needed to support an organization's strategy (Kaplan & Norton, 1992).

In the public sector, these objectives enable government to align its human resources and information technology with its strategy. Governments must continually improve CVSO skills for better understanding citizen or veteran expectations. Therefore, the CVSOs will be able to teach and transfer the service delivery process to veterans. It is critical while improving veteran Web sites to understand clearly veterans' expectations in order to empower them (Cetiner & Ryan, 2004). This suggests the following hypothesis (see Figure 1):

H1: *The Web site-supported learning and innovation perspective will have a positive effect on e-government service delivery performance.*

Web Site-Supported Internal Process Perspective

The internal business process ensures the highest quality of services. This perspective demonstrates how organizations identify performance indicators via their business unit internal capabilities. When the internal business processes are running well, the organization's services can meet the customers' needs and requirements in order to create value for them (Pandey, 2005).

Once government has a clear image of its veterans, it can then determine the ways by which it will achieve the productivity improvements to reach its performance objectives. The internal process perspective captures these public efforts and activities by deepening relationships with veterans.

Usability of a Web site can enhance internal operations by enabling process automation (Damsgaard, 2002). For example, CVSOs who do not use a Web site may spend a significant amount of time on assisting veterans. If CVSOs can automate service delivery to veterans, this may free up CVSOs to pursue other government activities. By automating the online support to veterans, CVSOs can perform more productive functions. Thus, we propose the hypothesis below (see Figure 1):

H2: *The Web site-supported internal process perspective will have a positive effect on e-government service delivery performance.*

Web Site-Supported Veteran Value Proposition Perspective

A large, satisfied and loyal customer base means more revenue and improved financial performance. This perspective shows how organizations use performance indicators and outcome measures of their business units to create value for current and future customers. The customer perspective requires a company to know how it should create value for its customers if it is to succeed (Pandey, 2005).

In the public sector, every interaction on the Web site with a veteran should take place in the context of all previous online interactions with that veteran. An online conversation should pick up where the last one left off, whatever the duration between the previous and the current conversations. A government unit will be more successful in creating veteran value if its veteran perspective is built on the basis of direct veteran interaction.

CVSOs can offer direct Web site access to veterans so each time a veteran visits a government Web portal is an occasion to deliver additional services or provide a path for other government units that want to reach that veteran. CVSOs add new value for their veterans by supplying them with regulatory information about such subjects as waste disposal at no fee; this strengthens relationships with veterans, but it weakens the business of private companies that provide such information for a fee. Accordingly, it is hypothesized that (see Figure 1).

H3: *The Web site-supported veteran value proposition perspective will have a positive effect on e-government service delivery performance.*

Web Site-Supported Financial Perspective

The financial perspective provides a valuable way for analyzing organizations' performance and profitability. This perspective indicates how organizations take into account measures of their business units' past performance. But the financial dimension alone is not enough to guide performance in creating value; it depends on non-financial measures for providing the bottom-line score (Pandey, 2005).

Usability of a Web site by CVSOs will reduce cost and increase veteran participation; therefore, increased electronic service delivery performance will follow. Good electronic service encourages veterans to use the Web site more often, which means they become more likely to use it for transactions and support. Electronic service thus lowers the cost of serving veterans.

Veterans should visit a Web site because they have heard that this specific Web site will serve their interests and needs. If the Web site fulfills the expectations of the veterans, they will start using the Web services more intensively. The veterans will invest time and effort in developing relationships with other veterans and in creating, for example, a personalized interface. Therefore, the veterans are likely to have high switching costs if they change the Web site, and this increases the probability of locking in the veterans. Therefore, we propose the following hypothesis (see Figure 1):

H4: *The Web site-supported financial perspective will have a positive effect on e-government service delivery performance.*

METHODOLOGY: MULTIPLE CASE STUDIES

This research focuses on government Web sites that deliver electronic services to veterans. The Web site of the North Dakota Government Rural Outreach (GRO) Initiative has been selected as the sample U.S. government Web site for this research. That Web site has been chosen because it has a component dedicated to veterans and CVSOs.

The procedure used to process data was based on the guidelines suggested by Churchill (1979). In order to follow these guidelines, multiple case studies have been conducted. In total, 19 CVSOs of 17 counties in North Dakota were asked to participate in the multiple case studies. The interviews were held individually with each CVSO for one hour. The interviews were tape-recorded, and the texts of the interviews were transcribed and entered into Windows Word software. To help conduct the interviews, an interview guide was used.

The study used content analysis to analyze the data obtained from a sample of CVSOs to test the hypotheses. The common response is obtained by using the generalized answer from all CVSOs. The procedure for data analysis was based on totaling the overall counts of messages coded in each category and the analysis of the messages. Each research variable has played the role of category. Occurrences of the messages in each category have been analyzed to determine relative importance. Higher relative counts reflect the significance of the particular message within the specific category.

Chi-square is used as a statistical test in this article because the data are in the form of raw frequency counts of things in two or more mutually exclusive and exhaustive categories (Connor-Linton, 2003). In this study, we converted interview message frequencies into percentages. Any count of messages collected from the interview of the CVSOs fall into only one category or value on each variable. For example, our data are counts of positive or negative messages expressing a preference for independent and dependent variables used in this study. Each message is counted only once. In other words, the interview contents have been transformed in quantitative data by counting the negative and positive messages about each category of our research design. The dependent variable (e-government service delivery performance) and the independent variables (the Web site-supported learning and innovation perspective, the Web site-supported internal process perspective, the Web site-supported veteran value proposition perspective, and the Web site-supported financial perspective) have been used as categories. Because chi-square is a lot like percentage, it extrapolates a population characteristic (a parameter) from the sampling characteristic (a statistic) similarly to the way percentage standardizes a frequency to a total column N of 100 (Connor-Linton, 2003). We converted the number of messages counted into percentages as if there were 100 messages in each category of the independent or dependent variable for comparability.

Finally, a chi-square statistical test has been used because according to Connor-Linton (2003), for the chi-square approximation to be adequate, the total number of subjects should be at least 10. In accordance with Eisenhardt (1989, cited in Chan & Ngai, 2007), the sample for a qualitative field study implying the use of chi-square should be at least 10 organizations, subjects or observations participating to the preliminary interview. The results obtained in this study are based on a sample of 19 interviews. The guidelines established by Connor-Linton (2003) for the utilization of a chi-square statistical test may be applied to the total number of CVSOs participating in our interview. This is considered very satisfactory. Also, the use of a content analysis approach does not require a specific number of observations (Churchill, 1979).

RESULTS AND FINDINGS

Procedures for Testing the Hypotheses

In this research, Chi-square, a nonparametric test of statistical significance for bivariate tabular analysis, was used to test the hypotheses and analyze data. The test of the hypotheses determined whether or not two different samples (of people or texts) are significantly different in some aspect of their behaviors. Then it can be generalized from the samples that the populations are also different in the behaviors (Connor-Linton, 2003).

Chi-square was selected in this research because it is a rough estimate of confidence; it accepts weaker, less accurate data as input than do parametric tests (like t-test, f-fisher, and analysis of variance) (Connor-Linton, 2003). Also, it does not require the sample data to be more or less normally distributed in the population from which the sample is drawn (Connor-Linton, 2003). But Chi-square requires that data are in the form of raw frequency counts of phenomena in two or more mutually exclusive and exhaustive categories (Connor-Linton, 2003).

Hypothesis Testing (H1)

Table 1 summarizes the number, percent, and chi-square of messages related to the impact of the Web site-supported learning and innovation perspective on e-government service delivery performance.

Table 1 presents the results of chi-square tests carried out to determine whether positive messages about the Web site-supported learning and innovation perspective predominate over negative messages about the Web site-supported learning and innovation perspective. The results show that positive messages were significantly greater than negative messages (82.23 at $p<0.001$). Thus, Hypothesis 1, which states that *Web site-supported*

Table 1. Number, percent, and chi-square of learning and innovation

Coded Messages	Number	Percentage (%)	Number of CVSOs	Chi-Square
IVSS	41	77	14	15.86***
IVNS	12	23	5	
Total	53	100	19	
YVCB	37	71	11	9.30***
YVNC	15	29	7	
Total	52	100	18	
TVSI	29	81	14	13.44***
TVNI	7	19	3	
Total	36	100	17	
HTST	41	87	16	43.08***
HTNT	6	13	3	
Total	47	100	19	
TFEE	20	44	7	0.55
TFNE	25	56	12	
Total	45	100	19	
The total chi-square value for Table 1 is				82.23***

$^{***}P<0.001$

learning and innovation perspective will have a positive effect on e-government service delivery performance, is supported.

Hypothesis Testing (H2)

Table 2 summarizes the number, percent, and chi-square of messages related to the impact of the Web site-supported internal process perspective on e-government service delivery performance.

Table 2 presents the results of chi-square tests carried out to determine whether positive messages about the Web site-supported internal process perspective predominate over negative messages about the Web site-supported internal process perspective. The results show that positive messages were significantly greater than negative messages (30.76 at p<0.001). Thus, Hypothesis 2, which states that Web site-supported internal process perspective *will have a positive effect on e-government service delivery performance*, is supported.

Hypothesis Testing (H3)

Table 3 summarizes the number, percent, and chi-square of messages related to the impact of the Web site-supported veteran value proposition perspective on e-government service delivery performance.

Table 3 presents the results of chi-square tests carried out to determine whether positive messages about the Web site-supported veteran value proposition perspective predominate over negative messages about the Web site-supported veteran value proposition perspective. The results show that positive messages were significantly greater than negative messages (37.83 at p<0.001). Thus, Hypothesis 3, which states Web site-supported veteran value proposition perspective *will have a positive effect on e-government service delivery performance*, is supported.

Table 2. Number, percent, and chi-square of internal process

Coded Messages	Number	Percentage (%)	Number of CVSOs	Chi-Square
IFOF	25	45	10	0.64
IFNF	31	55	9	
Total	56	100	19	
IAFV	32	84	17	17.78***
IANV	6	16	2	
Total	38	100	19	
VCWR	29	78	15	11.91***
VCNR	8	22	4	
Total	37	100	19	
VCUI	31	54	11	0.43
VCNI	26	46	8	
Total	57	100	19	
The total chi-square value for Table 2 is				30.76***

****p< 0.001*

Table 3. Number, percent, and chi-square of veteran value proposition

Coded Messages	Number	Percentage (%)	Number of CVSOs	Chi-Square
DWVO	29	59	10	1.65
DWNO	20	41	9	
Total	49	100	19	
IPVV	42	71	11	10.59***
IPNV	17	29	8	
Total	59	100	19	
IFUV	33	77	14	12.30***
IFNV	10	23	4	
Total	43	100	18	
IICV	36	77	12	13.29***
IINV	11	23	5	
Total	47	100	17	
The total chi square value for Table 3 is				37.83***

***$p<0.001$

Hypothesis Testing (H4)

Table 4 summarizes the number, percent and chi-square of messages related to the impact of the Web site-supported financial perspective on e-government service delivery performance.

Table 4 presents the results of chi-square tests carried out to determine whether there are any significant differences between:

- the proportion of increase of initial cost and decrease of initial cost messages, and

Table 4. Number, percent, and chi-square of financial perspective

Coded Messages	Number	Percentage (%)	Number of CVSOs	Chi-Square
				5.91**
DCAB	21	34	8	
DCNB	40	66	9	
Total	61	100	17	
				5.58**
ATCD	38	65.50	10	
ATND	20	34.50	9	
Total	58	100	19	

**$P<0.025$

- the proportion of increase of cost over time and decrease of cost over time messages.

The results show that decrease of cost over time messages were significantly greater than increase of cost over time messages. Thus, Hypothesis 4, which states that the Web site supported financial perspective *will have a positive effect on e-government service delivery performance,* is supported (5.58 at $p<0.025$). This finding suggests that, as time goes by, a different group or generation of veterans might be comfortable with using the Internet. However, increase of initial cost messages were significantly greater than decrease of initial cost messages (5.91 at $p<0.025$). The result also shows that the Web site-supported financial perspective has a negative impact on e-government service delivery performance at the earliest stage of the Web site's use to deliver service to veterans because of the digital divide among various generations of veterans.

DISCUSSION

The results show that the different aspects of the relations between CVSOs and veterans fit with the four interrelated Balanced Scorecard factors. The results also show that three of the four Web site-supported Balanced Scorecard dimensions (learning and innovation, internal process and veteran value proposition) have a positive impact on e-government service delivery performance. However, the impact of the fourth Web site-supported Balanced Scorecard perspective (financial) on e-government service delivery performance is different because of the digital divide among the various generations of veterans.

As time goes by, that finding could change because the high cost associated with the earliest stage of the Web site's use by CVSOs to serve veterans will decrease. This decrease will result from a different group of veterans who might be comfortable using the Internet. Therefore, the digital divide gap between veterans will be small in terms of using the Internet and make them more self-sufficient as far as using technology. The results also show that to increase e-government service delivery performance, veteran self-service needs to be promoted on Web sites. Then, e-government service delivery will free up CVSOs and allow them to focus less on routine tasks that could be easily handled by Web sites.

The digital divide is frequently referred to as the gap between technology "haves" and "have nots" (Kvasny & Keil, 2006; Sipior, Ward, Volonino, & Marzec, 2004). The digital divide is the separation between those who have access to digital technology and those who do not (Dewan & Riggins, 2005). It has been found that faster growth of use of technology occurs when a community has a more well-developed telecommunications infrastructure, more competition in the technology market, and lower technology network access costs (Kauffman & Techatassanasoontorn, 2005). Some factors and indicators that previously may have been expanding the digital divide among veterans with earlier technologies are narrowing the gap as the Internet becomes the defining technology of the information age. Community members like CVSOs can run programs with assistance from volunteers to reduce the digital divide gap. The programs will allow veterans to learn computer skills and their discomfort will decline (Sipior et al., 2003). Counties' initiatives can take a form of community technology centers where veterans can come to get exposure to the Internet and learn something about computers and their applications (Kvasny & Keil, 2006). Counties can also take a very different approach such as providing free internet access to the veterans via a digital cable configuration (Kvasny & Keil, 2006).

Online veteran participation is the most important CVSOs-to-veteran relationship but has received the least attention. Some authors argue that e-service should not be primarily about participation or direct access to government information. First and foremost, it should be about self-help. The opportunity is the construction of a civic cyber-space where citizens talk to each other rather than to the state because the digital divide allows, in the United States for instance, white government service receivers to have more access to the Internet than minority government service receivers. E-government service delivery offers some advantages because citizens can receive their services from home, work, or wherever they have Internet access. The economic costs of e-government service delivery may be reduced. In the presence of the digital divide, both the traditional aspect and the electronic aspect of service provision should be available, negatively affecting the impact of IT introduction. Internet service delivery should not eliminate the traditional service provision, but rather complement it by improving the efficiency and effectiveness of the delivery process.

Governments need to overcome some issues before e-government service delivery will be a reality. First, due to the advent of the digital divide, governments face legal challenges and claims that Internet service delivery effectively discriminates against ethnic minorities. Second, remote Internet service delivery, such as in rural areas, is made from a wide variety of personal computers and Web browsers. Although most combinations of computers and browsers may be compatible with the e-government service delivery system, some older browser versions may not be able to support the security and privacy features required by the government system. For this specific issue, free distribution of the latest versions of the most common browser may be a temporary solution.

In the presence of the digital divide, it is also very important to point out the negative impact of the use of wireless technologies on e-government service delivery performance. In fact, some characteristics of wireless technologies, distinguishing them from PCs and laptops, may have significant implications on citizen privacy and security. Wireless devices are personal devices, carried by government employees and citizens on their person. The successful usage of wireless devices by government agencies will depend on the technological readiness of the government constituents. There are three main issues that need to be taken into account when government adopts wireless technology to deliver service to citizens. First, the wireless device is one that is always handy and available to an individual. Second, a wireless device is distinctly personal, and its usage can be tracked to an individual rather than to a household, as is the case with PCs and other devices. Third, wireless technology is location aware. That means it is easy to track where the user is physically as long as the wireless device is on.

The first main issue is that of accessibility. A wireless device enables a citizen to access government services at any time and from any place. It also allows government employees to access each other at any time and from any place. The second issue is that a citizen can be targeted and reached by a government agency because a wireless device is associated with a particular individual. The third issue is location awareness which implies that a government can pinpoint where the citizen is using the wireless device. It may be advantageous in an emergency situation; however it implies a violation and invasion of citizen privacy.

At this point of wireless development, and due to limited security and privacy issues, as well as incompatible standards, it is more suitable for government to employee applications rather than government to citizen applications.

The findings of this study also demonstrate the importance of usability of a Web site. The International Organization for Standardization defines usability as the extent to which a service can be used by specified users to achieve specified goals with effectiveness, efficiency, and satisfaction in a specified context of use (Cetiner & Ryan, 2004). Therefore, veterans can benefit from the usability functionalities of Web sites to receive e-government service to achieve goals with satisfaction.

The findings and analysis of this research also show that veterans would embrace the e-government service deliveries if more initiatives were put in place to facilitate their online interaction with CVSOs.

CONCLUSION

Theoretical and Managerial Implications

The results of this study will allow Internet and information systems experts to inform government agencies, veteran service and benefit providers, and CVSOs about the impact of Internet network use on e-government service delivery performance. These experts have to show state governments reliable methods of fixing e-government service delivery performance that involve increasing the level of Balanced Scorecard interrelated four factors when CVSOs use Web sites to provide electronic services to veterans.

This study added to the understanding and applicability of the Internet to the public sector. The findings of this research should be used as a diagnostic tool for government constituent managers, including e-government managers. The results of this study show that the challenge for a portal is to offer services that lock-in the government agencies, CVSOs, and veteran service providers.

Rare is the existing research that has developed a research model dedicated to the combined Balanced Scorecard and e-government service delivery performance. The literature review above lets us understand that since the late 1980s, a growing number of studies on e-government models and frameworks developments have been carried out. However, many studies have failed to develop a research model examining the impact of usability of Web portals supported Balanced Scorecard factors on e-government service delivery performance. This study expands the Balanced Scorecard theory and applies it to the e-government service delivery performance.

Limitations

There were three notable limitations with the study. Firstly, the study focuses on only one state (CVSOs in North Dakota) and it remains to be seen if the results apply to other states. This may restrict the generalizability of the findings. Secondly, CVSOs are in the public sectors where the confidentiality and the rules to release personal and professional information are strict. Consequently, it is often fruitless to ask for some information related to the relationship between CVSOs and individual veterans. Even if a small percentage of CVSOs maintain such information, it is unlikely that it is readily available to researchers due to suspicions about the intended use. Finally, there were limitations of the questionnaire used, in particular, to the profile of respondents. From certain statements, we became aware that some respondents did not necessarily have a general knowledge about using the Internet. In addition, it will be more interesting if individual veterans participate in the study.

Future Directions

Even though the data we collected for this study focused on CVSOs, the research approach could be applied to any group of citizens. Quantitative analysis will be needed to perform more tests with the hypotheses developed in this study. Future research is necessary because this research model must be tested further. E-government technology has attracted significant attention among researchers and development communities. Many exciting research issues are being addressed and some are yet to be addressed, and we hope that this article inspires others to do future research by expanding or enhancing this research model.

REFERENCES

Carter, L., & Belanger, F. (2005). The utilization of e-government services: Citizen trust, innovation and acceptance factors. *Information Systems Journal, 15*, 5-25.

Cetiner, M., & Ryan, T. (2004). How to improve the usability of government Web sites. In *Proceedings of the Seventh Conference of the Southern Association for Information Systems, 1*(1), 65-68.

Chan, S.C.H., & Ngai, E.W.T. (2007). A qualitative study of information technology adoption: How ten organizations adopted Web-based training. *Information Systems Journal, 17*, 289-315.

Churchill, G. A. (1979). A paradigm for developing better measures of marketing constructs. *Journal of Marketing Research, 16*(1), 64-73.

Connor-Linton, J. (2003). *Chi-square tutorial.* Retrieved November 14, 2007, from http://www.georgetown.edu/faculty/ballc/webtools/web_chi_tut.html

Dewan S., & Riggins, F. J. (2005). The digital divide: Current and future research directions. *Journal of the Association for Information Systems, 6*(12), 215-235.

Dufner, D., Holley, L., & Reed, B. (2002). Can private sector strategic information systems planning techniques work for the public sector? *Communications of the Association for Information Systems, 8*(27), 413-431.

Gant, D. B., & Gant, J. P. (2002). State Web portals: Delivering and financing e-service. *E-government Series* (pp. 1- 60).

Griffith, T., Sawyer, J., & Neale, M. (2003). Virtualness and knowledge in teams: Managing the love triangle of organizations, individuals, and information technology. *MIS Quarterly, 2*(27), 265-287.

Hsu, K-H. (2005). Using balanced scorecard and fuzzy data envelopment analysis for multinational R & D project performance assessment. *The Journal of American Academy of Business, Cambridge, 1*(27), 189-196.

Huang, D. C., & Hu, Q. (2004). Integrating Web services with competitive strategies: The balanced scorecard approach. *Communications of the Association for Information Systems, 1*(13), 57-80.

Jain, A. (2004). IT penetration in the government workforce: A resource based analysis of the 50 U.S. state governments. In *Proceedings of the Tenth Americas Conference on Information Systems. (Online)*

Kaplan, R., & Norton, D. P. (1992). The balanced scorecard–measures that drive performance. *Harvard Business Review*, 71-79.

Kauffman, R. J., & Techatassanasoontorn, A. A. (2005). Is there a global digital divide for digital wireless phone technologies. *Journal of the Association for Information Systems, 6*(12), 289-316.

Kvasny, L., & Keil, M. (2006). The challenges of redressing the digital divide: A tale of two U.S. cities. *Information Systems Journal, 16*, 23-53.

Levenburg, N. M., & Klein, H. A. (2006). Delivering customer services online: Identifying best practices of medium-sized enterprises. *Information Systems Journal, 16*, 135-155.

Moon, M. J. (2002). State government e-procurement in the information age: Issues, practices, and trends. *E-government series* (pp. 1-65).

Moon, M. J., & Norris, F. D. (2005). Does managerial orientation matter? The adoption of reinventing government and e-government at the municipal level. *Information Systems Journal, 15*, 43-60.

Pandey, I. M. (2005). Balanced scorecard: Myth and reality. *The Journal for Decision Makers, 1*(30), 51-66.

Park, J. H., Lee, J. K., & Yoo J. S. (2005). A framework for designing the balanced supply chain scorecard. *European Journal of Information Systems*, *14*, 335-346.

Ramesh, B., & Tiwana A. (2001). E-services: Models and methods for design, implementation, and delivery. In *Proceedings of the 34th Hawaii International Conference on System Sciences*, 1 page.

Sipior, J. C., Ward, B. T., Volonino, L., & Marzec, J. Z. (2004). A community initiative that diminished the digital divide. *Communications of the Association for Information Systems*, *13*(1), 29-56.

Stowers, G. N. L. (2004). Measuring the performance of e-government. *E-government series* (pp.1-52).

Wright, W., Smith, R., Jesser, R., & Stupeck, M. (1999). Information technology, process reengineering and performance measurement: A balanced scorecard analysis of Compaq Computer Corporation. *Communications of the Association for Information Systems*, *8*(1), 1-47.

APPENDIX

Learning and Innovation Perspective

Coded Messages	Definition of Messages
IVSS	Increase of veteran self-service (veterans use GRO Web site more)
IVNS	Decrease of veteran self-service (veterans visit more CV-SOs)
YVCB	Young veterans (young veterans who come back from Iraq) are computer literate.
YVNC	Old veterans (old veterans who came back from past wars) are not computer literate
TVSI	Teaching veterans the way to secure and to release information on the Internet
TVNI	Not teaching veterans the way to secure and to release information on the Internet
HTST	Having CVSO training sessions twice a year
HTNT	Not having CVSO training sessions twice a year
TFEE	There's formal evaluation of Web services provided to veterans
TFNE	There's no formal evaluation of Web services provided to veterans

Internal Process Perspective

Coded Messages	Definition of Messages
IFOF	I fill out the forms for veterans online
IFNF	I help veterans to fill out the forms online
IAFV	I'm the advocate for the veteran's online activities
IANV	I'm not the advocate for the veteran's online activities
VCWR	The veterans choose on the Internet the NSO (national service organization) they want to have as their representative
VCNR	I choose on the Internet an NSO to represent the veterans
VCUI	Veterans were confused by using the Internet and they called me for help
VCNI	Veterans were not confused by using the Internet

Coded Messages	Definition of Messages
DWVO	I deal with veterans online on a one-on-one basis
DWNO	I deal with veterans online through other people in the community
IPVV	I pay visits to veterans face-to-face
IPNV	I pay visits to veterans online
IFUV	I follow up with all veterans online
IFNV	I follow up with some veterans online
IICV	I initiate contact with veterans online
IINV	Veterans initiate contact with me online

Veteran Value Proposition Perspective

Financial Perspective

Coded Messages	Definition of Messages
DCAB	Decrease cost at the beginning of Internet use
DCNB	Increase cost at the beginning of Internet use
ATCD	Over time, cost of Internet use will decrease
ATND	Over time, cost of Internet use will increase

This work was previously published in Journal of Electronic Commerce in Organizations, Vol. 6, Issue 2, edited by M. Khosrow-Pour, pp. 11-28, copyright 2008 by IGI Publishing (an imprint of IGI Global).

Chapter 17
Development and Testing of an E-Commerce Web Site Evaluation Model

Luiz Antonio Joia
Brazilian School of Public and Business Administration of Getulio Vargas Foundation and Rio de Janeiro State University, Brazil

Luiz Cláudio Barbosa de Oliveira
Brazilian School of Public and Business Administration of Getulio Vargas Foundation, Brazil

ABSTRACT

The scope of this research is to develop and test a model for evaluating B2C e-commerce Web sites. Consequently, this study seeks to investigate the relationship between the Web site interface of B2C e-commerce and virtual customer behavior, concentrating on purchasing attitude and intention. The objective of this article is, therefore, to research which features of a virtual store effectively influence the user purchasing process via the Internet. In order to achieve this aim, the technology acceptance model was adapted for pertinent theoretical references about consumer behavior on the Web. The model developed was then tested and validated through a structural equation modeling approach. An empirical analysis of the CD e-retailing industry in Brazil was then conducted. The study concludes that, for the electronic commerce of CDs, ease of use, trust, pleasure, and attractiveness as perceived by visitors to CD e-retailing Web sites are key issues for understanding customer attitude and purchasing intention.

INTRODUCTION

In a virtual enterprise, most of the contact with customers hinges on the interaction between the customer and the Web site. However, most B2C e-commerce Web sites were not developed to cater to this aspect, thus creating obstacles to winning over user trust in order to make the customer feel sufficiently comfortable to conduct a commercial transaction (Zhang, Small, von Dran, & Barcellos, 2000). Indeed, some Web sites effectively discourage the purchase, producing results that are diametrically opposed to those desired (Nielsen, 2001).

Nonetheless, the online retailing industry has experienced sustained growth due to the Internet. This digital channel has increased market-share in the retailing industry as a whole (Parente, 2000). However, very little is yet known about how this virtual environment affects customer behavior, that is, customer attitude and intention throughout the course of the purchasing process. Moreover, it is important to stress that in this specific scenario, all interaction between the customer and the digital company is developed via the Web site. So, its characteristics are linked to the subjective and objective elements that will influence the purchase, which makes careful planning of the Web site of paramount importance, as this will either lead to the success or failure of a virtual enterprise (Nielsen, 2001).

Hence, the scope of this article is to build a B2C e-commerce Web site evaluation model that makes it possible to pinpoint the Web site characteristics that contribute to improving the virtual customer's attitude and purchasing intention. This model will then be applied to the Brazilian CD e-retailing industry. Similar studies regarding B2C e-commerce Web site evaluation have already been undertaken by several authors (Lohse & Spiler, 1999; Turban & Gehrke, 1999; Zhang et al., 2000; Zhang & Dran, 2001; Schubert & Selz, 1999; Ledered, Mauin, Sena, & Zhuang, 2000; Heidjen, 2001, to name but a few), though all of them use different methodologies and pursue varied objectives.

In this research, an attempt was made to adapt the different aspects of existing models, distinguishing clearly between the external aspects of Web sites, namely those related to the idiosyncrasies of the user, from their internal aspects, that is, those related to their design. In this way, relationships between the different factors under scrutiny were sought, in order to understand their contribution to the closure of an online sale.

This article is structured as follows. First, the theoretical background used to develop the model is presented. The technology acceptance model (TAM) adapted to B2C e-commerce, as well as the classification of online customers, is then explained. Using this bibliographical review, the research methodology adopted, the hypotheses to be tested in the article, and the model developed to evaluate the influence of the e-retailing Web site on the increase of customer attitude and purchasing intention are set forth. The model developed is then tested, validated, and reviewed through its application to the Brazilian CD e-retailing industry. Finally, the contributions and limitations of the study are presented, in addition to recommendations for further research addressing this area.

BIBLIOGRAPHICAL REVIEW

Consumer Behavior on the Web

According to Hoffman and Novak (1995), the relationship between clients and companies changes with the passing of time, thanks to the new possibilities that the Internet has brought to marketing for: i) seeking new forms of response on the part of consumers, ii) increasing penetration in target segments, and iii) increasing client reach. By means of the Internet, organizations seek to increase the influence of their power, without losing the quality and personal touch in attendance.

For Brondmo (2001), relationships between the client and the company on the Web fall into a predictable cycle of four stages, namely (i) awareness, (ii) permission, (iii) involvement, and (iv) loyalty.

In order to get the client's attention, it is necessary to implement promotional actions that are aligned with the marketing plan of the company and based on the type of relationship that one seeks to develop with the client (Brondmo, 2001).

By facilitating dialog between company and client, the Internet makes it even more important to draw the client's attention. This is due to the fact that on the Web it is the clients who choose which site to visit among an almost inexhaustible variety of options (Godin, 2001). The content of the site should be sufficiently attractive to lead people to want to visit, highlighting the importance of making relevant information available to clients (Turban, Lee, King, & Chung, 2000).

Once the company has aroused the client's awareness, it should request permission for communication. The goal at this stage is to stimulate the interest of potential clients in establishing a relationship with the company (Brondmo, 2001).

At this stage, the initiative for a commercial contact with the client should come from the company that will transmit content to this prospective client (Turban et al., 2000). One example of the application of this type of technology is e-mail, when used for sending publicity material. For this type of publicity to be more effective, the message must be of interest to consumers and must have their prior permission for transmission. Request for permission from the consumer for the transmission of information helps to build up a relationship of trust, which the consumer wants and values (Hoffman & Bateson, 1999).

In the process of involvement with the client, the diversity of information existing on the product becomes relevant in that the Internet is a means that does not stimulate impulse purchases (Reedy, Schullo, & Zimmerman, 2001). The client needs to be convinced of the usefulness of the product in order to acquire it. Therefore, supplying plentiful information and offering product samples, such as for example the first chapter of a book and helping to locate information, are important factors that can influence the consumer's purchasing decision (Reedy et al., 2001).

For most clients, choice is influenced by value, which is based mainly on perceptions of quality, price, and purchasing convenience (Cao, Zhang, & Seydel, 2005; Constantinides, 2004; Rust, Zeithaml, & Lemon, 2001).

When potential clients make their first purchase, they are no longer mere listeners and become *de facto* clients. At this stage, a sound basis for a relationship should be built by means of offering information, promotions, and entertainment that are in line with the interests of the client. It is through this relationship that involvement arises (Brondmo, 2001).

Lastly, to progress from the stage of involvement to that of loyalty there must be a sense of commitment between the parties. At this stage, information about the client must be constantly gathered, stored, and analyzed, for the purpose of increasing the level of individualized communications and tailoring the products and services offered to clients (Brondmo, 2001). Thus, one of the indicators of client loyalty is the frequency with which the latter purchases from the same company.

In addition to this, the rate of conversion of visitors into clients becomes the crucial point in the construction of the relationship of a Web site with its users. In this respect, understanding the link between how consumers use the Internet and their own lifestyles and predisposition to purchase via the digital realm are topics discussed by various authors.

Moe and Fader (2001) identified two patterns of customer behavior on the Web, namely exploratory search or objective search. Turban et al. (2000) classified these behavior modes as hedonist or utilitarian, whereas Wolfinbarger and Gilly (2001) defined these two forms of behavior as being experience- or objective-based. Whatever the case, it is possible to identify two different profiles in Internet users, which is a relevant fact for the evaluation of a B2C Web site.

On the other hand, in their research into lifestyle and purchasing profile, Bellman, Lohse, and Johnson (1999) concluded that a group of indicators exists—associated with how wired the users are—that makes it possible to evaluate their predisposition to make purchases online. According to the authors, demographic indicators have no influence on the purchasing behavior of consumers online. Once on the Internet, the best indicator for the tendency to purchase is the search for information about products and the time of Internet use. As a result, the authors concluded that major purchasers via the Internet have a very similar lifestyle, known as the wired style (Bellman et al., 1999).

Kim, Cho, and Rao (2000) extended the concept of predisposition to purchase online, to include the perceived trust and benefit in the purchase as being crucial factors in any transaction. Furthermore, they corroborated the ideas of Bellman et al. (1999), to the extent that they consider that Internet-wired consumers perceive greater benefits in an online purchase. They further concluded that consumers who are more time-oriented also perceive greater benefits in purchasing goods and services online, which tallies with the utilitarian-hedonist behavior proposed by Moe and Fader (2001), Wolfinbarger and Gilly (2001), and Turban et al. (2000).

Technology Acceptance Model on the Web

Davis (1989) developed the original version of the technology acceptance model. His intention was to explain the use of information technology by users, adopting a causal relationship based on beliefs, perceptions, attitudes, intentions, and behavior. According to the model, these concepts should be linked to the theory of reasoned action developed by Fishbein and Ajzen (van der Heidjen, 2000a), which contends that consumers' perceptions are responsible for their attitudes towards a given object. These attitudes, in turn, are thought to define consumers' intentions, thereby modifying their behavior.

The TAM model has been applied as a way of explaining customer satisfaction in relation to a specific Web site by several researchers, including van der Heidjen (2000b), Ledered et al. (2000) and Venkatesh (2000), to name but a few. Each researcher has different objectives, placing varying emphasis on each of the factors involved and including or excluding explanatory components.

According to the model proposed by Davis (1989), which was further adapted for the Web by van der Heidjen (2000b), perceived utility and ease of use are the main factors influencing a person's attitude in the use of information systems. These factors might change the user's behavior, generating either positive or negative motivation. The user's final behavior therefore determines the success or failure of information system use. The model assumes that all the factors can be measured and are positively associated with the use of the information system, that is, the more users perceive the utility of the information system, the more they use it, and vice-versa.

Perceived utility was defined as the expectation people have about how a particular information system can improve their performance. The perceived ease of use, on the other hand, would be the time spent by users to learn how to use a specific information system. The attitude related to the use of the information system is represented by the users' evaluation about their desires to employ the technology (Ledered et al., 2000). Moreover, the behavior accrued from use would be the attitude in action, that is, the accomplishment of the desire to use the technology (van der Heidjen, 2000b).

On the other hand, Zhang et al. (2000) classified the attributes associated with acceptance of a Web site into hygienic and motivating factors. This distinction was conducted by using a research sample. The authors identified four categories as being eminently motivational, namely simplicity, cognitive responses, credibility, and appearance. In addition to this, three categories were considered predominantly hygienic: privacy and security, navigation, and technical aspects. One of the interesting conclusions of this work was to identify that for acceptance of a Web site, certain (hygienic) characteristics must always exist, while other (motivational) interface characteristics contribute purely to user satisfaction, such as simplicity and appearance. These factors only come into play if the hygienic factors are already in place.

The Importance of Trust on the Web

Normally, consumers purchase from retail outlets of well-known brand names even though these may not offer the product they are seeking at the best price. This would indicate that other factors such as trust might be very important in electronic commerce (Flavián, Guinalíu, & Gurrea, 2006; Greenwald & Kephart, 1999).

The importance of trust in the virtual market can also be perceived in consumer behavior research on the Web. In the majority of cases, lack of security and perceived untrustworthiness are singled out as motives why some people do not make purchases online (Hanson, 2000).

In as much as the Internet creates purchasing habits in which both customers and retailers are spatially and temporarily separated, and the parties involved are generally anonymous during the transaction phase, trust in online transactions is becoming a key factor in the success of Internet activities (Smith, Bailey, & Brynjolfsson, 1999; Urban, Sultan, & Qualls, 2000).

When consumers purchase an item in a physical store, the transaction is normally concluded simultaneously, that is, they pay for the merchandise and receive it at the same moment. In the majority of cases in a virtual store, there is a time lag between the consummation of the order, payment, and receipt of the product. More often than not, the consumer will never even speak directly to any employee of the company (Reedy et al., 2001).

The lack of physical presence and the inherent anonymity of the milieu make client perceptions about the company all the more important in business conducted via the Internet (Loebbecke & Hornig, 2001).

Hoffman, Novak, and Peralta (1998) and Jarvenpaa, Tractinsky, and Vitale (2000), among others, tackle the challenge of measuring the relation of purchasing in a digital environment and trust in a Web site, detecting a marked causal relationship. In addition to this, they concluded that the concern of users with trustworthiness in a Web site depends on their idiosyncratic characteristics, namely style and orientation in the digital world.

THE PROPOSED MODEL

The proposed model for this research draws on the theoretical references presented earlier. It links the characteristics of trustworthiness proposed by Hoffman et al. (1998) and Jarvenpaa et al. (2000), with the characteristics presented by models for evaluation of the success of electronic commerce Web sites. One of those used was the technological acceptance model (TAM), modified for the Web by Heidjen (2001), and Web site design evaluation models developed by Turban and Gehrke (1999), Turban et al. (2000), Zhang et al. (2000), Cao et al. (2005), and Abdinnour-Helm (2005). In order to complement the study, it is necessary to consider the characteristics of the users and their lifestyle (Ahn, Ryu, & Han, 2004; Bellman et al., 1999), as well as their search and purchase behavior on the Internet (Constantinides, 2004; Wolfinbarger & Gilly, 2001; Moe & Fader, 2000; Turban et al., 2000).

The model is based on existing correlations between issues associated with the user, namely external factors not controlled by the graphical interface, and issues related to the Web site design, that is, internal factors controlled by the virtual enterprise. Therefore, user style and user behavior orientation are factors considered external to the graphical interface (antecedents to the model). On the other hand, the Web site design influences the consumer's perceptions in relation to four aspects: i) perceived utility, ii) ease of use, iii) perceived pleasure and attractiveness, and iv) trust associated with the purchase, all considered internal factors related to the design of the Web site. Hence, the combination of these two main tracks (internal and external factors) would determine user attitudes related to their virtual consumption in a specific store, that is, the attitudes that would lead the customer to a behavior mode inclined to make a purchase and, finally, to an intention to buy.

External Factors

With respect to the external factors associated with commercial transactions on the Web, customers can be classified according to their browsing objectives, their search for sales, their lifestyles, and so forth. They have unique needs and desires and behave according to their past purchasing experiences (Kotler & Armstrong, 1999).

In this manner, based on the theoretical references presented, the hypotheses relating to the external aspects of the Web site to be statistically tested are divided up into groups and outlined.

In the first group, wired user style is evaluated, that is, the more wired the style of the users, the greater their perception of the Web site design characteristics. Consequently, the hypotheses presented (H1 to H4) were formulated:

H1: *A wired user's style is associated positively with his/her idea of the perceived utility of a Web site.*

H2: *A wired user's style is associated positively with his/her idea of the perceived ease of use of a Web site.*

H3: *A wired user's style is associated positively with his/her idea of the perceived pleasure and attractiveness of a Web site.*

H4: *A wired user's style is associated positively with his/her idea of the perceived trust in a Web site.*

In the second group, the hypotheses relating to user orientation are formulated, that is, the more the user's orientation is based on objectives, the greater the perception of the aspects of utility, ease of use, pleasure, and attractiveness of the Web site. On the other hand, the perception of trust in the Web site will be diminished. Consequently, the hypotheses presented (H5 to H8) were formulated:

H5: *A user's orientation based on objectives is associated positively with his/her idea of the Web site's perceived utility.*

H6: *A user's orientation based on objectives is associated positively with his/her idea of the Web site's perceived ease of use.*

H7: *A user's orientation based on objectives is associated positively with his/her idea of the perceived pleasure and attractiveness associated with the Web site.*

H8: *A user's orientation based on objectives is associated negatively with his/her idea of the perceived trust in a Web site.*

For measurement of the latent external factors: "user style" and "user orientation," the following indicators presented in Table 1 were adopted.

Internal Factors

The analysis of the key success factors of a CD e-retailing Web site is the main scope of this research. In this manner, based on the theoretical references presented, hypotheses relating to the

Table 1. Indicators associated with the external factors of the model

Variable of the Model	Indicators
User Style (Wired/Not Wired)	Frequency of purchase of products via the Internet
	Frequency of search for information via the Internet
	Time of use of the Internet
	Daily number of e-mails exchanged
	Time of availability of the Internet at work
User/Consumer Orientation (Hedonist/Utilitarian)	Use of the Internet for recreation purposes
	Use of the Internet to visit specific Web sites
	Use of the Internet to save time
	Use of the Internet for a defined purpose
	Search for information on specific products

Source: Ahn et al. (2004); Constantinides (2004); Wolfinbarger & Gilly (2001); Moe & Fader (2000); Turban et al. (2000); Bellman et al. (1998).

internal aspects of the Web site are outlined (H9 to H12) for them to be statistically tested:

H9: *The user's idea of the perceived utility and relevance of information is associated positively with his/her attitude and intention to purchase CDs on an e-retailing Web site.*

H10: *The user's idea of the perceived ease of use is associated positively with his/her attitude and intention to purchase CDs on an e-retailing Web site.*

H11: *The user's idea of the perceived pleasure and attractiveness is associated positively with his/her attitude and intention to purchase CDs on an e-retailing Web site.*

H12: *The user's idea of the perceived trust is associated positively with his/her attitude and intention to purchase CDs on an e-retailing Web site.*

As stated, the proposed model combines/includes certain characteristics of the technology acceptance model (TAM), making it easier to

Table 2. Indicators associated with the internal factors of the model

Variable of the Model	Indicators
Perceived Utility	Content
	Velocity of information
	Navigation
Perceived Ease of Use	Organization
	Technical aspects
Pleasure and Attractiveness	Appearance
	Relaxation
Trust	Privacy
	Transaction security
	Attitude
Attitude and Intention to Purchase	Intention to return
	Intention to purchase

Source: Cao et al. (2005); Abdinnour-Helm (2005); Ahn et al. (2004); Heidjen (2001); Zhang et al. (2000); Jarvenpaa et al.(2000); Turban et al.(2000); Turban & Gehrke (1999); Hoffman et al. (1998).

Figure 1. Proposed model for Web site evaluation

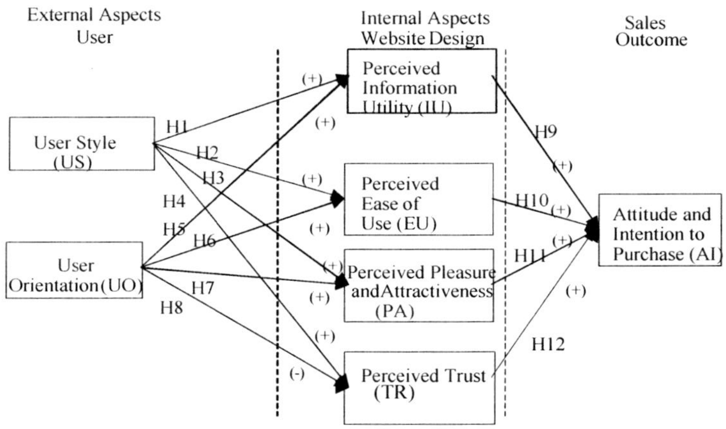

Source: Cao et al. (2005); Abdinnour-Helm (2005); Ahn et al. (2004); Constantinides (2004); Hoffman et al. (1998); Jarvenpaa et al. (2000); Bellman et al. (1998); Moe & Fader (2000); Wolfinbarger & Gilly (2001); Turban et al. (2000); Turban & Gehrke (1999); Zhang et al. (2000); Heidjen (2001).

Table 3. Sampling procedure data

Parameter	Description	Value
N	Number of cases in the sample group	2000
N'	Number of cases randomly selected	1500
N	Number of cases in the sample	496
F	Sample Fraction—conversion rate	33.07%

Source: Trochim (2001)

understand the critical success factors associated with a graphical interface.

For measurement of the latent internal factors, namely perceived utility, perceived ease of use, pleasure and attractiveness, trust, and attitude, and intention to purchase, the following indicators presented in Table 2 were adopted.

However, all these elements must be tested according to the data collection and sample evaluation. Only after that, is it possible to verify whether or not the model tallies fully with the hypotheses presented.

In Figure 1, the model is depicted as a whole, through the relationships among the hypotheses already presented above.

Sampling Procedures

The population of this study encompasses all Brazilians who have purchased a CD at least once using the Internet as their transaction/information channel. A sample of 2,000 subjects was taken from Web-based community groups, discussion groups on the Internet, and other digital sources. From this sample, a group of 1,500 subjects was randomly selected. From this group, 496 subjects agreed to participate in this research (33.07% conversion rate), as presented in Table 3.

In the case in question, a form to be filled out online was used for standard response. An e-mail was sent to users, encouraging them to participate in the research. After reading it, a hyperlink directed the users to the research Web site. Maximum care was taken in this process to avoid the impression of direct mail (or spam), which would not correspond to the initial objective. Since this survey deals with electronic commerce via the Internet, it was understood that utilization of this channel to receive the data would not cause any bias in the sample. After all, every online purchaser is a Web surfer and could therefore access the survey. A Web site was created for this purpose and all the data necessary for the research gathered between May and August 2005, duly explaining the reasons behind the study, ensuring the privacy of the respondent, and the seriousness of the work.

In this research, the simple random sampling method proposed by Trochim (2001) was used, based on the list of potential respondents obtained. Use of an online questionnaire, with response validation by means of the use of Javascript, avoided the respondent leaving any question unanswered. The questionnaire was prepared following the recommendations of Taylor (2000) on e-surveys, that is, easy and quick to answer, in order to facilitate the data tabulation process.

The techniques used for data analysis in this research presuppose that the sample is both normal and random. In order to assess the former premise, tests to evaluate the sample's kurtosis and skewness were conducted. In addition to this, an analysis of the normality of the sample was carried out (Sharma, 1996) as well as the chi-square goodness of fit test (Snedecor & Cochran, 1989). In order to test whether the sample was random, the Run Test (Wonnacott & Wonnacott, 1972) was used. None of the tests indicated that the premises were violated.

The questionnaire used a five-point Likert scale, in which all of the variables were considered ordinals for statistical purposes. In this kind of measurement, the variables can be adequately ordered, though no inferences can be drawn about the distances between them. The scale used is discussed in the next section.

Data Measurement and Collection

One of the greatest challenges in collecting data for research is the creation and formatting of a questionnaire. In order to avoid misinterpretations in this endeavor, a preliminary questionnaire was initially distributed to a pilot group of 14 subjects. This group not only answered the questionnaire, but could also suggest changes in fuzzy points and in the order of the questions. This process had a great impact on the original document, narrowing the scope of the questions and eliminating redundant and/or useless questions. This initiative solved problems of scale, the ordering of items and questions, as well as making it easier to fill out the form.

The modified questionnaire was then submitted to a pre-test involving the same pilot group, in order to conduct initial data analysis. In this way, it was possible to estimate the sample size and verify whether the statistical tools to be applied complied with the research objectives.

The definitive questionnaire was developed using the contributions of this pilot group, as well as the contributions made by the following authors: Hoffman et al. (1998); Jarvenpaa et al. (2000); Bellman et al. (1998); Moe & Fader (2000); Wolfinbarger & Gilly (2001); Turban & Gehrke (1999); Turban et al. (2000); Zhang et al. (2000); Heidjen (2001).

As stated earlier, a Web site was created for data gathering and between May and August 2005 the questionnaires, duly filled out by the respondents, were collected.

Statistical Analysis Processing

In order to analyze the data accrued from this research, statistical methods and multivariate analysis techniques were used. The former are useful for analyzing each variable graphically, identifying outliers, and confirming the necessary premises. The latter are useful for validating the hypotheses and the proposed model as a whole, as well as the relationships among the constructs. In order to achieve this, the SAS 8.2 and LISREL 8.51 statistical packages were used to perform factorial analysis and structural equation modeling (SEM).

The use of SEM gives flexibility to the data analysis, making it possible to confirm the existence of relationships among constructs (confirmatory model strategy). Furthermore, SEM permits the development of a more complex structure of both observed variables (indicators) and latent variables (constructs) (Hair et al., 1998). The analysis process starts with the evaluation of the reliability of the measures and the validation (convergent and discriminant) of the indicators in relation to the constructs. In order to achieve this, factorial analysis was used to verify the reliability and validity of the constructs. After this analysis, the structural model itself was investigated, whereby its goodness of fit, as well as the proposed relationships among the latent variables, was duly established.

DATA ANALYSIS

The data analysis followed the procedure suggested by Hair et al. (1998). First, the coherence of the estimates and the goodness of fit of the model were evaluated. This was followed by an analysis of the measurement model, that is, the relationships between the indicators and the constructs (observed and latent variables). Finally, the results of the structural model analysis were reported, as well as the evaluation of the hypotheses used to test the relationships among the constructs and considerations about the relationship levels among them.

Estimates and Goodness of Fit of the Model

The structural equation model can be analyzed using either the correlation matrix or the covariance matrix. In order to achieve the proposed outcomes, the correlation matrix was used in this research. The first obstacle to determining this matrix lies in the use of ordinal variables. The correlation among continuous variables is measured by the Pearson coefficient. However, when dealing with correlations among ordinal variables, the polyserial/polychoric correlation matrix must be used (Jöreskog & Sörbom, 1993). Furthermore, it is necessary to calculate the asymptotic covariance matrix, to be used as a weighting matrix associated with the correlation measures. Besides, according to Jöreskog and Sörbom (1993), the model must be estimated by the minimum weight least square method (MWLS), instead of the traditional maximum likelihood estimator method (MLE).

The first analysis to be made addressed the coherence of the estimates. In order to achieve this, three items needed to be verified: i) negative error variance (also known as "Heywood case" variance), ii) standard coefficients greater than 1.0, and iii) very high standard error. After these analyses, it was verified that the model was stable without any offending estimate or identification error.

The next step was to evaluate the model validity, which was achieved by analysis of several goodness of fit indexes, as presented in Table 4.

The model is well adjusted to the data collected, as the NFI, TLI, and CFI indexes are above the suggested minimum of 0.9, according to Klem (1995). Also, the AGFI calculated was close to 0.86, again confirming the goodness of fit of the model, as the closer the value is to 1.0, the better the model. The residuals seem to be small, as the value of the RMSR (root mean square residual) was around 5%, confirming yet again the goodness of fit of the model according to Hair et al. (1998). The value of the adjusted residual index

Table 4. Results of the analysis of the model's goodness of fit

Goodness of Fit Indexes of the Model	Value	Goodness of Fit Indexes of the Model	Value
Final Fit Function	0.302	Normed Fit Index (NFI)	0.9788
Goodness of Fit Index (GFI)	0.9246	Tucker-Lewis Index (TLI)	0.9962
GFI Adjusted for df (AGFI)	0.8573	RMSEA Estimate	0.05815
Root Mean Square Residual (RMSR)	0.0526	RMSEA 90% Lower Confidence Limit	0.0331
Chi-Square	198.39	RMSEA 90% Upper Confidence Limit	0.0832
Chi-Square DF	170	Bentler's Comparative Fit Index (CFI)	0.9351
Pr > Chi-Square	0.07	Hoelter's Critical N	52

Source: SAS 8.2 Output, the CALIS procedure

(RMSEA) must lie between 0.00 (perfect fit) and 0.08 (at a 5% level of significance), which occurs with the calculated value of 0.0582. The sample size is also adequate, as it is greater than the critical value presented in the Hoelter index (Hair et al., 1998). The chi-square test alone does not support the validity of the model. The index accrued from this statistic is a measure of the badness of fit of the model, measuring the unexplained variation of the residuals. However, as argued by Jöreskog and Sörbom (1993), the calculation of this measure is based on the sample size (N-1 times the minimum value of the fit function, where N is the size of the sample), being overestimated for samples having more than 200 observations (such as this one). In order to avoid this kind of problem, it is better to use measures that offset the size of the sample, such as AGFI, RMSEA, NFI, and CFI.

Hence, it can be concluded that the model cannot be rejected as invalid for explaining the observed data.

Measurement Model

Analysis of the measurement model starts with the assessment of the significance level of the indicators, followed by verification of the scale validation. In the t-test applied to the coefficients of the indicators, at a 1% level of significance, all the indicators were considered statistically significant in relation to their associated constructs. After that, it is necessary to evaluate the reliability of each construct. A construct is said to be reliable when its indicators explain most of its variation. Three independent methods are usually used to evaluate the reliability of a construct: Cronbach's α-value, composite reliability, and analysis of the extracted variance (Sharma, 1996). The extracted variance is also used to verify the discriminant validity (Pedhazur & Schmelkin, 1991).

According to Nunnally (1978), Cronbach's α-value must be higher than 0.7, whereas the composite reliability must be above 0.6. Another important measure is the extracted variance. This measure is the existing rate of dispersion among the indicators for a specific construct. According to Klem (1995), this measure must be higher than 0.5, for a level of significance of 5%. The three measures quoted are useful for explaining how the indicators confirm the construct. Higher values of these measures demonstrated that the construct was well measured (Sharma, 1996).

Table 5 depicts Cronbach's α-value, the composite reliability, and the extracted variance for the sample used in this research.

Convergent validity analysis was also performed, though space does not permit inclusion of the results thereof in this article. All of the constructs obtained sufficient scores in convergent validity, which measures how the items of a construct are positively correlated to each other (Malhotra, 2001).

Table 5. Reliability of the indicators

Construct	Cronbach's α-value	Composite Reliability	Extracted Variance
US	0.70	0.62	0.51
UO	0.77	0.70	0.58
IU	0.75	0.70	0.54
EU	0.81	0.77	0.59
PA	0.89	0.84	0.69
TR	0.92	0.87	0.72
AI	0.98	0.91	0.80

Source: SAS 8.2 Output, the CALIS procedure

Figure 2. Path diagram: final solution coefficients

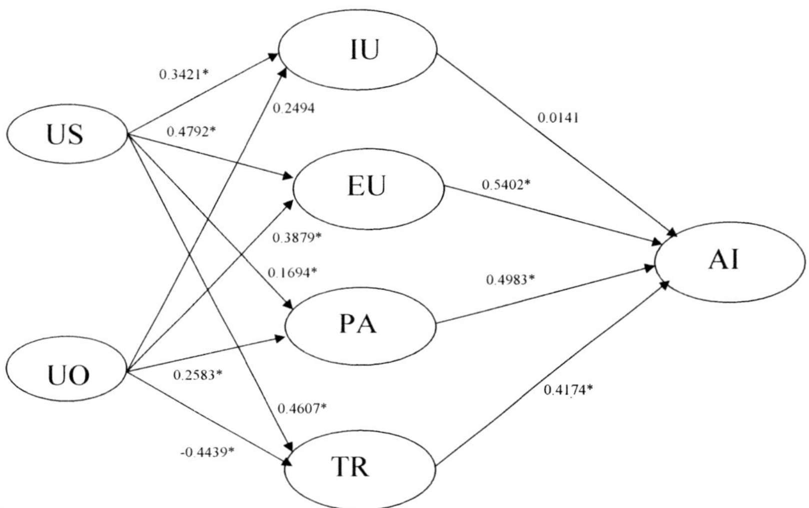

(*) *Significant at 1% level*
Source: SAS 8.2 Output, the CALIS procedure

Then there is also discriminant validity, which assesses the degree to which a concept and its indicators differ from another concept and its indicators. In order to perform discriminant validity analysis, the chi-square difference analysis proposed by Bagozzi, Yi, and Phillips (1991) was used. This statistical procedure involves four steps: a confirmatory factor analysis (CFA) model is defined for the constructs; a null model where the covariance of constructs is equal to 1 is defined; an alternative model, in which the covariance (ϕ) is freely estimated, is defined; and the chi-square difference with one degree of freedom is used to check the null hypothesis that the model's fit is the same. All constructs passed the discriminant validity analysis test.

Hypothesis Test

Once the goodness of fit values of the model had been calculated and the measurement model evaluated, it was possible to test the relationships between the constructs and their related hypoth-

Table 6. Results of the structural model

Structural Path	Hypothesis	Coefficients	t-Test	Decision
User Style -> Information Utility	H1	0.3421	6.0533	Supported
User Style -> Ease of Use	H2	0.4792	8.7907	Supported
User Style -> Pleasure and Attractiveness	H3	0.1694	3.6932	Supported
User Style -> Trust	H4	0.4607	7.1701	Supported
User Orientation -> Information Utility	H5	0.2494	2.2857	Not Supported
User Orientation -> Ease of Use	H6	0.3879	4.0621	Supported
User Orientation -> Pleasure and Attractiveness	H7	0.2583	2.7272	Supported
User Orientation -> Trust	H8	-0.4439	-3.5794	Supported
Information Utility -> Attitude and Intention	H9	-0.0141	0.1731	Not Supported
Ease of Use -> Attitude and Intention	H10	0.5402	6.5388	Supported
Pleasure and Attractiveness -> Attitude and Intention	H11	0.4983	8.3099	Supported
Trust -> Attitude and Intention	H12	0.4174	10.0312	Supported

Source: SAS 8.2 Output, the CALIS procedure

eses. The following path diagram (Figure 2) is used to depict the values calculated.

Based on the results obtained by SEM, the hypotheses were tested (at a 1% level of significance) and the coefficients obtained are depicted in Table 6.

Ten out of the twelve proposed relationships revealed statistical significance, confirming the hypothesis presented. Two hypotheses were not supported (H5 and H9). According to Table 6, the user's orientation (UO) was not supported as related to the user's perception of information utility (IU). Similarly, information utility (IU) does not seem to affect the user's attitude and intention to purchase (AI).

According to the tests, the ease of use, the user's perceived pleasure, and the trust perceived by the user in the Web site elicit a positive attitude, increasing the user's intention to purchase CDs via the Internet. In other words, the greater the user's perceived ease of use (EU), perceived pleasure and attractiveness (PA), and perceived trust (TR) in the Web site, the better the user's attitude in relation to the Web site and the greater the user's intention to purchase/return to the virtual store (AI).

Based on the data analysis conducted by the authors and the information accrued from the structural model, it can be concluded that virtual stores should concentrate on continuous enhancement of their Web sites, so as to increase both traffic and sales. In order to accomplish this, the research showed which indicators are more relevant for each construct, as well as the effects of each valid construct within the model (including the priority among them), so as to improve the user's attitude and intention to purchase CDs on an e-retailing Web site.

CONCLUSION, RESEARCH LIMITATIONS, AND FURTHER STUDIES

Based on the theoretical background outlined, a causal model was proposed and tested, blending several characteristics of other models currently in use. Each part of the proposed model was supported by components of other models, which explains how and why all these parts are consolidated in this new enlarged model. These elements were divided into internal and external facets of a Web site. Naturally, this research seeks to evaluate the existing features of a virtual store that improve the user's attitude towards the Web site, as well as his/her intention to purchase CDs. Indeed, as explained earlier, it is of paramount importance to consider the external aspects of this virtual arena, so as not to miss important explanations about the Web site's performance. Moreover, the user's characteristics must be established, in order to better understand his/her behavior in a digital environment. Several models fail to take these issues into consideration, which is why this research offers original contributions for understanding this environment.

The goodness of fit analysis of the model showed that it was adequate, with a well-dimensioned sample size. The indexes calculated (NFI, TLI, AGFI, RMSEA, and CFI) indicated a very good match between the model and the collected data of the sample. This goodness of fit was even greater than that obtained by Heidjen (2001) in his proposed model. Furthermore, the reliability of the measurement scale was also adequate, as Cronbach's α-values were all greater than 0.7 (Nunnally, 1978), reflecting the goodness of fit between the scale and the constructs measured.

The SEM structural equation supported the contention that user style (US) and user orientation (UO) are relevant to understanding the behavior of an on-line customer. The data proved that the user's attitude and intention is modified by the way he/she perceives the interface. As explained earlier, the ease of use (EU) related to the Web site presented the highest relationship coefficient along with the user's attitude and intention to purchase (AI), as this factor results in a 54% increase in this scale (AI). Another important factor in this research is the pleasure and attractiveness (PA) perceived by the user, which is also an important component in the purchase pattern of a virtual consumer. This construct contributes with a weight of nearly 50% in the user's attitude and intention to purchase (AI) via the Web site. Lastly, the relationship between perceived trust (TR) and user attitude and intention to purchase (AI) achieves a value of 42%. Thus, a digital store's investment in user-friendliness, a visually pleasing layout, and a safe transaction process results in improved expectations related to sales, as the user's attitude and intention to purchase are increased. Moreover, depending on the user's style and orientation, his/her perceptions of the Web site characteristics are different.

Although supported by theory, only two relationships were not supported by the data collated, namely that between user orientation (UO) and perceived information utility (IU), and that between information utility (IU) and user attitude and intention to purchase (AI). Despite the existing theory (Turban et al., 2000, Zhang et al., 2000, Heidjen, 2001), the data collected did not support these relationships. Possibly, a rapid change in virtual customer characteristics, with customers tending to search for useful and relevant information in the easiest and swiftest manner available, can render these constructs and their associated relationships redundant to the respondents. Nevertheless, further research addressing this specific issue is required.

The contributions resulting from this study must be assessed in light of the limitations existing in the method used. One obstacle was associated with the definition of the sample. The sample selection process does not exclude the voluntary/intentional nature of the respondent in participating in the research. After all, not all of the consumers invited to participate actually did so, as the conversion rate was 33.07%. This may lead to conclusions that are only valid for a profile of people who do not object to responding to questions contained in online questionnaires.

Similarly, these consumers were all Brazilians. The study did not seek to analyze the influence of the characteristics of the country in the attitude and intention to purchase via the Web. In other words, this is not a cross-cultural study, though one should not overlook a potential cultural influence associated with the propensity of Internet users to purchase via the Web.

Furthermore, one cannot neglect the relationship between the type of product sold and its propensity to being sold by electronic means, as was proven by Hassanein and Head (2004). Therefore, a study of the attitude and intention of virtual consumers associated purely with the purchase of CDs must also be seen as a research constraint, since this is one of the most homogenous and predictable products to be offered for sale via the Web (Hassanein & Head, 2004).

In this context, it was assumed that the theoretical references lead to a nomological network capable of supporting the model depicted in Figure 1 (Trochim, 2004). For this reason, exploratory factor analysis to verify whether the consumers can be classified as hedonistic or utilitarian was not conducted. However, it is possible that some clients feature both of these characteristics. This is a limitation that could be mitigated in future studies addressing this realm.

Another limitation of the method adopted is related to the use of the attitude and intention (AI) variable as a single construct. Based on the studies of Fishbein and Ajzen (van der Heidjen, 2000a), this construct encompasses different theoretical concepts. However, if this construct were divided in two, there would not be sufficient indicators to assess discriminant validities, which would lead to measurement errors. Thus, the option chosen by the authors was to measure them jointly.

Refinement of measurement of these constructs, by performing similar analysis addressing other segments of B2C e-commerce, is a recommended area for further research. This would permit verification of whether or not the same pattern occurs in other industries. Furthermore, incorporating new theoretical concepts to the model, as well as fine-tuning the measurement of constructs, could confirm the external validity of this model, making it possible to extend the conclusions drawn in this article to all areas of the e-commerce realm.

A significant improvement in this study would be the inclusion of the CD digital store's markup as a dependent variable. This would permit not only validation of the model *vis-à-vis* the Web site's actual financial performance, but also measurement of the return in profits on CD sales associated with each change in website features. However, barriers involved in obtaining such data may hamper future researchers in their attempt to achieve this.

Another important type of study would be longitudinal analysis of the parameters. This would show to what extent the coefficients of the relationship between the concepts change over time. According to Zhang and Dran (2001), some characteristics of the Web site tend to alter with time. Therefore, what is now seen as an additional element will, in future, become a premise that all virtual stores will need to include. For this reason, the type of analysis performed in this research must be continually refined. This is because there is always the risk, inherent to the nature of the business, of a change in consumer behavior, in the information technologies and in the competition itself.

As a last recommendation, the authors suggest a control study in virtual undertakings. By controlling the variables and making adjustments to the interface, it would be possible to check if the result (attitude and intention to purchase) alters as expected/predicted in the model. In this way, it would be possible to establish the quantitative relationship between past behavior and the prediction made by the model.

The analysis of past behavior and the test of future behavior are very significant. To create in-market experiments deliberately (for example, altering certain characteristics of the Web site) makes it possible to attribute differences in the responses to quite complex combinations. By creating a rapid test-analysis cycle, it is possible to adapt constantly what is tested to what is learnt by the organization. The ongoing learning process conducted in this manner offers two benefits: 1) it provides rapid feedback from the market, enabling the company to detect changes in client behavior and the effect of actions by the competition and 2) it reveals new ways of generating economic value by means of fine-tuning the consumer's experience, moving from opportune actions to the generation of sustainable value. However, this type of coordination and cooperation for conducting analysis demands a degree of integration that is normally still difficult to be encountered within companies.

REFERENCES

Abdinnour-Helm, S. F., Chaparro, B. S., & Farmer, S. M. (2005). Using the end-user computing satisfaction (EUCS) instrument to measure satisfaction with a Web site. *Decision Sciences, 36*(2), 341-364.

Ahn, T., Ryu, S., & Han, I. (2005). The impact of the online and offline features on the user acceptance of Internet shopping malls. *Electronic Commerce Research and Applications, 3*, 405-420.

Bagozzi, R. P., Yi, Y., & Phillips, L. W. (1991). Assessing construct validity in organizational research. *Administrative Science Quarterly, 36*(3), 421-458.

Bellman, S., Lohse, G.L., & Johnson E. J. (1999). Predictors of online buying behavior. *Communications of ACM, 42*(12), 32-38.

Brondmo, H. P. (2001). *The engaged customer: The new rules of Internet direct marketing.* Harper Business.

Cao, M., Zhang, Q., & Seydel, J. (2005). B2C e-commerce Web site quality: an empirical examination. *Industrial Management & Data Systems, 105*(5), 645-661.

Constantinides, E. (2004). Influencing the online consumer's behavior: the Web experience. *Internet Research, 14*(2), 111-126.

Davis, F. D. (1989). Perceived usefulness, perceived ease of use, and user acceptance of information technology. *MIS Quarterly, 13*(2), 319-340.

Flavián, C., Guinalíu, M., & Gurrea, R. (2006). The role played by perceived usability, satisfaction and consumer trust on Web site loyalty. *Information and Management, 43*(1), 1-14.

Godin, S. (2001). *Unleashing the idea virus: Stop marketing at people! Turn your ideas into epidemics by helping your customers do the marketing thing for you.* Hyperion.

Greenwald, A. R., & Kephart, J. O. (1999). Shopbots and pricebots. In *Proceedings of the International Joint Conference on Artificial Intelligence.*

Hair, J. F., Jr., Anderson, R. E., Tatham, R. L., & Black, W. C. (1998). *Multivariate data analysis* (5th. ed.). New Jersey: Prentice Hall.

Hanson, W. (2000). *Principles of Internet marketing.* South-Western College Publishing.

Hassanein, K., & Head, M. (2004, June 21-23). The influence of product type on online trust. In *Proceedings of the 17th Electronic Commerce Conference.* Bled, Slovenia.

Hoffman, D., & Bateson, J. E. G. (1999). *Essentials of services marketing.* International Thomson Publishing.

Hoffman, D. L., & Novak, T. P. (1995). *Marketing in hypermedia computer mediated environment: Conceptual foundations.* Owen Graduate School of Management and Vanderbilt University. Retrieved on May 2003, from http://www2000.ogsm.vanderbilt.edu

Hoffman, D. L., Novak, T. P., & Peralta M. (1998). *Building consumer trust in online environments: The case for information privacy.* Owen Graduate School of Management and Vanderbilt University. Retrieved on May 2003, from http://www2000.ogsm.vanderbilt.edu

Jarvenpaa, S. L., Tractinsky, N., & Vitale M. (2000). Consumer trust in an Internet store. *Information Technology & Management, 1*(1), 45-71.

Jöreskog, K. G., & Sörbom, D. (1993). *LISREL 8: structural equation modeling with SIMPLIS command language.* Mooresville, IN: Scientific Software.

Kim, D. J., Cho, B., & Rao, H. R. (2000, December 10-13). Effects of consumer lifestyles on purchasing behavior on the Internet: a conceptual framework and empirical validation. In *Proceedings of the Twenty First International Conference on Information Systems* (pp. 688-695). Brisbane, Australia.

Klem, L. (1995). *Path analysis: Reading and understanding multivariate statistics.* In L. G. Grimm & P. R. Yarnold (Eds.), Washington, D.C.: American Psychological Association.

Kotler, P., & Armstrong, G. (1999). *Princípios de marketing* (7th ed.). Rio de Janeiro: LTC.

Ledered, A. L., Mauin, D. J., Sena, M. P., & Zhuang, Y. (2000). The technology acceptance model and the World Wide Web. In *Proceedings of the Conference of Decision Support Systems* (Vol. 29, pp. 269-282).

Loebbecke, C., & Hornig, V. (2001, June 25-26). Increasing trust in e-commerce: Concepts and examples of insurance solutions. In *Proceedings of the 14th Bled Electronic Commerce Conference.* Bled, Slovenia.

Lohse, G. L., & Spiller, P. (1999). Electronic shopping: The effect of customer interface on traffic and sales. *Communications of ACM, 41*(7), 81-87.

Malhotra, N. K. (1993). Marketing research: An applied orientation. New Jersey: Prentice Hall, International Edition.

Moe, W. W., & Fader, P. S. (2001). *Modeling online store visit patterns as a measure of customer satisfaction.* Retrieved May 2003, from http://216.107.131.155/archive/papers/

Nielsen, J. (2001). *Projetando Web Sites—design Web usability* (2nd ed.). Rio de Janeiro: Editora Campus.

Nunnally, J. C. (1978). Psychometric theory (2nd ed.). New York: McGraw Hill.

Parente, J. (2000). *Varejo no Brasil.* São Paulo: Editora Atlas.

Pedhazur, E., & Schmelkin, L. P. (1991). *Measurement, design and analysis: An integrated approach.* Hillsdale: Lawrence Erlbaum Associates.

Reedy, J., Schullo, S., & Zimmerman, K. (2001). *Electronic marketing: Integrating electronic resources into the marketing process.* South-Western.

Rust, R. T., Zeithaml, V. A., & Lemon, K. N. (2001). *Driving customer equity: How customer lifetime value is reshaping corporate strategy.* The Free Press.

Schubert, P., & Selz, D. (1999). Measuring the effectiveness of electronic commerce sites going beyond traditional marketing paradigms. In *Proceedings of the 32nd Hawaii International Conference on System Sciences.*

Sharma, S. (1996). *Applied multivariate techniques.* New York: John Willey & Sons.

Smith, M. D., Bailey, J., & Brynjolfsson, E. (1999). *Understanding digital markets: review and assessment.* Retrieved May 2003, from http://www.ecommerce.mit.edu/papers/ude

Snedecor, G. W., & Cochran, W. G. (1989). *Statistical methods.* Iowa State University Press.

Taylor, H. (2000). Does Internet research work? *Journal of the Market Research Society, Winter,* 13.

Trochim, W. M. K. (2001). *The research methods knowledge base* (2nd ed.). Retrieved May 2003, from at http://www.trochim.human.cornell.edu/kb/index.htm

Trochim, W. M. K. (2004). *The nomological network* (2nd ed.). Retrieved August 2004, from http://www.socialresearchmethods.net/kb/nomonet.htm

Turban, E., & Gehrke, D. (1999). Determinants of successful Web site design: Relative importance and recommendations for effectiveness. In *Proceedings of the 32nd Hawaii International Conference on System Sciences.*

Turban, E., Lee, J., King, D., & Chung H. M. (2000). *Electronic commerce—a managerial perspective.* Prentice Hall.

Urban, G. L., Sultan, F., & Qualls, W. J. (2000). Placing trust at the center of your Internet strategy. *Sloan Management Review, Fall.*

van der Heijden, H. (2000a). *E-TAM: A revision of the technology acceptance model to explain Web site revisits. Research Memorandum, 29.*

van der Heijden, H. (2000b). *Using the technology acceptance model to predict Web site usage: Extension and empirical test. Research Memorandum, 25.*

van der Heijden, H. (2001, June). Factors influencing the usage of Web sites: The case of a generic portal in the Netherlands. In *Proceedings of the 14th Bled Electronic Commerce Conference: e-everything: e-commerce, e-government, e-household, e-democracy.* Bled, Slovenia.

Venkatesh, V. (2000). *Determinants of perceived ease of use: Integration control, intrinsic motivation and emotion into the technology acceptance model.* Retrieved May 2003, from http://isr.commerce.ubc.ca/Abstract/11-4-Venkatesh.html

Wolfinbarger, M., & Gilly, M.. (2001). *Shopping online for freedom, control and fun.* California State University Long Beach: College of Business Administration.

Wonnacott, T., & Wonnacott, R. (1972). *Introductory statistics for business and economics* (2nd ed.). New York: John Wiley & Sons, Inc.

Zhang, P., Small, R. V., von Dran, G. M., & Barcellos, S. (2000). A two factor theory for Web site design. In *Proceedings of the 33rd Hawaii International Conference on System Sciences.*

Zhang, P., & von Dran, G. (2001). Expectations and rankings of Web site quality features: Results of two studies on user perceptions. In *Proceedings of the 34th Hawaii International Conference on System Sciences.*

This work was previously published in Journal of Electronic Commerce in Organizations, Vol. 6, Issue 3, edited by M. Khosrow-Pour, pp. 37-53, copyright 2008 by IGI Publishing (an imprint of IGI Global).

Compilation of References

Aalto, L., Gothlin, N., Korhonen, J., & Ojala, T. (2004). Bluetooth and WAP push based location-aware mobile advertising system. In *Proceedings of the 2nd International Conference on Mobile Systems, Applications, and Services* (pp. 49-58). ACM.

Abdinnour-Helm, S. F., Chaparro, B. S., & Farmer, S. M. (2005). Using the end-user computing satisfaction (EUCS) instrument to measure satisfaction with a Web site. *Decision Sciences*, *36*(2), 341-364.

Abdul-Gader, K., & Kozar, K. A. (1995). The Impact of Computer Alienation on Information Technology Investment Decisions: An Exploratory Cross-National Analysis. *MIS Quarterly*, *19*(4), 535–559. doi:10.2307/249632

Abolhasan, M., Wysocki, T., & Dutkiewicz, E. (2004). A review of routing protocols for mobile ad hoc networks. *Ad Hoc Networks*, *2*, 1-22.

Adams, D., Nelson, R., & Todd, P. (1992). Perceived Usefulness, Ease of Use, and Usage of Information Technology: A Replication. *MIS Quarterly*, *16*(2), 227–247. doi:10.2307/249577

Adebanjo, D. (2003). Classifying and selecting e-CRM applications: an analysis-based proposal. *Management Science*, *41*(6), 570-577.

Adomavicious, G., & Tuzhilin, A. (1999). User profiling in personalization applications through rule discovery and validation. In *Proceedings of the ACM Fifth International Conference on Data Mining and Knowledge Discovery (KDD'99)* (pp. 377-381).

Ahn, T., Ryu, S., & Han, I. (2005). The impact of the online and offline features on the user acceptance of Internet shopping malls. *Electronic Commerce Research and Applications*, *3*, 405-420.

Ajzen, I. (1991). The Theory of Planned Behavior. *Organizational Behavior and Human Decision Processes*, *50*(2), 179–211. doi:10.1016/0749-5978(91)90020-T

Ajzen, I., & Driver, B. (1992). Application of the Theory of Planned Behaviour to Leisure Choice. *Journal of Leisure Research*, *24*(3), 207–224.

Ajzen, I., & Fishbein, M. (1980). The prediction of behavior from attitudinal and normative variables. *Journal of Experimental Social Psychology*, *6*(4), 466-487.

Ajzen, I., & Fishbein, M. (1980). *Understanding Attitudes and Predicting Social Behavior*. London: Prentice-Hall, Englewood Cliffs.

Ajzen, I., & Madden, T. (1986). Prediction of Goal-Directed Behavior: Attitudes, Intentions, and Perceived Behavioral Control. *Journal of Experimental Social Psychology*, *22*, 453–474. doi:10.1016/0022-1031(86)90045-4

Åkesson, M., & Ihlström, C. (2006a). Towards an ubiquitous media environment-adding the e-newspaper channel. In *Proceedings of TAGA 2006*, Vancouver, Canada.

Åkesson, M., & Ihlström, C. (2006b). Designing and evaluating the calm e-newspaper. In *Proceedings of ECIS 2006*, Göteborg, Sweden.

Al-Gahtani, S., & King, M. (1999). Attitudes, Satisfaction and Usage: Factors Contributing to Each in the Acceptance of Information Technology. *Behaviour & Information Technology*, *18*(4), 277–297. doi:10.1080/014492999119020

Al-Hajri, S. (2005). *Internet Technology Adoption in the Banking Industry*. Victoria University, Melbourne.

Al-Naeem, T., Rabhi, F. A., Benatallah, B., & Ray, P. K. (2004). Systematic Approaches for Designing B2B

Applications. *International Journal of Electronic Commerce*, *9*(2), 41–70.

Alt, R. (2008). *Überbetriebliches Prozessmanagement - Gestaltungsmodelle und Technologien zur Realisierung integrierter Prozessportale*. Berlin: Logos.

Alt, R., & Fleisch, E. (2000). Business Networking Systems - Characteristics and Lessons Learned. *International Journal of Electronic Commerce*, *5*(2), 7–27.

Alt, R., Reichmayr, C., & Zurmühlen, R. (2000). Developing eCommerce within Business Networks - The Case of ETA SA. In *Proceedings of the 13th Bled Electronic Commerce Conference, Bled*.

Alt, R., Reichmayr, C., Cäsar, M. A., & Zurmühlen, R. (2002). Evolution of Electronic Catalogs to Customer Process Portals - A structured Approach at ETA S.A. In *Proceedings of the 15th Bled Electronic Commerce Conference, Bled*.

Al-Wohaibi, M., Masoud, F., & Edwards, H. (2002). Fundamental Risk Factors in Deploying IT/IS Projects in Omani Government Organisations. *Journal of Global Information Management*, *10*(4), 1–22.

Amazon (2008). *Kindle: Amazon's new wireless reading device*. Retrieved January 14th, 2008, from http://www.amazon.com/Kindle-Amazons-Wireless-Reading-Device/dp/B000F173MA

Amit, R., & Zott, C. (2001). Value creation in e-business. *Strategic Management Journal*, *22*, 493-520.

An introduction to theory and research. Reading, MA: Addison-Wesley.

Anderson, R. E., & Srinivasan, S. S. (2003). E-satisfaction and e-loyalty: A contingency framework. *Psychology and Marketing*, *20*, 123-138.

Anderson, R., & Srinivasan, S. (2003). E-Satisfaction and E-Loyalty: A Contingency Framework. *Psychology and Marketing*, *20*(2), 123–138. doi:10.1002/mar.10063

Andreou, A., Chrysostomou, C., Leonidou, C., Mavromoustakos, S., Pitsillides, A., Samaras, G. (2002). *Mobile commerce applications and services: A design development approach*. The First International Conference on Mobile Business M-Business2002, July 8-9, Athens, Greece, 2002.

Androutsellis-Theotokis, S., & Spinellis, D. (2004). A survey of peer-to-peer content distribution technologies. *ACM Computing Surveys*, *36*(*4*), 335-371.

Ansari, A., & Mela, C.F. (2003). E-customization. *Journal of Marketing Research*, *40*, 131-145.

Anton, J. (1996). *Customer relationship management: Making hard decisions with soft numbers*. Englewood Cliffs, NJ: Prentice-Hall.

Apache Software Foundation. (n.d.). *Apache HTTP Server Project*. Retrieved June 21, 2008, from http://httpd.apache.org/

Arbuckle, J. L., & Wothe, W. (2000). *Amos 4.0 user's guide*. Small Waters Corporation.

Archer, N., & Yufei, Y. (2000). Managing business-to-business relationships throughout the e-commerce procurement life cycle. *Internet Research: Electronic Networking Applications and Policy*, *10*(5), 385–395. doi:10.1108/10662240010349390

Ariely, D. (2000). Controlling the information flow: Effects on consumers' decision making and preferences. *Journal of Consumer Research*, *27*(2), 233-248.

Athanassopoulos, A. (2000). Customer Satisfaction Cues to Support Market Segmentation and Explain Switching Behavior. *Journal of Business Research (Bangladesh)*, *47*, 191–207. doi:10.1016/S0148-2963(98)00060-5

Athanassopoulos, A. D. (2000). Customer satisfaction cues to support market segmentation and explain switching behavior. *Journal of Business Research*, *47*(3), 191-207.

Ayadi, A. (2007). Business Challenges of Online Banking Portals. In A. Tatnall (Ed.), *Encyclopaedia of Portal Technology and Applications* (Vol. 1, pp. 102-105). Hershey, PA: Information Science Reference.

Babcock, B., Babu, S., Datar, M., Motwani, R., & Widom, J. (2002, June). Models and issues in data stream systems. In *Proceedings of the 21st ACM SIGACT-SIGMOD-SIGART Symposium on Principles of Database Systems* (PODS'02) (pp. 1-16).

Bagozzi, R. P., Yi, Y., & Phillips, L. W. (1991). Assessing construct validity in organizational research. *Administrative Science Quarterly*, *36*(3), 421-458.

Bagozzi, R., Davis, F., & Warshaw, P. (1992). Development and Test of a Theory of Technological Learning and Usage. *Human Relations*, *45*(7), 659–670. doi:10.1177/001872679204500702

Bagozzi, R., Wong, N., Abe, S., & Bergami, M. (2000). Cultural and Situational Contingencies and the Theory of Reasoned Action: Application to Fast Food Restaurant Consumption. *Journal of Consumer Psychology*, *9*(2), 97–106. doi:10.1207/S15327663JCP0902_4

Bahl, P., & Padmanabhan, V. N. (2000). *RADAR: Proceedings of IEEE INFOCOM: An in-building RF based user location and tracking system.*

Bakos, J. Y. (1991). A strategic Analysis of Electronic Marketplaces. *MIS Quarterly*, *15*(3), 295–310. doi:10.2307/249641

Balasubramanian, S., Peterson, R.A., & Jarvnpaa, S.L. (2002). Exploring the implications of m-commerce for markets and marketing. *Journal of the Academy of Marketing Science*, *30*(4), 348-361.

Banavar, G., Beck, J., Gluzberg, E., Munson, J., Sussman, J., & Zukowski, D. (2000). Challenges: An application model for pervasive computing. In *Proceedings of the 6th Annual International Conference on Mobile Computing and Networking* (pp. 266-274).

Bansal, H. S., & Taylor, S. F. (1999). The service provider switching model (SPSM): A model of consumer switching behavior in the service industry. *Journal of Service Research*, *2*(2), 200-218.

Barnes, D., Hinton, M., & Mieczkowska, S. (2005). Enhancing Customer Service Operations in E-Business: The Emotional Dimension. *Journal of Electronic Commerce in Organizations*, *3*(2), 17–32.

Barney, J. (1991). Firm resources and sustained competitive advantage. *Journal of Management*, *17*, 99-120.

Basant, R., Commander, S. J., Harrison, R., & Menezes-Filho, N. (2006). ICT Adoption and Productivity in Developing Countries: New Firm Level Evidence from Brazil and India. IZA Discussion Paper No. 2294

Basselier, G., Reich, B.H., & Benbasat, I. (2001). Information technology competence of business managers: A definition and research model. *Journal of Management Information Systems*, *17*(4), 159-182.

Bayne, K.M. (2002). *Marketing without wires*. Wiley.

Bebko, C. P. (2000). Service intangibility and its impact on consumer expectations of service quality. *Journal of Services Marketing*, *14*(1), 9-26.

Beck, L., & Ajzen, I. (1991). Predicting Dishonest Actions Using the Theory of Planned Behavior. *Journal of Research in Personality*, *25*(3), 285–301. doi:10.1016/0092-6566(91)90021-H

Bellman, S., Lohse, G.L., & Johnson E. J. (1999). Predictors of online buying behavior. *Communications of ACM*, *42*(12), 32-38.

Benbasat, I., Goldstein, D. K., & Mead, M. (1987). The Case Research Strategy in Studies of Information Systems. *MIS Quarterly*, *11*(3), 369–386. doi:10.2307/248684

Benedicto, J., Dinwiddy, S. E., Gatti, G., Lucas, R., & Lugert, M. (2000) *GALILEO: Satellite System Design and Technology Developments*. European Space Agency.

Berry, R. W. (1981). The employee as customer. *Journal of Retail Banking*, *3*(March), 33-40.

Besley, T., & Case, A. (1993). Modeling Technology Adoption in Developing Countries. *The American Economic Review*, *83*(2), 396–402.

Bhappu, A. D., & Schultze, U. (2006). The role of relational and operational performance in B2B customers' adoption of self-service technology. *Journal of Service Research*, *8*(4), 372–385. doi:10.1177/1094670506286571

Bharadwaj, A.S., Sambamurthy, V., & Zmud, R.W. (2000). IT capabilities: Theoretical perspectives and empirical operalization. In *Proceedings of the 21st International Conference on Information Systems*, Brisbane, Australia, (pp. 378-385).

Bieberstein, N., et al. (2005). *Service-oriented architecture compass: Business value, planning, and enterprise roadmap.* Englewood Cliffs, NJ: Prentice Hall.

Bitner, M. J. (1990). Evaluating service encounters: The effects of physical surroundings and employee responses. *Journal of Marketing*, *54*(2), 69-81

Bitner, M. J., Brown, S. W., & Meuter, M. L. (2000). Technology Infusion in Service Encounters. *Journal of the Academy of Marketing Science*, *28*(1), 138–149. doi:10.1177/0092070300281013

Bobbitt, L., & Dabholkar, P. (2001). Integrating Attitudinal Theories to Understand and Predict Use of Technology-Based Self-Service: The Internet as an Illustration. *International Journal of Service Industry Management*, *12*(5), 423–450. doi:10.1108/EUM0000000006092

Bolton, L., Warlop, L., & Alba, J. W. (2003). Consumer perceptions of price (un)fairness. *Journal of Consumer Research*, *29*(4), 474-491.

Bolton, R. N., & Lemon, K. N. (1999). A dynamic model of customers' usage of services: Usage as an antecedent and consequence of satisfaction. *Journal of Marketing Research*, *36*(2), 171-186.

Bolton, R. N., Kannan, P. K., & Gramlett, M. D. (2000). Implications of loyalty program membership and service experiences for customer retention and value. *Journal of the Academy of Marketing Science*, *28*(1), 95-108.

Bonfield, E. (1974). Attitude, Social Influence, Personal Norm and Intentions as Related to Brand Purchase Behavior. *JMR, Journal of Marketing Research*, *11*(Nov), 379–389. doi:10.2307/3151284

Brondmo, H. P. (2001). *The engaged customer: The new rules of Internet direct marketing*. Harper Business.

Brown, J., Bovey, D., & Chen, X. (1997). Context-aware applications: From the laboratory to the marketplace. *IEEE Personal Communications*, *4*(5), 58-64.

Brown, R.B. (1994). Reframing the competency debate. *Management Learning*, *25*(2), 289-299.

Brucks, M., Zeithaml, V. A., & Naylor, G. (2000). Price and brand name as indicators of quality dimensions for consumer durables. *Journal of Academy of Marketing Science*, *28*(3), 359-374.

Buchanan, T., & Smith, J. L. (1999). Using the Internet for psychological research: Personality testing on the World-Wide Web. *British Journal of Psychology*, *90*, 125-144.

Bucklin, C. B., DeFalco, S. P., DeVincentis, J. R., & Levis, J. P. I. (1996). Are you tough enough to manage your channels? *McKinsey Quarterly*, *1*, 104–114.

Bulander, R., Decker, M., & Schiefer, G. (2005). Comparison of different approaches for mobile advertising. In *Proceedings of the Second IEEE International Workshop on Mobile Commerce and Services* (pp. 174-182).

Bunduchi, R. (2005). Business relationships in Internet-based Electronic Markets: The role of goodwill trust and transaction costs. *Information Systems Journal*, *15*, 321–341. doi:10.1111/j.1365-2575.2005.00199.x

Business Week. (2001, December 3). *It's all about trust*.

Caldow, J. (2001). *E-gov goes wireless: From Palm to shining Palm*. Institute of Electronic Government, IBM Corporation.

Callon, M., Courtial, J. P., Turner, W. A., & Bauin, S. (1983). From Translations to Problematic Networks: An Introduction to Co-Word Analysis. *Social Sciences Information. Information Sur les Sciences Sociales*, *22*(2), 191–235. doi:10.1177/053901883022002003

Campbell, A. (2004). *Technology levels the playing field*. Retrieved January 2, 2008, from http://trendtracker.blogspot.com/2004/04/trend-technology-levels-playing-field.html

Canada IT. (2006). *TMW systems acquires Maddocks Systems Inc*. Retrieved September 2006 from http://www.canadait.com/cfm/index.cfm?It=106&Id=24394&Se=2&Lo=2

Candace, D.P. (2005). *E-commerce and m-commerce technologies*. IRM Press.

Cao, M., Zhang, Q., & Seydel, J. (2005). B2C e-commerce Web site quality: an empirical examination. *Industrial Management & Data Systems*, *105*(5), 645-661.

Capra, L., Emmerich, W., & Mascolo, C. (2003). CARISMA: Context - Aware Reflective Middleware System for Mobile Applications. *IEEE Transactions on Software Engineering*, *29*(10), 929-945.

Carlsson, C., Carlsson, J., Hyvönen, K., Puhakainen, J., & Walden, P. (2006). Adoption of mobile devices/services – searching for answers with the UTAUT. In *Proceedings of the 39th Hawaii International Conference on System Sciences*, Hawaii. (CD-ROM). Computer Society Press.

Carter, J., Bitting, E., & Ghorbani, A. A. (2002). Reputation formalization within information sharing multiagent architectures. *Computational Intelligence, 18*(4), 515-534.

Carter, L., & Belanger, F. (2005). The utilization of e-government services: Citizen trust, innovation and acceptance factors. *Information Systems Journal, 15*, 5-25.

Carver, C. S., & Scheier, M. F. (1990). Origins and functions of positive and negative affect: A control-process view. *Psychological Review, 97*(1), 19-35.

Castro, P., Chiu, P., Kremenek, T., & Muntz, R. (Eds. Abowd G. D., Brumitt B., & Shafer S.A.N.). (2001). *Proceedings of Ubicomp-2001: A Probabilistic Room Location Service for Wireless Networked Environments.* Springer-Verlag.

Cenfetelli, R. T., Benbasat, I., & Al-Natour, S. (2008). Addressing the What and How of Online Services: Positioning Supporting-Services Functionality and Service Quality for Business-to-Consumer Success. *Information Systems Research, 19*(2), 161–181. doi:10.1287/isre.1070.0163

Cerami, E. (2002). *Web services essentials.* New York: O'Reilly.

Cetiner, M., & Ryan, T. (2004). How to improve the usability of government Web sites. In *Proceedings of the Seventh Conference of the Southern Association for Information Systems, 1*(1), 65-68.

Chan, A. (2004). *Mediacorp and Singapore press holdings merge their TV and free newspaper operations.* Retrieved November 15, 2007, from Singapore press holdings, from http://www.sph.com.sg/news/archive/press_040917_003.html

Chan, A.T., & Chuang, S.-N. (2003). MobiPADS: A reflective middleware for context-aware mobile computing. *IEEE Transactions on Software Engineering, 29*(10), 1072-1085.

Chan, S.C.H., & Ngai, E.W.T. (2007). A qualitative study of information technology adoption: How ten organizations adopted Web-based training. *Information Systems Journal, 17*, 289-315.

Chandon, J.-L., Leo, P.-Y., & Philippe, J. (1997). Service encounter dimensions-a dyadic perspective: Measuring the dimensions of service encounters as perceived by customers and personnel. *International Journal of Service Industry Management, 8*(1), 65-86.

Chang, E., West, M., & Hadzic, M., (2006, August). A digital ecosystem for extended logistics enterprises. In *Proceedings of the 11th International Workshop on Telework.* Fredericton, Canada:

Chang, M. (1998). Predicting Unethical Behavior: A Comparison of the theory of Reasoned Action and the Theory of Planned Behavior. *Journal of Business Ethics, 17*, 1825–1834. doi:10.1023/A:1005721401993

Chang, V., & Uden, L. (2007, November). The three e's: E-learning, e-knowledge and e-ecosystems for SMEs. In Proceedings of the Eighth International We-B (Working for E-Business) Conference. Melbourne Australia:

Chaston, I., & Mangles, T. (2002). E-commerce in small UK manufacturing firms: A pilot study on internal competencies. Journal of Marketing Management, *18*(3-4), 341-360.

Chatterjee, D., Grewal, R., & Sambamurthy, V. (2002). Shaping up for e-commerce: Institutional enablers of the organizational assimilation of Web technologies. *MIS Quaterly, 26*(2), 65-90.

Chen, E., Vahidov, R., & Kersten, G. E. (2004). Agent-supported negotiations in the e-marketplace. *International Journal of Electronic Business, 3*(1), 28-49.

Chen, G., & Kotz, D. (2000). *A survey of context-aware mobile computing research.* Technical Report TR2000-381. Department of Computer Science, Dartmouth College.

Chen, H., Finin, T., & Joshi, A. (2003). An ontology for context-aware pervasive computing systems. *Knowledge Engineering Review, 18*(3), 197-207.

Chen, Q., & Chen, H.-M. (2004). Exploring the success factors of eCRM strategies in practice. *Database Marketing & Customer Strategy Management, 11*(4), 333–343. doi:10.1057/palgrave.dbm.3240232

Cheng, V. S. Y., Hung, P. C. K., & Chiu, D. K. W. (2007). Enabling Web services policy negotiation with privacy preserved using XACML. In *Proceedings of the 40th Annual Hawaii International Conference on System Science (HICSS 2007)* (p. 33a).

Cheung, S. C., Hung, P. C. K., & Chiu, D. K. W. (2003). On the e-negotiation of unmatched logrolling views. In *Proceedings of the 36th Annual Hawaii International Conference on System Science (HICSS 2003)* (p. 29).

Chiou, J.-S. (2004). The antecedents of consumers' loyalty toward Internet service providers. *Information & Management, 41*, 685-695.

Chiu, D. K. W., Cheung, S. C., & Till, S. (2003). A three-layer architecture for e-contract enforcement in an e-service environment. In *Proceedings of the 36th Annual Hawaii International Conference on System Science (HICSS 2003)* (p. 74a).

Chiu, D. K. W., Cheung, S. C., Hung, P. C, K., Chiu, S. Y. Y., & Chung, A. K. K. (2005). Developing e-negotiation support with a meta-modeling approach in a Web services environment. *Decision Support Systems, 40*(1), 51-69.

Chiu, D. K. W., Cheung, S. C., Hung, P. C. K., & Leung, H. F. (2005). Facilitating e-negotiation processes with semantic Web technologies. In *Proceedings of the 38th Annual Hawaii International Conference on System Science (HICSS 2005)* (p. 36a).

Chiu, D. K. W., Poon, J. K. M., Lam, W. C., Tse, C. Y., Sui, W. H. T., & Poon, W. S. (2005). How ontologies can help in an e-marketplace. In *Proceedings of the 13th European Conference on Information Systems, Information Systems in a Rapidly Changing Economy (ECIS 2005)* (p. 57).

Chlamtac, I., Conti, M., & Liu, J. J.-N. (2003). Mobile ad hoc networking: Imperatives and challenges. *Ad Hoc Networks, 1*(1), 13-64.

Chozam Inc. (2005). *Technology information center.* Retrieved August 30, 2005, from http://www.chozamtech.com/software/industry/transportation/transportation_software.html

Churchill, G. A. (1979). A paradigm for developing better measures of marketing constructs. *Journal of Marketing Research, 16*(1), 64-73.

Chyi, H. I. (2002). No one would pay for it? Web content as inferior goods. In *Proceedings of the Fifth World Media Economics Conference*, Turku, Finland.

Clark, J., & Soliman, F. (1999). A Graphical Method for Assessing Knowledge-Based Systems Investments. *Logistics Information Management, 12*(1-2), 63–77. doi:10.1108/09576059910256277

Clemons, E. K., Gu, B., & Lang, K. R. (2002, January 7-10). Newly vulnerable markets in an age of pure information products: An analysis of online music and online news. In *Proceedings of the 35th Annual Hawaii International Conference on System Sciences* (CD-ROM). Computer Society Press.

Cole M., O'Keefe, R. M., & Siala, H. (1999). From the user interface to the consumer interface. *Information Systems Frontiers, 1*(4), 349-361.

commitment, and membership behaviors in professional associations. *Journal of Marketing, 64*(July), 34-49.

Connor-Linton, J. (2003). *Chi-square tutorial.* Retrieved November 14, 2007, from http://www.georgetown.edu/faculty/ballc/webtools/web_chi_tut.html

Constantinides, E. (2004). Influencing the online consumer's behavior: the Web experience. *Internet Research, 14*(2), 111-126.

Constantiou, I., Damsgaard, J., & Knutsen, L. (2004). Investigating mobile services' adoption and diffusion in the Danish market. In *Proceedings of the 27th Information Systems Research Seminar in Scandinavia*, Sweden.

Cristobal, E., Flavian, C., & Guinaliu, M. (2007). Perceived e-service quality (PeSQ): Measurement validation and effects on customer satisfaction and web site loyalty. *Managing Service Quality, 17*(3), 317–140. doi:10.1108/09604520710744326

Cronin, J., Brady, M. K., & Hult, G. T. (2000). Assessing the effects of quality, value, and customer satisfaction on consumer behavioral intentions in service environments. *Journal of Retailing, 76*(2), 193-218.

Crosby, L. A., & Johnson, S. L. (2004). Making the intangible...tangible. *Marketing Management, 13*(3), 12-13.

Crosby, L. A., Evans, K. R., & Cowles, D. (1990). Relationship quality in services selling: An interpersonal influence perspective. *Journal of Marketing, 54*, 68–81.

Cullen, A. J., & Webster, M. (2007). A model of B2B e-commerce, based on connectivity and purpose. *International Journal of Operations & Production Management, 27*(2), 205–225. doi:10.1108/01443570710720621

Daft, R., & Lengel, R. (1986). Organizational information requirements, media richness, and structural design. *Management Science*, *32*(5), 554–571. doi:10.1287/mnsc.32.5.554

Dahlstrom, R., & Nygaard, A. (1999). An Empirical investigation of ex post transaction costs in franchised distribution channels. *Journal of Marketing Research*, *36*(2), 160-171.

Daniel, E. M., & White, A. (2005). The future of inter-organisational system linkages: findings of an international Delphi study. *European Journal of Information Systems*, *14*(2), 188–203. doi:10.1057/palgrave.ejis.3000529

Das, T. K., & Teng, B.-S. (1998). Between trust and control: Developing confidence in partner cooperation in alliances. *Academy of Management Review*, *23*(3), 491-512.

Davenport, T. H., & Brooks, J. D. (2004). Enterprise systems and the supply chain. *Journal of Enterprise Information Management*, *17*(1), 8–19. doi:10.1108/09576050410510917

Davis, F. D. (1986). *A Technology Acceptance Model for Empirically Testing New End-User Information Systems: Theory and Results*. MIT, Boston.

Davis, F. D. (1989). Perceived usefulness, perceived ease of use, and user acceptance of information technology. *MIS Quarterly*, *13*(2), 319-340.

Davis, F. D., & Venkatesh, V. (1996). A Critical Assessment of Potential Measurement Biases in the Technology Acceptance Model: Three Experiments. *International Journal of Human-Computer Studies*, *45*(1), 19–45. doi:10.1006/ijhc.1996.0040

Davis, F. D., & Venkatesh, V. (2004). Toward Preprototype User Acceptance Testing of New Information Systems: Implications for Software Project Management. *IEEE Transactions on Engineering Management*, *51*(1), 31–46. doi:10.1109/TEM.2003.822468

Davis, F. D., Bagozzi, R., & Warshaw, P. (1989). User Acceptance of Computer Technology: A Comparison of Two Theoretical Models. *Management Science*, *35*(8), 982–1003. doi:10.1287/mnsc.35.8.982

Day, G. S., & Hubbard, K. J. (2003). Customer Relationships Go Digital. *Business Strategy Review*, *14*(1), 17–26. doi:10.1111/1467-8616.00240

de Bruijn, J., Fensel, D., Keller, U., & Lara, R. (2005). Using the Web Service Modeling Ontology to Enable Semantic E-Business. *Communications of the ACM*, *48*(12), 43–47. doi:10.1145/1101779.1101807

Dedrick, J., Gurbaxani, V., & Kraemer, K. (2003). Information Technology and Economic Performance: A Critical Review of the Empirical Evidence. *ACM Computing Surveys*, *35*(1), 1–28. doi:10.1145/641865.641866

Dewan S., & Riggins, F. J. (2005). The digital divide: Current and future research directions. *Journal of the Association for Information Systems*, *6*(12), 215-235.

Dey, A., & Abowd, G. (1999). *Towards a better understanding of context and context-awareness*. Technical Report 99-22. Georgia Institute of Technology.

Dey, A.K. (2001). Understanding and using context. *Personal and Ubiquitous Computing*, *5*(1), 4-7.

Dias, C. (2001). Corporate Portals: a literature review of a new concept in information management. *International Journal of Information Management*, *21*, 269–287. doi:10.1016/S0268-4012(01)00021-4

Dierickx, I., & Cool, K. (1989). Asset stock accumulation and sustainability of competitive advantage. *Management Science*, *35*(12), 1504-1511.

Doney, P. M., & Cannon, J. P. (1997). An Examination of the Nature of Trust in Buyer-Seller Relationships. *Journal of Marketing*, (April): 35–51. doi:10.2307/1251829

Drucker, P.F. (1995). *Managing the non-profit organisation: Practices and principles*. Butterworth-Heinemann.

Dufner, D., Holley, L., & Reed, B. (2002). Can private sector strategic information systems planning techniques work for the public sector? *Communications of the Association for Information Systems*, *8*(27), 413-431.

Durkin, M.G., & McGowan, P. (2001). "Net Effect"-views from the periphery: Exploring the role and importance of the Internet in marketing activity in entrepreneurial firms. *Irish Marketing Review*, *14*, 15-25.

Durlacher Research Ltd. (2000). *Mobile commerce report*. Retrieved from http://www.durlacher.com/fr-research-reps.htm.

Dwyer, F. R., Schurr, P. H., & Oh, S. (1987). Developing buyer-seller relationships. *Journal of Marketing, 51*, 11-27.

Dyer, J. H., & Chu, W. C. (2000). The Determinants of trust in Supplier Auto-maker relationships in the U.S., Japan and Korea. *Journal of International Business Studies, 31*(2), 259–185. doi:10.1057/palgrave.jibs.8490905

East, R. (1993). Investment Decisions and Theory of Planned Behaviour. *Journal of Economic Psychology, 14*, 337–375. doi:10.1016/0167-4870(93)90006-7

Easton, J. (2002). *Going wireless: Transform your business with mobile technology*. Harper Business.

Eikebrokk, T.R., & Olsen, D.H. (2002). Understanding e-business competencies in SMEs. *Seeking success in e-business—multidisciplinary approach* (pp. 575-602). Kluwer Academic Publishers.

E-ink Corporation. (2007). *Electronic displays*. Retrieved November 15, 2007, from http://www.eink.com/products/customers.html

Eisenhardt, K. M. (1989). Building Theories from Case Study Research. *Academy of Management Review, 14*(4), 532–550. doi:10.2307/258557

Eisenhardt, K. M. (1989). Building theories from case study research. *Academy of Management Review, 14*(4),532-550.

Ej-Techonologies. (2009). Jprofiler. Retrieved from http://www.ej-technologies.com/products/jprofiler/overview.html

Elliot, S., & Fowell, S. (2000). Expectations versus reality: A snapshot of customer experiences with Internet retailing. *International Journal of Information Management, 20*, 323-336.

Engel, J., Blackwell, R., & Miniard, P. (1995). *Consumer behaviour* (8th ed.). Fort Worth, Texas: Dryden Press.

Ephrati, E., & Rosenschein, J. S. (1992). Constrained intelligent action: Planning under the influence of a master agent. In *Proceedings of the Tenth National Conference on Artificial Intelligence*, San Jose, California, USA, (pp. 263-268).

Esmahi, L., Dini, P., & Bernard, J. C. (1999, June). Towards an open intelligent market place for mobile agents. In *Proceedings of the Eighth IEEE International Workshops on Enabling Technologies: Infrastructure for Collaborative Enterprises (WET-ICE'99)*, Stanford University, California, USA, (pp. 279-286).

Esmahi, L., Dini, P., & Bernard, J. C. (2000). MIAMAP: A virtual market place for intelligent agents. In *Proceedings of the 33rd Hawaii International Conference on System Sciences (HICSS-33)* (Vol. 8, p. 8018).

Europe's Information Society. (2004a). *User interaction*. Retrieved from http://europa.eu.int/ information_society

European Commission. (2004). *Digital business ecosystems*. Retrieved April 12, 2008, from http://www.digital-ecosystems.org/

Fahy, P., & Clarke, S. (2004, June). CASS - Middleware for mobile context-aware applications. In *Proceedings of the 2nd ACM SIGMOBILE International Conference on Mobile Systems, Applications and Services* (MobiSys'04).

Faniel, I. M. (2005). Enhancing perceived value of cross-departmental knowledge on innovation via KM technology features. *Academy of Management Proceedings*, 1-16.

Fickas, S., Korteum, G., & Segall, Z., (1997). Software organization for dynamic and adaptable wearable systems. In *Proceedings of the 1st IEEE International Symposium on Wearable Computers* (pp. 56-63).

Fidler, R. (1997). *Mediamorphosis: Understanding new media*. Thousand Oaks, CA: Pine Forge Press.

Fikes, R., Engelmore, R., Farquhar, A., & Pratt, W. (1995). *Network-based information broker*. Stanford University: Knowledge System Laboratory. Retrieved December 15, 2006, from: http://www-ksl.stanford.edu/KSL_Abstracts/KSL-95-13.html

Fishbein, M. (1967). *Readings in attitude theory and measurement*. New York: Wiley.

Fishbein, M., & Ajzen, I. (1975). *Belief, Attitude, Intention, and Behavior: An Introduction to Theory and Research*. Reading: Addison-Wesley.

Fishman, G. S. (1996). *Monte Carlo Concepts, Algorithms and Applications*. Springer Verlag.

Flanagan, J. (1954). The critical incident technique. *Psychological Bulletin, 51*(4), 327-358.

Flavian, C., Guinaliu, M., & Gurrea, R. (2006). The role played by perceived usability, satisfaction and consumer trust on Website loyalty. *Information & Management, 43*, 1-14.

Fornell, C., & Wernerfelt, B. (1987). Defensive marketing strategy by customer complaint management: A theoretical analysis. *Journal of Marketing Research,* 24(4), 337-346.

Fornell, C., Johnson, M. D., Anderson, E. W., Cha, J., & Bryant, B. E. (1996). The American customer satisfaction index: Nature, purpose, and findings. *Journal of Marketing, 60*(October), 7-18.

Fredberg, T. (2003). *Interface strategies: Internet and the business of large Swedish daily newspapers*. Gothenburg: Institute for Management of Innovation and Technology.

Freed, L. (2006). *Best Practices for Measuring the Website Contribution to Multi-Channel Purchase Behavior – Introducing the Multi-Channel Value Index* (Forsee Results White Paper, Winter, 1-10).

Frolick, M.N., Chen, L.-D. (2004, Spring). Assessing m-commerce opportunities. *Information Systems Management*, pp. 53 -61.

Fuglsang, L. (2007). Critical incident Teknikken i Teknikker i Samfundsvidenskaberne, In L. Fuglsang, P. Hagedorn-Rasmussen, P. Bitch Olsen (Eds.), *Roskilde universitetsforlag (*pp. 260-277).

Fuglsang, L., & Sundbo, J. (2006). Flow and consumers in e-services self-services: New provider-consumer relations. *The Service Industries Journal, 26*(4), 361-379.

Gallivan, M, J. (2001). Organizational adoption and assimilation of complex technological innovations: Development and application of a new framework. *Database for Advances of Information Systems, 32*(3), 51-86.

Ganesan, S. (1994). Determinants of long-term orientation in buyer-seller relationships. *Journal of Marketing, 58*, 1–19. doi:10.2307/1252265

Gant, D. B., & Gant, J. P. (2002). State Web portals: Delivering and financing e-service. *E-government Series* (pp. 1- 60).

Garbarino, E., & Johnson, M. S. (1999). The different roles of satisfaction, trust, and commitment in customer relationships. *Journal of Marketing, 63*(April), 70–87.

Garvin, D. A. (1983). Quaity on the line. *Harvard Business Review, September-October*, 65-75.

Gebauer, J., Gribbins, J., & Shaw, J. (2006). *Task-technology fit for mobile information systems.* Working Paper, 06-0107. University of Illinois at Urbana-Champaign, College of Business.

Gefen, D. (2000). E-commerce: The role of familiarity and trust. *Omega, 28*(6), 725-737.

Gefen, D. (2002). Customer Loyalty in e-Commerce. *Journal of the Association for Information Systems, 3*(2), 27–51.

Gefen, D. (2002). Customer loyalty in e-commerce. *Journal of the Association for Information Systems, 3*, 27–51.

Gefen, D., & Straub, D. (1997). Gender Differences in Perception and Adoption of E-Mail: An Extention to the Technology Acceptance Model. *MIS Quarterly, 21*(4), 389–400. doi:10.2307/249720

Gefen, D., Karahanna, E., & Straub, D. W. (2003). Trust and TAM in online shopping: An integrated model. *MIS Quarterly, 27*(1), 51-90.

Geihs, K. (2001). Middleware challenges ahead. *IEEE computer, 34*(6), 24-31.

Genesereth, M. R., Ginsberg, M. L., & Rosenschein, J. S. (1986, August). Cooperation without communication. In *Proceedings of the Fifth National Conference on Artificial Intelligence*, Philadelphia, PA, (pp. 51-57).

George, J. (2002). Influences on the Intent to Make Internet Purchases. *Internet Research: Electronic Networking Applications and Policy, 12*(2), 165–180. doi:10.1108/10662240210422521

Germanakos, P., Mourlas, C., & Samaras, G. (2005c). Considering the new user requirements for apt mobile Internet services delivery. In *Proceedings of the IADIS*

International Conference on WWW/Internet 2005 (pp. 148-152).

Germanakos, P., Samaras, G., & Christodoulou, E. (2005a). Multi-channel delivery of services—the road from eGovernment to mGovernment: Further technological challenges and implications. In *Proceedings of the First European Conference on Mobile Government* (Euro mGov 2005) (pp. 210-220).

Germanakos, P., Tsianos, N., Lekkas, Z., Mourlas, C., & Samaras, G. (2006). Capturing essential intrinsic user behaviour values for the design of comprehensive Web-based personalized environments. *Computers in Human Behavior Journal*, Special Issue on Integration of Human Factors in Networked Computing. (accepted)

Germanakos, P., Tsianos, N., Mourlas, C., & Samaras, G. (2005b). New fundamental profiling characteristics for designing adaptive Web-based educational systems. In *Proceeding of the IADIS International Conference on Cognition and Exploratory Learning in Digital Age (CELDA2005)* (pp. 10-17).

Geyskens, I., Steenkamp, J., Steenkamp, E. M., Scheer, L. K., & Kumar, N. (1996). The effects of trust and interdependence on relationship commitment: a trans-atlantic study. *International Journal of Research in Marketing, 13*, 303-317.

Ghosh, S. (1998). Making Business Sense of the Internet. *Harvard Business Review*, (March-April): 126–133.

Giaglis, G. (2004). Κινητό και Ασύρματο Επιχειρείν. Εκδόσεις Ι. ΣΙΔΕΡΗΣ. ISBN 960-08-0331-5.

Giaglis, G.M., Kourouthanassis, P., & Tsamakos, A. (2003). Towards a classification framework for mobile location services. In B.E. Mennecke & T.J. Strader (Eds.), *Mobile Commerce: Technology, Theory, and Applications*. Hershey, PA: Idea Group Publishing. Retrieved from http://www.eltrun.gr/papers/mBook-classification.pdf..

Gibson, A., & Frakes, A. (1997). Truth or Consequences: A Study of Critical Issues and Decision Making in Accounting. *Journal of Business Ethics, 16*, 161–171. doi:10.1023/A:1017914713375

Giorgetti, M., & Louarn, M. L. (2004). *Technologies for digital ecosystems—Supporting regional growth and innovation*. Retrieved February, 2007, from http://www.ve-forum.org/apps/pub.asp?Q=1270

Glaser, B. G., & Strauss, A. L. (1967). *The discovery of grounded theory: Strategies for qualitative research*. New York: Aldien de Gruyter,.

Godin, S. (2001). *Unleashing the idea virus: Stop marketing at people! Turn your ideas into epidemics by helping your customers do the marketing thing for you*. Hyperion.

Goldenhar, L., & Connell, C. (1993). Understanding and Predicting Recycling Behavior: An Application of the Theory of Reasoned Action. *Journal of Environmental Systems, 22*, 91–103.

Google. (n.d.). *Google Maps for Mobile*. Retrieved March 12, 2009, from http://www.google.com/gmm/

Grandon, E., & Pearson, J.M. (2003). Strategic value and adoption of electronic commerce: An empirical study of Chilean small and medium businesses. *Journal of Global Information Technology Management, 6*(3).

Grant, I., & O'Donohoe, S. (2007). Why young consumers are not open to mobile marketing communication. *International Journal of Advertising, 26*(2), 223-246.

Greenwald, A. R., & Kephart, J. O. (1999). Shopbots and pricebots. In *Proceedings of the International Joint Conference on Artificial Intelligence*.

Gregor, S. (2006). The Nature of Theory in Information Systems. *MIS Quarterly, 30*(3), 611–642.

Grewal, D. Levy, M., & Lehmann, D. R. (2004). Retail branding and customer loyalty: an overview. *Journal of Retailing, 80*(4), 249-263.

Grewal, D., & Sharma, A. (1999). Value-based marketing strategies. In *Proceeding of the Annual Conference of the Academy of Marketing Science*. Coral Gables, FL.

Grewal, D., Gopalkrishnan, R., Iyer, & Levy, M. (2004). Internet retailing: Enablers, limiters, and market consequences. *Journal of Business Research, 57*(7), 703–713.

Grewal, D., Monroe, K. B., & Krishnan, R. (1998). The effects of price comparison advertising on buyers' perceptions of acquisition value and transaction value. *Journal of Marketing, 62*, 46-59.

Griffith, T., Sawyer, J., & Neale, M. (2003). Virtualness and knowledge in teams: Managing the love triangle of organizations, individuals, and information technology. *MIS Quarterly, 2*(27), 265-287.

Grönroos, C. (1987). Developing the service offering–a source of competitive advantage. In C.F. Surprenant (Ed.), *Add value to your service*. Chicago, IL: American Marketing Association.

Gruen, T. W., Summers, J. O., & Acito, F. (2000). Relationship marketing activities,

GT Nexus. (2005). GT Nexus' Web site. Retrieved August 30, 2005, from http://www.gtnexus.com/en/delivery_model/

Gu, T., Pung, H.-K., & Zhang, D.-Q. (2005). A service-oriented middleware for building context-aware services. *Journal of Network and Computer Applications, 28*, 1-18.

Guan S. U., & Zhu, F. (2002). Agent fabrication and its implementation for agent-based electronic commerce. *International Journal of Information Technology and Decision Making, 1*(3), 473-490.

Gulati, R. (1995). Does Familiarity Breed Trust? The Implications of Repeated Ties for Contractual Choice in Alliances. *Academy of Management Journal, 38*(1), 85–112. doi:10.2307/256729

Haas, L.M., Kossmann, D., Wimmers, E.L., & Yang, J. (1997, August). Optimizing queries across diverse data sources. In *Proceedings of 23rd International Conference on Very Large Data Bases* (VLDB'97) (pp. 276-285).

Hagel, J. III, & Brown, J. S. (2001). Your Next IT Strategy. *Harvard Business Review, 79*(October), 105–113.

Haig, M. (2002). *Mobile marketing: The message revolution*. Kogan Page.

Hair, J. F., Black, B., Babin, B., Anderson, R. E., & Tatham, R. L. (2005). *Multivariate data analysis*. Upper Saddle River, NJ: Prentice Hall.

Hair, J. F., Jr., Anderson, R. E., Tatham, R. L., & Black, W. C. (1998). *Multivariate data analysis* (5th. ed.). New Jersey: Prentice Hall.

Hallen, G., & Latino, R. J. (2003, June). Eastman Chemical's success story. *Quality Progres*, 50-54.

Hammond, K. (2001). B2C E-commerce 2000-2010: What experts predict. *Business Strategy Review, 12*, 43-50.

Hansmann, U., Merk, L., Nicklous, M., & Stober T. (2003). *Pervasive computing*. Berlin, Heidelberg, New York: Springer-Verlag.

Hanson, W. (2000). *Principles of Internet marketing*. South-Western College Publishing.

Harrell, G., & Bennett, P. (1974). An Evaluation of the Expectancy Value Model of Attitude Measurement for Physician Prescribing Behavior. *JMR, Journal of Marketing Research, 11*(3), 269–278. doi:10.2307/3151142

Harrison, D., Mykytyn, P., & Riemenschneider, C. (1997). Executive Decisions About Adoption of Information Technology in Small Business: Theory and Empirical Tests. *Information Systems Research, 8*(2), 171–195. doi:10.1287/isre.8.2.171

Hassanein, K., & Head, M. (2004, June 21-23). The influence of product type on online trust. In *Proceedings of the 17th Electronic Commerce Conference*. Bled, Slovenia.

Haynes, J.L. (1979). A new look at managerial performance: The AMA model of worthy performance. *Management Review, 2-3.*

Hedman, J., & Kalling, T. (2002). *IT and business models: Theories and concepts*. Copenhagen: CBS Press.

Hedman, J., & Kalling, T. (2003). The business model concept: Theoretical underpinnings and empirical illustrations. *European Journal of Information Systems, 12*, 49-59.

Held, J., & Bowers, J. (2001). *Securing E-Business Applications and Communications*. Boca Raton, FL: Auerbach.

Hellier, P., Geursen, G., Carr, R., & Rickard, J. (2003). Customer repurchase intention: A general structural equation model. *European Journal of Marketing, 37*(11), 1762-1800.

Heman, R.D., & Renz, D.O. (1999). Theses on nonprofit organisational effectiveness. *Nonprofit and Voluntary Sector Quarterly, 28*(2), 107-126.

Hendrickson, A., & Collins, M. (1996). An Assessment of Structure and Causation of IS Usage. *The Data Base for Advances in Information Systems*, *27*(2), 61–67.

Herzberg, A. (2003). Payments and banking with mobile personal devices. *Communications of the ACM*, *46*(5), 53-58.

Hesmondhalgh, D. (2002). *The cultural industries*. SAGE.

Hevner, A. R., March, S. T., & Park, J. (2004). Design science in information systems research. *MIS Quarterly*, *28*(1), 75-105.

Hightower, J., & Borriello, G. (2001). Location Systems for Ubiquitous Computing. *IEEE Computer Magazine*, *34*(8), 57–66.

Hirschman, A. O. (1970). *Exit voice and loyalty: Responses to decline in firms, organizations, and states*. Cambridge, MA: Harvard University Press.

Hoffman, D. L., & Novak, T. P. (1995). *Marketing in hypermedia computer mediated environment: Conceptual foundations*. Owen Graduate School of Management and Vanderbilt University. Retrieved on May 2003, from http://www2000.ogsm.vanderbilt.edu

Hoffman, D. L., Novak, T. P., & Peralta M. (1998). *Building consumer trust in online environments: The case for information privacy*. Owen Graduate School of Management and Vanderbilt University. Retrieved on May 2003, from http://www2000.ogsm.vanderbilt.edu

Hoffman, D., & Bateson, J. E. G. (1999). *Essentials of services marketing*. International Thomson Publishing.

Hogg, L. M., & Jennings, N. R. (1997, November 8-10). Socially rational agents. In *Proceedings of the AAAI Fall symposium on Socially Intelligent Agents*, Boston, Mass., (pp. 61-63).

Holmquist, L., Hook, K., Juhlin, O., & Persson P. (2002). Challenges and opportunities for the design and evaluation of mobile applications. In *Proceedings of the Workshop on Main Issues in Designing Interactive Mobile Services*.

Holtzblatt, K. (2005). Customer-centered design for mobile applications. *Personal and Ubiquitous Computing*, *9*(4), 227-237.

Homburg, C., Hoyer, W. D., & Koschate, N. (2005). Customers' reactions to price increases: Do customer satisfaction and perceived motive fairness matter? *Journal of the Academy of Marketing Science*, *33*(1), 36-49.

Hong, I. (2002). A new framework for interorganisational systems based on the linkage of participants' roles. *Information & Management*, *4*(2), 261–270. doi:10.1016/S0378-7206(01)00095-7

Hsu, K-H. (2005). Using balanced scorecard and fuzzy data envelopment analysis for multinational R & D project performance assessment. *The Journal of American Academy of Business, Cambridge*, *1*(27), 189-196.

Hu, W.-C., Lee, C.-w., & Yeh, J.-h. (2004). Mobile commerce systems. In S. Nansi (Ed.), *Mobile Commerce Applications* (pp. 1-23). Hershey, PA: Idea Group Publishing.

Huang, D. C., & Hu, Q. (2004). Integrating Web services with competitive strategies: The balanced scorecard approach. *Communications of the Association for Information Systems*, *1*(13), 57-80.

Huber, G., & Daft, R. 1987. The information environment of organizations. In F.Jablin et al. (Ed.), *Handbook of Organizational Communication* (pp. 130-164). Thousand Oaks, CA: Sage Publications.

IFRA. (2006). *eNews initiative*. Retrieved November 15, 2007, from http://www.ifra.com/website/website.nsf/html/CONT_ENEWS?OpenDocument&ENW&

Igbaria, M., Guimaraes, T., & Davis, G. (1995). Determinants of Microcomputer Usage Via a Structural Equation Model. *Journal of Management Information Systems*, *11*(4), 87–114.

Ihlström Eriksson, C., & Kalling, T. (2007, June 7-9). Proposing a business model framework for the e-newspaper. In *Proceedings of the 15th European Conference on Information Systems, ECIS 2007*, St. Gallen.

Ihlström Eriksson, C., & Svensson, J. (2007, May 31-June 2). The mobile e-newspaper innovation-examining the pre-adoption phase. *LA Global Mobility Round Table*, Los Angeles.

Ihlström Eriksson, C., Åkesson, M., & Hakeröd, J. (2007, August 11-14). Advertising in ubiquitous media environments. In *Proceeding of IRIS'30*, Tampere, Finland.

Ihlström, C. (2005). The e-newspaper innovation-converging print and online. In *Proceedings of the International Workshop on Innovation and Media: Managing changes in Technology, Products and Processes*, Stockholm, Sweden.

Ihlström, C., & Henfridsson, O. (2005). Online newspapers in Scandinavia: A longitudinal study of genre change and interdependency. *IT & People*, *18*(2), 172-192.

Ihlström, C., & Palmer, J. (2002). Revenues for online newspapers: Owner and user perceptions. *Electronic Markets: The International Journal of Electronic Commerce & Business Media*, *12*(4), 228-236.

Ihlström, C., Åkesson, M., & Nordqvist, S. (2004). From print to Web to e-paper-the challenge of designing the e-newspaper. In *Proceedings of ICCC 8th International Conference on Electronic Publishing, ELPUB 2004*, Brasilia, (pp. 249-260).

Ihlström, C., Svensson, J., & Åkesson, M. (2005). How would you like your e-newspaper? Converging the best of two worlds. In *Proceedings of HCI International 2005*, Las Vegas. CD-ROM.

Interchange of Data Between Administrations. (2004). *Multi-channel delivery of eGovernment services*. Retrieved from http://europa.eu.int/idabc/.

Ioannidis, A., Spanoudakis, M., Sianas, P., Priggouris, I., Hadjiefthymiades, S., & Merakos, L. (2004). Using XML and related standards to support Location Based Services. In *Proceedings of SAC'2004, 19th ACM Symposium on Applied Computing, Web Technologies and Applications Special Track, Nicosia, Cyprus*.

iRex Technologies. (2006). *The iLiad*. Retrieved November 15, 2007, from http://www.irextechnologies.com

Issarny, V., Sacchetti, D., Tartanoglou, F., Sailhan, F., Chibout, R., Levy, N., & Talamona, A. (2005). Developing ambient intelligence systems: A solution based on Web services. *Journal of Automated Software Engineering*, *12*(1), 101-137.

Jackson, R., & Wang, P. (1995). *The convergence of database marketing and interactive media networks*. In E. Forrest & R. Mizerski (Eds.), *Interactive marketing: The future present* (pp. 355-372). NTC Publishing Group.

Jain, A. (2004). IT penetration in the government workforce: A resource based analysis of the 50 U.S. state governments. In *Proceedings of the Tenth Americas Conference on Information Systems. (Online)*

Jarvenpaa, S. L., & Todd, P. A. (1997). Consumer reactions to electronic shopping on the world wide Web. *International Journal of Electronic Commerce*, *1*(2), 59-88.

Jarvenpaa, S. L., & Tractinsky, N. (1999). Consumer trust in an Internet store: A cross-cultural validation. *Journal of Computer Mediated Communication*, *5*(2), 1-35.

Jarvenpaa, S. L., Tractinsky, N., & Vitale, M. (2000). Consumer trust in an Internet store. *Information Technology and Management*, *1*, 45–71. doi:10.1023/A:1019104520776

Jayawardhena, C. (2007). Outcomes of service encounter quality in a business-to-business context. *Industrial Marketing Management*, *36*(5), 575-588.

Jennings, N. R., Faratin, P., Lomuscio, A. R., Parsons, S., Wooldridge, M. J., & Sierra, C. (2001). Automated negotiations: Prospects, methods and challenges. *Group Decision and Negotiation*, *10*(2), 199–215.

Johnson, J. T., Barksdale, H. C., Jr., & Boles, J. S. (2001). The strategic role of the salesperson in reducing customer defection in business relationships. *Journal of Personal Sell and Sales Management*, *21*(Spring), 123–134.

Johnston, H. R., & Vitale, M. R. (1988). Creating competitive advantage with Interorganizational Information systems. *MIS Quarterly*, *12*(2), 153–165. doi:10.2307/248839

Jöreskog, K. G., & Sörbom, D. (1993). *LISREL 8: structural equation modeling with SIMPLIS command language*. Mooresville, IN: Scientific Software.

Julian, C., & Ramaseshan, B. (1994). The Role of Customer-Contact Personnel in the Marketing of a Retail Bank's Services. *International Journal of Retail and Distribution Management*, *22*(5), 29–34. doi:10.1108/09590559410067316

Jungk, R., & Müllert, N. (1996). *Future workshops-how to create desirable futures*. London: Institute for Social Inventions.

Kaasinen, E. (2003). User needs for location-aware mobile services. In *Personal and Ubiquitous Computing* (Vol. 7).

Kalafatis, S. P., & Miller, H. (1997). A re-examination of the commitment-trust theory. In H. G. Gemünden, T. Ritter, & W. Achim (Eds.), *Relationships and networks in international markets*.(pp. 213-227). Oxford, UK: Pergamon.

Kalakota, R., & Whinston, A. (1997). *Electronic Commerce: A manager's Guide*. Reading, Mass: Addison-Wesley.

Kaplan, R., & Norton, D. P. (1992). The balanced scorecard–measures that drive performance. *Harvard Business Review*, 71-79.

Kaplan, S., & Sawhney, M. (2000). E-Hubs: The New B2B Marketplaces. *Harvard Business Review*, *78*(3), 97–103.

Karahanna, E., & Limayem, M. (2000). E-mail and V-mail Usage: Generalizing Across Technologies. *Journal of Organizational Computing and Electronic Commerce*, *10*(1), 49–66. doi:10.1207/S15327744JOCE100103

Karahanna, E., Straub, D., & Cherveny, N. (1999). Information Technology Adoption Across Time: A Cross - Sectional Comparison of Pre-Adoption and Post-Adoption Believes. *MIS Quarterly*, *23*(2), 192–213. doi:10.2307/249751

Kauffman, R. J., & Techatassanasoontorn, A. A. (2005). Is there a global digital divide for digital wireless phone technologies. *Journal of the Association for Information Systems*, *6*(12), 289-316.

Keen, P.G.W. (2000, March 13). Ensuring E-trust. *Computerworld*.

Keidl, M., & Kemper, A. (2004, March). A framework for context-aware adaptable Web services. In *Proceedings of 9th International Conference on Extending Database Technology* (EDBT '04) (pp. 826-829).

Kemper, H.G., & Wolf, E. (2002). Iterative process models for mobile application systems: A framework. In *Proceedings of the Twenty-Third International Conference on Information Systems* (pp. 401-413).

Kersten, G. E., & Lai, H. (2005). *Satisfiability and completeness of protocols for electronic negotiations* (Research paper INR01/05). Retrieved January 4, 2007, from http://interneg.concordia.ca/interneg/research/papers/2005/01.pdf

Kettinger, W., Grover, V., Guha, S., & Segars, A. (1994). Strategic Systems Revisited: A Study in Sustainability and Performance. *MIS Quarterly*, *18*(1), 31–58. doi:10.2307/249609

Khalfan, A. (2004). Information Security Considerations in IS/IT Outsourcing Projects: A Descriptive Case Study of Two Sectors. *International Journal of Information Management*, *24*(1), 29–42. doi:10.1016/j.ijinfomgt.2003.12.001

Khedro, T., & Genesereth, M. R. (1994). Progressive negotiation for resolving conflicts among distributed heterogeneous cooperating agents. In *Proceedings of the 12th National Conference on Artificial Intelligence*, Seattle, Washington, USA, (pp. 381-386).

Kim, D. J. Ferrin, D.L. & Rao, H.R. (2008). Trust and Satisfaction, Two Stepping Stones for Successful E-Commerce Relationships. *Information Systems Research, Articles in Advance* (pp. 1-21). Informs.

Kim, D. J., Cho, B., & Rao, H. R. (2000, December 10-13). Effects of consumer lifestyles on purchasing behavior on the Internet: a conceptual framework and empirical validation. In *Proceedings of the Twenty First International Conference on Information Systems* (pp. 688-695). Brisbane, Australia.

King, W., & Teo, T. (1996). Key Dimensions of Facilitators and Inhibitors for the Strategic Use of Information Technology. *Journal of Management Information Systems*, *12*(4), 35–53.

Kleijnen, M., Wetzels, M., & de Ruyter, K (2004). Consumer acceptance of wireless finance. *Journal of Financial Services Marketing*, *8*(3), 206-217.

Klem, L. (1995). *Path analysis: Reading and understanding multivariate statistics*. In L. G. Grimm & P. R. Yarnold (Eds.), Washington, D.C.: American Psychological Association.

Klemperer, P. (1999). Auction theory: A guide to the literature. *Journal of Economic Surveys*, *13*, 227-286.

Klemperer, P. (2002). What really matters in auction design. *Journal of Economic Perspectives*, *16*(1), 169-189.

Klobas, J., & Clyde, L. (2000). Adults Learning to Use the Internet: A Longitudinal Study of Attitudes and Other Factors Associated with Intended Internet Use.

Library & Information Science Research, 22(1), 5–34. doi:10.1016/S0740-8188(99)00038-9

Knell, S. (2003). The shape of things to come: Museums in the technological landscape. *Museum and Society, 1*(3), 132-146.

Kohler, A., & Gruhn, V. (2004). Analysis of mobile business processes for the design of mobile information systems. In *Proceedings of the 5th International Conference on Electronic Commerce and Web Technologies* (3182) (pp. 238–247).

Kolos-Mazuryk, L., Poulisse, G., & van Eck, P. (2005). Requirements engineering for pervasive services. In *Proceedings of the Second Workshop on Building Software for Pervasive Services* (pp. 18-22).

Koning, J. L. (2002). A review on the interaction issues in agent-based marketplaces. *International Journal of Information Technology and Decision Making, 1*(3), 457-472.

Kotler, P., & Armstrong, G. (1999). *Princípios de marketing* (7th ed.). Rio de Janeiro: LTC.

Kourouthanassis, P., Spinellis, D., Roussos, G., & Giaglis, G. (2002). Intelligent Cokes and diapers: MyGROCER ubiquitous computing environment. In *Proceedings of the 1st International Mobile Business Conference* (pp. 150-172).

Kraus, S., Wilkenfeld, J., & Zlotkin, G. (1995). Multiagent negotiation under time constraints. *Artificial Intelligence, 75*(2), 297-345.

Krogstie, J., Lyytinen, K., Ophahl, A., Pernici, B., Siau, K., & Smolander K. (2003). *Mobile information systems – Research challenges on the conceptual and logical level.* Lecture Notes in Computer Science – Advanced Conceptual Modeling Techniques (Vol. 2784) (pp. 124-135). Berlin, Heidelberg: Springer-Verlag, .

Krueger, C. C., van der Beek, K., & Swatman, P. (2004). New and emerging business models for online news: A survey of 10 European countries. In *Proceedings of the 17th Bled eCommerce Conference*, Slovenia.

Kubicek, H. (1992). The Organization Gap in Large-Scale EDI Systems. In R. J. Streng, C. F. Ekering, E. van Heck & J. F. H. Schultz (Eds.), *Scientific Research on EDI "Bringing Worlds Together"* (pp. 11-41). Amsterdam: Samsom.

Kurihara, Y., Takaya, S., Harui, H., & Kamae, H. (Eds.). (2008). *Information Technology and Economic Development.* Hershey: Idea Group Reference.

Kurkovsky, S., & Harihar, K. (2006). Using ubiquitous computing in interactive mobile marketing. *Personal and Ubiquitous Computing Journal*, 227-240.

Kurland, N. (1995). Ethical Intentions and the Theories of Reasoned Action and Planned Behavior. *Journal of Applied Social Psychology, 25*(4), 297–313. doi:10.1111/j.1559-1816.1995.tb02393.x

Kvasny, L., & Keil, M. (2006). The challenges of redressing the digital divide: A tale of two U.S. cities. *Information Systems Journal, 16*, 23-53.

Lager, M. (2005). *Breaking Down the Silos, Customer Relationship Management.* Retrieved from http://www.destinationCRM.com

Lamparter, S., Ankolekar, A., Studer, R., Oberle, D., & Weinhardt, C. (2006). A policy framework for trading configurable goods and services in open electronic markets. In *Proceedings of the 8th International Conference on Electronic Commerce: The New E-commerce: Innovations for Conquering Current Barriers, Obstacles and Limitations to Conducting Successful Business on the Internet* (pp. 162-173).

Lane, C., & Bachmann, R. (1996). The social constitution of trust: Supplier relations in Britain and Germany. *Organization Studies, 17*(3), 365-395.

Lassar, W., Mittal, B., & Sharma, A. (1995). Measuring customer-based brand equity. *Journal of Consumer Marketing, 12*(4), 11-19.

Latour, B. (1986). The Powers of Association. In J. Law (Ed.), *Power, Action and Belief. A New Sociology of Knowledge? Sociological Review monograph 32* (pp. 264-280). London: Routledge & Kegan Paul.

Latour, B. (1993). *We Have Never Been Modern* (C. Porter, Trans.). Hemel Hempstead: Harvester Wheatsheaf.

Latour, B. (1996). *Aramis or the Love of Technology.* Cambridge, Ma: Harvard University Press.

Lavender, M. (2004). Maximizing customer Relationships and Minimizing Business Risk. *International Journal of Bank Marketing*, *22*(4), 291–296. doi:10.1108/02652320410542563

Law, J., & Callon, M. (1988). Engineering and Sociology in a Military Aircraft Project: A Network Analysis of Technological Change. *Social Problems*, *35*(3), 284–297. doi:10.1525/sp.1988.35.3.03a00060

Lawton, G. (2005). LAMP lights enterprise development efforts. *IEEE Computers*, *38*(9), 18–20.

Ledered, A. L., Mauin, D. J., Sena, M. P., & Zhuang, Y. (2000). The technology acceptance model and the World Wide Web. In *Proceedings of the Conference of Decision Support Systems* (Vol. 29, pp. 269-282).

Lederer, A., Maupin, D., Sena, M., & Zhuang, Y. (2000). *The Role of Ease of Use, Usefulness and Attitude in the Prediction of World Wide Web Usage.* Paper presented at the ACM SIGCPR Conference on Computer Personnel Research, Boston, MA.

Lee, D. R. (2005). Agricultural Sustainability and Technology Adoption: Issues and Policies for Developing Countries. *American Journal of Agricultural Economics*, *87*(5), 1325–1334. doi:10.1111/j.1467-8276.2005.00826.x

Lee, D.M.S., & Trauth, E.M. (1995). Critical skills and knowledge requirements of IS professionals: A joint academic/industry investigation. MIS Quarterly, *19*(3), 313-340.

Lee, K. K. K., Chiu, D. K. W., & Hung, P. C. K. (2006). Web-service based information integration for e-financial planning system matchmaking decision support. In *Proceedings of the IEEE/WIC/ACM International Conference on Web Intelligence and International Agent Technology Workshops, (WI-IAT 2006)* (pp. 259-262).

Lee, M., & Turban, E. (2001). Trust model for consumer Internet shopping. *International Journal of Electronic Commerce*, *6*(1), 75-91.

Legner, C., & Vogel, T. (2008). Leveraging Web Services for Implementing Vertical Industry Standards: A Model for Service-Based Interoperability. *Electronic Markets*, *18*(1), 39–52. doi:10.1080/10196780701797623

Leick, A. (1995). GPS *Satellite Surveying*, Second Edition, John Wiley & Sons.

Lemphers, D. (2007, November). *Communitize–Doing business with ourselves!* Keynote address at the Eighth International We-B (Working for E-Business) Conference, Melbourne Australia.

Levenburg, N. M., & Klein, H. A. (2006). Delivering customer services online: Identifying best practices of medium-sized enterprises. *Information Systems Journal*, *16*, 135-155.

Lewicki, R. J., & Bunker, B. B. (1996). Developing and Maintaining Trust in Work Relationships. In R.M. Kramer, & T.R. Tyler (Eds), *Trust in Organizations: Frontiers of Theory and Research* (pp. 114-139). Thousand Oaks, CA: Sage Publications.

Lewis, B.R., & Entwistle, T.W. (1990). Managing the service encounter: A focus on the employee. *International Journal of Service Industry Management*, *1*(3), 41-52.

Lewis, J. D., & Weigert, A. (1985). Trust as a social reality. *Social Forces, 63*, 967-985.

Li, C., . Xu, Y., & Li, H. (2005). An empirical study of dynamic customer relationship management. *Journal of Retailing and Consumer Services, 12*, 431-441.

Liao, S., Shao, Y., Wang, H., & Chen, A. (1999). The Adoption of Virtual Banking: An Empirical Study. *International Journal of Information Management*, *19*(1), 63–74. doi:10.1016/S0268-4012(98)00047-4

Linthicum, D. S. (2000). *Enterprise Application Integration.* Upper Saddle River, NJ: Addison-Wesley.

Lippert, M., Wolf, H., & Züllighoven, H. (2001). Domain Services for Multi-Channel Application Software. In *Proceedings of the 34th Annual Hawaii International Conference on System Sciences (HICSS-34), Maui.*

Liu, C., & Arnett, K. P. (2000). Exploring the factors associated with Web site success in the context of electronic commerce. *Information and Management, 38*(October), 23-34.

Loch, K., & Conger, S. (1996). Evaluating Ethical Decision Making and Computer Use. *Communications of the ACM*, *39*(7), 77–83. doi:10.1145/233977.233999

Locke, E., Fredrick, E., Lee, C., & Bobko, P. (1984). Effect of Self-Efficacy, goals, and Task Strategies on Task Performance. *The Journal of Applied Psychology*, *69*(2), 241–251. doi:10.1037/0021-9010.69.2.241

Loebbecke, C., & Hornig, V. (2001, June 25-26). Increasing trust in e-commerce: Concepts and examples of insurance solutions. In *Proceedings of the 14th Bled Electronic Commerce Conference*. Bled, Slovenia.

Lohse, G. L., & Spiller, P. (1999). Electronic shopping: The effect of customer interface on traffic and sales. *Communications of ACM, 41*(7), 81-87.

Loney, C., Jessup, L., & Valacich, J. (2004). Emerging mobile business models for mobile brokerage services. *Communications of the ACM, 47*(6), 71-77.

Lovelock, C. (1983, Summer). Classifying services to gain strategic marketing insights. *Journal of Marketing, 47*, 9-20.

Lovelock, C., & Wirtz, J. (2007). *Services marketing: People, technology, strategy* (6th ed.). Upper Saddle River, NJ: Pearson Prentice Hall.

Löwer, U. M. (2005). *Interorganisational Standards: Managing Web Services Specifications for Flexible Supply Chains*. Heidelberg: Physica-Verlag.

Lu, J., & Zhang, G. (2003). Cost benefit factor analysis in e-services. *International Journal of Service Industry Management, 14*(5), 570–595. doi:10.1108/09564230310500237

Lumayem, M., & Hirt, S. G. (2003). Force of habit and information systems usage: Theory and initial validation. *Journal of the Association for Information Systems, 4*, 65-95.

Lund, A.B. (2007). Media markets in Scandinavia: Political economy aspects of convergence and divergence. *Nordicom Review, Jubilee Issue 2007*, 121-134.

Lynch, J. G., & Ariely, D. (2000). Wine online: Search costs affect competition on price, quality and distribution. *Marketing Science, 19*(1), 83-103.

Maamar, Z., Mostefaoui, S., & Mahmoud, Q. (2005, January). Context for personalized Web services. In *Proceedings of 38th IEEE Hawaii International Conference on System Sciences* (HICSS'05) (pp. 166.2).

Macneil, R. (1981). *The new Social Contract*. New Haven, CT: Yale University Press.

Madhavan, J., Bernstein, P.A., Doan, A., & Halevy, A.Y. (2005, April). Corpus-based schema matching. In *Proceedings of the 21st International Conference on Data Engineering* (ICDE 2005) (pp. 57-68).

Madlberger, M. (2006). Exogenous and Endogenous Antecedents of Online Shopping in a Multichannel Environment: Evidence from a Catalog Retailer in the German-Speaking World. *Journal of Electronic Commerce in Organizations, 4*(4), 29–51.

Mahadevan, B. (2003). Making sense of emerging markets structures in B2B e-commerce. *California Management Review, 46*(1), 86–100.

Mahatanankoon, P., Wen, J.H., & Lim, B. (2005). Consumer-based m-commerce: Exploring consumer perception of mobile applications. *Computer Standards and Interfaces, 27*, 347-357.

Malhotra, N. K. (1993). Marketing research: An applied orientation. New Jersey: Prentice Hall, International Edition.

Mallat, N., Rossi, M., & Tuunainen, V.K. (2004). Mobile banking services. *Communications of the ACM, 47*(5), 43-46.

Mallat, N., Rossi, M., Tuunainen, V. K., & Öörni, A. (2006). The impact of use situation and mobility on the acceptance of mobile ticketing services. In *Proceedings of the 39th Hawaii International Conference on System Sciences*, Hawaii. (CD-ROM). Computer Society Press.

Malone, T. W., & Crowston, K. (1994). The interdisciplinary study of coordination. *ACM Computing Survey, 26*(1), 87-119.

March, S. T., & Smith, G. F. (1995). Design and natural science research on information technology. *Decision Support Systems, 15*(4), 251-266.

Marcolin, B., Compeau, D.R., Munro, M.C., & Huff, S.L. (2000). Assessing user competence: Conceptualization and measurement. *Information Systems Research, 11*(1), 37-60.

Markus, L. M., Majchrzak, A., & Gasser, L. (2002). A design theory for systems that support emergent knowledge processes. *MIS Quarterly, 26*(3), 179-212.

Marsh, S. (1992). *Trust and reliance in multiagent-systems: A preliminary report.* University of Stirling, Department of Computer Science and Mathematics,.

Marti, S., & Garcia-Molina, H. (2006). Taxonomy of trust: Categorizing P2P reputation systems. *Computer Networks, 50*, 472-484.

Mathieson, K. (1991). Predicting User Intentions: Comparing the Technology Acceptance Model with the Theory of Planned Behavior. *Information Systems Research, 2*(3), 173–191. doi:10.1287/isre.2.3.173

Mathwick, C., Malhotra, N. K., & Rigdon, E. (2002). The effect of dynamic retail experiences on experiential perceptions of value: An Internet and catalog comparison. *Journal of Retailing, 78*(1), 50-51.

McAfee, A. (2005). Will Web Services Really Transform Collaboration? *MIT Sloan Management Review, 46*(2), 78–84.

McGowan, P., Durkin, M.G., Allen, L., Dougan, C., & Nixon, S. (2001). Developing competencies in the entrepreneurial small firm for use of the Internet in the management of customer relationships. *Journal of Industrial Training, 25*(2-3-4), 126-136.

McGrath, R. G., & MacMillan, I. C. (2000). *The entrepreneurial mindset. Strategies for continuously creating opportunity in an age of uncertainty.* Cambridge: Harvard Business School Press.

McKitterick, D., & Dowling, J. (2003). *State of the art review of mobile payment technology.* Technical Report, TCD-CS-2003-24.

McKnight, D. H., & Chervany, N. L. (2001). What trust means in e-commerce customer relationships: an interdisciplinary conceptual typology. *International Journal of Electronic Commerce, 6*(2).

McKnight, D., & Chervany, N. (2002). What trust means in e-commerce customer relationship: an interdisciplinary conceptual typology. *International Journal of Electronic Commerce, 6*(2), 35-59.

Melone, N. P. (1990). A theoretical assessment of the user-satisfaction construct in information systems research. *Management Science, 36*(1), 76-81.

Merrilees, B., & Fenech, T. (2007). From catalog to Web: B2B multi-channel marketing strategy. *Industrial Marketing Management, 37*(1), 44–49. doi:10.1016/j.indmarman.2006.06.014

Merton, R. K. (1968). *Social theory and social structure.* New York: The Free Press.

Meuter, M.L., Ostrom, A.L., Roundtree, R., & Bitner, M. (2000). Self-service technologies: Understanding customer satisfaction with technology-based service encounters. *Journal of Marketing, 64*(3), 50-64.

Miceli, G., Ricotta, F., & Costabile, M. (2007). Customizing customization: A conceptual framework for interactive personalization. *Journal of Interactive Marketing, 21*(2), 6-25.

Michelsen, H., & Hartwich, F. (2004). *University-based Agricultural Research: A Comparative Study in Sub-Saharan Africa* Addis Ababa: International Service for National Agricultural Research

Microsoft, C. (n.d.a). *Internet Information Services.* Retrieved June 15, 2008, from http://www.microsoft.com/WindowsServer2003/iis/default.mspx

Microsoft, C. (n.d.b). *SQL Server2005.* Retrieved May 6, 2008, from http://www.microsoft.com/sql/default.mspx

Microsystems, S. Inc. (n.d.). *Sun Java Sysem Web Server.* Retrieved June 19, 2007, from http://www.sun.com/software/products/Web_srvr/home_Web_srvr.xml

Middleton, B., & Long, G. (1990). Marketing skills: Critical issues in marketing education and training. *Journal of Marketing Management, 5*(3), 325-43.

Miles, M. B., & Huberman, A. M. (1994). *Qualitative Data Analysis: An Expanded Sourcebook* (2. ed.). Thousand Oaks: Sage Publications.

Miles, M. B., & Huberman, A. M. (1994). *Qualitative data analysis.* Thousand Oaks, CA:Sage Publications.

Mirchandani & Motwani. (2001). Understanding small business electronic commerce adoption: An empirical analysis. *Journal of Computer Information Systems, Spring*, 70-73.

MISC. (2002). 6 Million Internet User Registration Milestone Reached, After 1st Quarter, As Banking Usage (Transactions) Grow 15% in Latest Quarter. Retrieved 17 December 2004, from www.marketintelligence.com.au

Mobile Electronic Transaction Ltd. (2003). MeT white paper on mobile ticketing. Retrieved from http://www.mobiletransaction.org/pdf/R200/white_papers/MeT_White_paper_on_mobile_ticketing_v1.pdf.

Moe, W. W., & Fader, P. S. (2001). *Modeling online store visit patterns as a measure of customer satisfaction*. Retrieved May 2003, from http://216.107.131.155/archive/papers/

Mohr, J., & Spekman, R. (1994). Characteristics of Partnership Success: Partnership attributes, Communication Behavior, and Conflict Resolution Techniques. *Strategic Management Journal*, *15*, 135–152. doi:10.1002/smj.4250150205

Molla, A., & Licker, P. S. (2005). Ecommerce Adoption in Developing Countries: a Model and Instrument. *Information & Management*, *42*(6), 877–899. doi:10.1016/j.im.2004.09.002

Montealegre, R. (1999). A Temporal Model of Institutional Interventions for Information Technology Adoption in Less-Developed Countries. *Journal of Management Information Systems*, *16*(1), 207–232.

Moon, M. J. (2002). State government e-procurement in the information age: Issues, practices, and trends. *E-government series* (pp. 1-65).

Moon, M. J., & Norris, F. D. (2005). Does managerial orientation matter? The adoption of reinventing government and e-government at the municipal level. *Information Systems Journal*, *15*, 43-60.

Moore, G., & Benbasat, I. (1991). Development of an Instrument to Measure the Perception of Adopting an Information Technology Innovation. *Information Systems Research*, *2*(3), 192–222. doi:10.1287/isre.2.3.192

Moore, W., Dean, D., Gerber, A., Wagenknecht, G., & Vanderheyden, P. (2004). Eclipse Development using the Graphical Editing Framework and the Eclipse Modeling Framework. Retrieved March 2005, from http://www.redbooks.ibm.com/abstracts/sg246302.html.

Moorman, C., Zaltman, G., & Deshpande, R. (1992). Relationships between providers and users of market research: The dynamics of trust within and between organizations. *Journal of Marketing Research*, *29*(3), 314-328.

Morgan, R. M., & Hunt, S. D. (1994). The commitment-trust theory of relationship marketing. *Journal of Marketing*, *58*, 20–38. doi:10.2307/1252308

Morlok, E., Nitzberg, B., & Balasubramaniam, K. (2000). *The parcel service industry in the US: Its Size and role in commerce* (Tech. Rep., University of Pennsylvania , School of Engineering and Applied Science, Systems Engineering Department). Available online from http://www.pti.psu.edu/mautc/docs/MA-III-0007.pdf

Mui, L., Halberstadt, A., & Mohtashemi, M. (2002). Notions of reputation in multiagents systems: A review. In *Proceedings of the First International Joint Conference on Autonomous Agents and Multiagent Systems* (pp. 280-287)l New York: ACM Press.

Mui, L., Mohtashemi, M., & Halberstadt, A. (2002). A computational model of trust and reputation. In *Proceedings of the 35th Hawaii International Conference on System Science (HICSS)* (p. 188). Big Island, HI:

Mukherjee, A., & Neth, P. (2003). A Model of Trust in Online Relationship Banking. *International Journal of Bank Marketing*, *21*(1), 5–15. doi:10.1108/02652320310457767

Muntermann, J., Rossnagel, H., & Rannenberg, K. (2004). Mobile financial information services, security and certification. In *Post-Proceedings of the CSES 2004 2nd International Workshop on Certification and Security in Inter-Organizational E- Services, IFIP 18th World Computer Congress 2004.*

Mykytyn, P., & Harison, D. (1993). The Application of the Theory of Reasoned Action to Senior Management and Strategic Information Systems. *Information Resources Management Journal*, *6*(2), 15–25.

Neumann, D., & Weinhardt, C. (2002). Domain-independent e-negotiation design: Prospects, methods, and challenges. In *Proceedings of the 13th International Workshop on Database and Expert Systems Applications (DEXA 2002)* (pp. 680-686).

Neumann, D., Gimpel, H., Holtmann, C., & Weinhardt, C. (2002). Agent-based bidding in electronic markets. In *Proceedings of the Second IFIP Conference on E-commerce, E-business, E-government (I3E 2002)* (pp. 553-567).

Neumann, D., Schnizler, B., Weber, I., & Weinhardt, C. (2007). Second-best combinatorial auctions—the case of the pricing-per-column mechanism. In *Proceedings of the 40th Annual Hawaii International Conference on System Science (HICSS 2007)* (p. 33b).

Newcomb, E., Pashley, T., & Stasko, J. (2003). Mobile computing in the retail arena. In *Proceedings of the ACM Conference on Human Factors in Computing Systems* (pp. 337-244). New York: ACM Press.

Newcomer, E. (2002). *Understanding Web services: XML, WSDL, SOAP, and UDDI.* Boston: Addison-Wesley Professional.

Newcomer, E., & Lomow, G. (2004). *Understanding SOA with Web Services.* Maryland: Addison-Wesley.

Ng, L. L. N., Chiu, D. K. W., & Hung, P. C. K. (2007). Tendering process model (TPM) implementation for B2B integration in a Web services environment. In *Proceedings of the 40th Annual Hawaii International Conference on System Science (HICSS 2007)* (p. 143b).

Nickerson, J., Hamilton, B., & Wada, T. (2001). Market position, resource profile, and governance: Linking Porter and Williamson in the context of international courier and small package services in Japan. *Strategic Management Journal, 22*(3), 251-273. Retrieved April 12, 2008, from http://cniss.wustl.edu/summerschool/nickersonpaper.pdf

Nielsen, J. (2001). *Projetando Web Sites—design Web usability* (2nd ed.). Rio de Janeiro: Editora Campus.

Noordewier, T. G., George, J., & Nevin, J. R. (1990). Performance outcomes of purchasing arrangements in industrial buy-vendor relationships. *Journal of Marketing Research, 54,* 80-93.

Nordhaug, O. (1998). Competence specificity in organizations. *International Studies of Management and Organizations, 28*(1), 8-29.

Normann, R. (1977). *Management for growth.* Chichester: John Wiley.

Normann, R. (1991). *Service management strategy and leadership in service businesses* (2nd ed.). Chichester: John Wiley and Sons.

Nour, M. A., & Fadlalla, A. (2000). A framework for Web marketing strategies. *Information Systems Management,* 41-50.

NTT DoCoMo, Inc. (2007). *i-mode.* Retrieved June 12, 2007, from http://www.nttdocomo.com/services/imode/index.html

Nunnally, J. C. (1978). Psychometric theory (2nd ed.). New York: McGraw Hill.

O'Connor, J., & Galvin, E. (2001). *Marketing in the digital age.* Prentice Hall.

OECD. (2000). *OECD small and medium size enterprise outlook.* Retrieved November 15, 2007, from http://www.oecd.org

Oliver, R. L. (1980). A cognitive model of the antecedents and consequences of satisfaction decisions. *Journal of Marketing Research, 17*(November), 460-469.

Oliver, R. L. (1997). *Satisfaction: A behavioral perspective on the consumer*: New York: McGraw-Hill.

Open Mobile Alliance Ltd. (n.d.). *WAP Forum.* Retrieved from June 13, 2008, from http://www.openmobilealliance.org/tech/affiliates/wap/wapindex.html

Oracle. (n.d.). *Oracle Databases.* Retrieved August 25, 2008, from http://www.oracle.com/database/index.html

Ortiz, S. Jr. (2000). Embedded databases come out of hiding. *IEEE Computer, 33*(3), 16–19.

Osborn, A. (1953). *Applied imagination.* New York: Scribner.

Österle, H. (1995). *Business in the Information Age - Heading for new Processes.* Berlin: Springer.

Österle, H. (2001). Enterprise in the Information Age. In H. Österle, E. Fleisch & R. Alt (Eds.), *Business Networking: Shaping Collaboration Between Enterprises* (Vol. 2, rev. and extended ed., pp. 17-54). Berlin: Springer.

Otim, S., & Grover, V. (2006). An empirical study on Web-based services and customer loyalty. *European*

Journal of Information Systems, *15*, 527–541. doi:10.1057/palgrave.ejis.3000652

Ozsu, T., & Valduriez, P. (1991). *Principles of distributed database systems*. Prentice-Hall.

Paarlberg, R. L. (2003). Technology Adoption in Developing Countries: The Case of Genetically Modified Crops. Retrieved April 2007, from http://www.au.af.mil/au/awc/awcgate/cia/nic2020/technology_adoption_nov6.pdf

Paavilainen, J. (2002). *Mobile business strategies: Understanding the technologies and opportunities*. Wireless Press.

Palmer, J. W. (2000). Electronic commerce in retailing: Convenience, search costs, delivery and price across retail formats. *Information Technology and Management*, *1*(1-2), 25–43. doi:10.1023/A:1019152403937

Pan, G., Pan, S. L., Newman, M., & Flynn, D. (2006). Escalation and de-escalation of commitment: a commitment transformation analysis of an e-government project. *Information Systems Journal, 16*, 3-21.

Pan, S. L., & Lee, J. (2003). Using e-CRM for a united view of the customer. *Communication of the ACM, 46*(4), 95-99.

Panayiotou, C., & Samaras, G. (2004). mPERSONA: Personalized portals for the wireless user: An agent approach. *Mobile Networks and Applications (MONET), Special Issue on Mobile and Pervasive Commerce*, *9*(6).

Panayiotou, C., & Samaras, G. (2006). Mobile user personalization with dynamic profiles: Time and activity. In *Proceedings of On the Move to Meaningful Internet Systems 2006: OTM 2006 Workshops* (PerSys 2006). Part II (pp. 1295-1304).

Pandey, I. M. (2005). Balanced scorecard: Myth and reality. *The Journal for Decision Makers, 1*(30), 51-66.

Papazafeiropoulos, G., Prigouris, N., Marias, I., Hadjiefthymiades, S., & Merakos, L. (2003). *Proceedings of the IST Mobile & Wireless Communications Summit: Retrieving Position From Indoor WLANs Through GWLC, Portugal.*

Parasuraman, A. (2000). Technology readiness index (TRI): A multiple item scale to measure readiness to embrace new technologies. *Journal of Service Research*, *2*(4), 307-320.

Parasuraman, A., & Grewal, D. (2000). The impact of technology on the quality-value-loyalty chain: A research agenda. *Journal of the Academy of Marketing Science, 28*(1), 168-174.

Parasuraman, A., Zeithaml, V. A., & Berry, L. L. (1988). SERVQUAL: A multiple item scale for measuring consumer perceptions of service quality. *Journal of Retailing, 64*(1), 12–40.

Parasuraman, A., Zeithaml, V. A., & Berry, L. L. (1994). Reassessment of expectations as a comparison standard in measuring service quality: Implications for further research. *Journal of Marketing, 58*, 111-124.

Parasuraman, A., Zeithaml, V., & Berry, L. (1985, Fall). A conceptual model of service quality and its implications for future research, *Journal of Marketing*, *49*, 41-50.

Parente, J. (2000). *Varejo no Brasil*. São Paulo: Editora Atlas.

Park, J. H., Lee, J. K., & Yoo J. S. (2005). A framework for designing the balanced supply chain scorecard. *European Journal of Information Systems*, *14*, 335-346.

Parlay – ETSI., (2001) *Open Service Access API. Part 6: Mobility SCF. ES 210 915-6 v.1.1.1.*

Parthasarathy, M., & Bhallachorjee, A. (1998). Understanding post-adoption behavior in the context of online service. *Information Systems Research, 9*(4), 362-378.

Pascoe, G. C. (1983). Patient satisfaction in primary health care: A literature review and analysis. *Eval Program Plann, 6*, 185-197.

Pascoe, J. (1998). Adding generic contextual capabilities to wearable computers. In Proceedings of the 2nd International Symposium on Wearable Computers (pp. 92-99).

Patton, M. Q. (2002). *Qualitative research & evaluation methods (3rd ed.)*. Thousand Oaks, CA: Sage Publications.

Payne, A., & Frow, P. (2004). The Role of Multichannel Integration in Customer Relationship Management. *Industrial Marketing Management*, *33*(6), 527–538. doi:10.1016/j.indmarman.2004.02.002

Pedersen, P E., Methlie, L B., & Thorbjørnsen, H. (2002). Understanding mobile commerce end-user adoption:

A triangulation perspective. In *Proceedings of the 35th Hawaii International Conference on System Sciences, Hawaii*. (CD-ROM). Computer Society Press.

Pedhazur, E., & Schmelkin, L. P. (1991). *Measurement, design and analysis: An integrated approach*. Hillsdale: Lawrence Erlbaum Associates.

Peteraf, M. A. (1993). The cornerstones of competitive advantage: A resource-based view. *Strategic Management Journal*, *14*(3), 179-191.

Petre, M., Minocha, S., & Roberts, D. (2006). Usability beyond the Website: An empirically-grounded e-commerce evaluation instrument for the total customer experience. *Behavior & Information Technology*, *25*(2), 189-203.

Phau, I., & Poon, S. M. (2000). Factors influencing the types of products and services purchased over the Internet. *Internet Research*, *10*(2), 102-113.

Piccoli, G., Brohman, M. K., Watson, R. T., & Parasuraman, A. (2004). Net-Based Customer Service Systems: Evolution and Revolution in Website Functionalities. *Decision Sciences*, *35*(3), 423–455. doi:10.1111/j.0011-7315.2004.02620.x

Pieterson, W., Ebbersa, W., & van Dijk, J. (2006). Personalization in the public sector: An inventory of organizational and user obstacles towards personalization of electronic services in the public sector. *Government Information Quarterly*, *24*(1), 148-164.

Plastic logic. (2007). *Plastic Logic raises $100 million to enable the first "take anywhere, read anywhere" electronic reader products*. Retrieved November 15, 2007, from http://www.plasticlogic.com/news-detail.php?id=300

Polymer Vision. (2007). *Press releases*. Retrieved November 15, 2007, from http://www.polymervision.com/News-Center/Press-Releases/Index.html

Porter, M. (2001). Strategy and the Internet. *Harvard Business Review*, (March): 63–78.

Porter, M. E. (1985). *Competitive advantage*. New York: Free Press.

Porter, M. E. (1991). Towards a dynamic theory of strategy. *Strategic Management Journal*, *12*, 95-119.

Postma, P. (1999). *The new marketing era: Marketing to the imagination in a technology-driven world*. McGraw-Hill.

Priggouris, G., Hadjiefthymiades, S., & Merakos, L. (2004). *An XML framework for multi-level access control in the enterprise domain, Proceedings of the 6th International Conference on Enterprise Information Systems, Porto, Portugal, April.*

Punie, Y., Burgelman, J-C., & Bogdanowicz, M. (2002). The future of online media industries. *Scenarios for 2005 and Beyond. IPTS*, *64*, 35-42.

Puschmann, T., & Alt, R. (2005). Developing an Integration Architecture for Process Portals. *European Journal of Information Systems*, *14*(2), 121–134. doi:10.1057/palgrave.ejis.3000527

Rai, A., Lang, S. S., & Welker, R. B. (2002). Assessing the validity of IS success models: An empirical test and theoretical analysis. *Information Systems Research*, *13*(1), 50-69.

Raihan, N., & Hamid, A. (2005). E-CRM: Are we there yet? *The Journal of American Academy of Business, March*, 51-57.

Raisinghani, M. (2000). *Electronic Commerce at the Down of the Third Millennium - Electronic Commerce: Opportunities and Challenges*. Thousand Oaks: Sage Publications.

Rajapakse, J., & Seddon, P. B. (2007). ERP Adoption in Developing Countries in Asia: A Cultural Misfit Unpublished Working paper. University of Melbourne.

Ramesh, B., & Tiwana A. (2001). E-services: Models and methods for design, implementation, and delivery. In *Proceedings of the 34th Hawaii International Conference on System Sciences*, 1 page.

Ramsey, J., & Smith, M. (1999). Managing Customer Channel Usage in the Australian Banking Sector. *Managerial Auditing Journal*, *14*(7), 329–338. doi:10.1108/02686909910289812

Randall, D., & Gibson, A. (1991). Ethical Decision Making in the Medical Profession: An Application of the Theory of Planned Behavior. *Journal of Business Ethics*, *10*(2), 111–122. doi:10.1007/BF00383614

Ranganathan, C., & Ganapathy, S. (2002). Key dimensions of business-to-user web sites. *Information & Management*, *39*(6), 457–465. doi:10.1016/S0378-7206(01)00112-4

Rappa, M. (2002). *Business models on the Web*. Retrieved November 15, 2007, from http://digitalenterprise.org/models/models.html

Rasmusen, E. (2001). *Games and information: An introduction to game theory* (3rd ed.). Blackwell Publishers.

Ratliff, J. M. (2002). NTT DoCoMo and its i-mode success: Origins and implications. *California Management Review*, *44*, 55-70.

Reedy, J., Schullo, S., & Zimmerman, K. (2001). *Electronic marketing: Integrating electronic resources into the marketing process*. South-Western.

Reichheld, F. F., & Schefter, P. (2000). E-loyalty: Your secret weapon on the Web. *Harvard Business Review, July-August*, 105-13.

Reimers, K. (2001). Standardizing the new e-business platform: Learning from the EDI experience. *Electronic Markets*, *11*(4), 231–237. doi:10.1080/101967801753405508

Rigby, D. K., Reichheld, F., & Schefter, P. (2002). Avoid the four perils of CRM. *Harvard Business Review, February*, 101-110.

Ring, P. S., & Van de Ven, A. H. (1994). Developing processes of cooperative inter-organizational relationships. *Academy of Management Review*, *19*, 90–118. doi:10.2307/258836

Ritter, P., Lorig, K., Laurent, D., & Matthews, K. (2004, July-September). Internet vs. mailed questionnaires: A randomized comparison. *Journal of Medical Internet Research*, *6*(3), e29.

Rogers, E. M. (1995). *Diffusion of Innovations* (4th ed.). New York: The Free Press.

Roman, M., Hess, C.K., Cerqueira, R., Ranganathan, A., Campbell, R.H., & Nahrstedt, K. (2002). Gaia: A middleware infrastructure to enable active spaces. *IEEE Pervasive Computing*, *1*(4), 74-83.

Romano, N. C., & Fjermesad, J. (2003). Electronic commerce customer relationship management: A research agenda. *Information Technology and Management, 4*, 233-258.

Rosenschein, J. S., & Genesereth, M. R. (1985, August). Deals among rational agents. In *Proceedings of the Ninth International Joint Conference on Artificial Intelligence*, Los Angeles, California, (pp. 91-99).

Rosenschein, J. S., & Zlotkin, G. (1994). *Rules of encounter*. MIT Press.

Roth, E. (2000). Game theory as a tool for market design. In F. Patrone, I. Garcia-Jurado, & S. Tijs (Eds.), *Game practice: Contributions from applied game theory* (pp. 7-18).

Roth, M.T., & Schwarz, P.M. (1997, August). Don't scrap it, wrap it! A wrapper architecture for legacy data sources. In *Proceedings of 23rd International Conference on Very Large Data Bases* (VLDB'97) (pp. 266-275).

Rousseau, D. M., Sitkin, S. B., Burt, R. S., & Camerer, C. (1998). Not so different after all: A cross-discipline view of trust. *Academy of Management Review*, *23*(3), 393–404.

Roy, J. (2001). Recent trends in logistics and the need for real-time decision tools in the trucking industry. In *Proceedings of the 34th Annual Hawaii International Conference* (p. 9). Retrieved April 12, 2008, from http://www.hicss.hawaii.edu/HICSS_34/PDFs/DTIST02.pdf

Rucci, A. J., Kirn, S. P., & Quinn, R. T. (1998). The rmployee-customer profit chain at SEARS. *Harvard Business Review, January/February*, 82-97.

Ruh, W. A., Maginnis, F. X., & Brown, W. J. (2001). *Enterprise Application Integration*. New York: John Wiley & Sons.

Rupnik, R., & Krisper, M. (2003). Mobile applications: A new application model in information systems. In *Proceedings of the 2nd International Conference on Mobile Business*.

Rust, R. T., & Lemon, K. N. (2001). E-service and the consumer. *International Journal of Electronic Commerce, 5*, 85–101.

Rust, R. T., Subramanian, B., & Wells, M. (1992). Making complaints a management tool. *Marketing Management*, *1*(3), 40-45.

Rust, R. T., Zeithaml, V. A., & Lemon, K. N. (2001). *Driving customer equity: How customer lifetime value is reshaping corporate strategy*. The Free Press.

Sääksjärvi, M., & Santonen, T. (2003). Customization as business model for online newspapers. In *Proceedings of the 16th Bled eCommerce Conference*, Bled, Slovenia.

Sabater, J., & Sierra, C. (2001). REGRET: A reputation model for gregarious societies. In *Proceedings of the Fourth Workshop on Deception Fraud and Trust in Agent Societies* (pp. 61-70). Montreal, Canada:

Sabater, J., & Sierra, C. (2002). Reputation and social network analysis in multiagent systems. In *Proceedings of the First International Joint Conference on Autonomous Agents and Multiagent Systems* (pp. 475-482). New York: ACM Press.

Sadeh, N. (2002). *M-commerce: Technologies, services, and business models* (pp. 177-179). New York: John Wiley & Sons.

Sako, M., & Helper, S. (1998). Determinants of trust in supplier relations: Evidence from the automotive industry in Japan and the United States. *Journal of Economic Behavior & Organization, 34*, 387–417. doi:10.1016/S0167-2681(97)00082-6

Salomann, H., Kolbe, L., & Brenner, W. (2006). Self-Services in Customer Relationships: Balancing High-Tech and High-Touch Today and Tomorrow. *e-Service Journal, 4*(2), 65-84.

Sambamurthy,V., & Zmud, R.W. (1994). *IT management competency assessment: A tool for creating business value through IT*. Morristown, NJ: Financial Executives Research Foundation.

Samtani, A., Leow, T.T., Lim, H.M., & Goh, P.G.J. (2003). Overcoming barriers to the successful adoption of mobile commerce in Singapore. *International Journal of Mobile Communications, 1*(1/2), 194-231.

Sandholm, T. W. (1993). An implementation of the contract net protocol based on marginal cost calculations. In *Proceedings 11th National Conference on Artificial Intelligence* (pp. 256-262).

Sandholm, T. W. (1996). *Negotiation among self-interested computationally limited agents*. Unpublished PhD dissertation, Department of Computer Science, University of Massachusetts. Retrieved February 15, 2006, from: http://www.cs.cmu.edu/~sandholm/dissertation.ps

Sandholm, T. W. (2000). Agents in electronic commerce: Component technologies for automated negotiation and coalition formation. *Autonomous Agents and Multi-Agent Systems, 3*(1), 73-96.

Sarker, S., & Wells, J. D. (2003). Understanding mobile handheld device use and adoption. *Communications of the ACM, Special Issue on mobile commerce opportunities and challenges, 46*(12), 35-40.

Sathye, M. (1999). Adoption of Internet Banking by Australian Consumers: An Empirical Investigation. *International Journal of Bank Marketing, 17*(7), 324–334. doi:10.1108/02652329910305689

Sawhney, M., & Zabin, J. (2002). Managing and measuring relational equity in the network economy. *Journal of the Academy of Marketing Science, 30*(4), 313–332. doi:10.1177/009207002236908

Schierholz, R., Ostrowski, A., Glissmann, S., Kolbe, L. M., & Brenner, W. (2007). Don't call us, we'll call you - Performance Management in Multi-Channel Environments. *Journal Information Science and Technology, 3*(2), 44–61.

Schilit, B., & Theimer, M. (1994). Disseminating active map information to mobile hosts. *IEEE Network, 8*(5), 22-32.

Schilit, B., Adams, N., & Want, R. (1994). Context-aware computing applications. In *Proceedings of the 1st International Workshop on Mobile Computing Systems and Applications* (pp. 85-90).

Schilling, V., & Sobotta, A. (1999). Prozesskostenrechnung der mittelständischen industrie. *Betriebswirtschaftliches Forschungszentrum für Fragen der mittelständischen Wirtschaft e.V.* Bayreuth, Germany.

Schmidt, A., & Laerhoven, K. (2001). How to build smart appliances. *IEEE Personal Communications, 8*(4), 66-71

Schneider, G. P. (2009). *Electronic Commerce* (8th ed.). Course Technology, Cengage Learning

Schoop, M., & Quix, C. (2001). DOC.COM: A framework for effective negotiation support in electronic marketplaces. *Computer Networks, 37*(2), 153-170.

Schubert, P., & Selz, D. (1999). Measuring the effectiveness of electronic commerce sites going beyond traditional marketing paradigms. In *Proceedings of the 32nd Hawaii International Conference on System Sciences.*

Schumpeter, J. A. (1934). *The theory of economic development.* Cambridge: Harvard University Press.

Scupola, A. (2003). The adoption of Internet commerce by SMEs in the south of Italy: An environmental, technological and organizational perspective. *Journal of Global Information Technology Management, 6*(1), 51-71.

Scupola, A. (Ed.). (2008). *Cases on managing e-service.* Hershey, PA: IGI Publishing (Forthcoming).

Searby, S. (2003). Personalisation—An overview of its use and potential. *BT Technology Journal, 21*(1), 13-19.

Sebastianell, R., & Tamimi, N. (2002). How product quality dimensions relate to defining quality. *International Journal of Quality & Reliability Management, 19*(4), 442-453.

Segars, A., & Grover, V. (1993). Re-Examining Perceived Ease of Use and Usefulness: A Confirmatory Factor Analysis. *MIS Quarterly, 17*(4), 517–525. doi:10.2307/249590

Sharma, N., & Patterson, P. G. (1999). The impact of communication effectiveness and service quality on relationship commitment in consumer professional services. *The Journal of Services Marketing, 13*(2), 151-170.

Sharma, S. (1996). *Applied multivariate techniques.* New York: John Willey & Sons.

Shifter, D., & Ajzen, I. (1985). Intention, Perceived Control and Weight Loss: An Application of the Theory Planned Behaviour. *Journal of Personality and Social Psychology, 49*, 843–851. doi:10.1037/0022-3514.49.3.843

Shimp, T., & Kavas, A. (1984). The Theory of Reasoned action Applied to Coupon Usage. *The Journal of Consumer Research, 11*(Dec), 795–809. doi:10.1086/209015

Shostack, G.L. (1985). Planning the service encounter. In J.A. Czepiel, M.R. Solomon, & C.F. Surprenant (Eds.), *The service encounter* (pp. 243-254). Lexington: Lexington Books.

Siau, K., & Shen, Z. (2003). Building customer trust in m-commerce. *Communications of the ACM, 46*(4), 91-94.

Silicon Valley Daily. (2006). *HP Regains Lead in Global PC Sales.* Retrieved June 11, 2007, from http://www.svdaily.com/gartner1.html

Sipior, J. C., Ward, B. T., Volonino, L., & Marzec, J. Z. (2004). A community initiative that diminished the digital divide. *Communications of the Association for Information Systems, 13*(1), 29-56.

Sirdeshmukh, D., Singh, J., & Sabol, B. (2002). Consumer trust, value, and loyalty in relational exchanges. *Journal of Marketing, 66*, 15-37.

Smith, M. D., Bailey, J., & Brynjolfsson, E. (1999). *Understanding digital markets: review and assessment.* Retrieved May 2003, from http://www.ecommerce.mit.edu/papers/ude

Smith, R. G. (1988). The contract net protocol: high-level communication and control in a distributed problem solver'. In A. H. Bond & L. Gasser (Eds.), *Readings in distributed artificial intelligence* (pp. 357-366). Morgan Kauffmann Publishers.

Smith, R. G., & Davis, R. (1981). Frameworks for cooperation in distributed problem solving. *IEEE Transactions on Systems, Man and Cybernetics, 11*(1), 61-70.

Snedecor, G. W., & Cochran, W. G. (1989). *Statistical methods.* Iowa State University Press.

Snoj, B., Korda, A. P., & Mumel, A. (2004). The relationships perceived quality, perceived risk and perceived product value. *Journal of Product & Brand Management. 13*(3), 156-167.

Sony Styles. (2007). *Sony reader.* Retrieved November 15, 2007, from http://www.sonystyle.com/webapp/wcs/stores/servlet/CategoryDisplay?catalogId=10551&storeId=10151&langId=-1&categoryId=16184&XID=O:ebook%20reader:corp_read_gglsrch

Sparks, P., & Shepherd, R. (1992). Self-Identity and the Theory of Planned Behaviour: Assessing the Role of Identification with Green Consumerism. *Social Psychology Quarterly, 55*(4), 388–399. doi:10.2307/2786955

Speier, C., Morris, M., & Bridggs, C. (1995). *Attitudes Toward Computers: The Impact of Performance.* Paper presented at the AIS, Pittsburgh, PA.

Stafford, M.R., & Faber, R.J. (2004). *Advertising, promotion and new media.* M.E. Sharpe.

Stanoevska-Slabeva, K. (2003). Towards a reference model for m-commerce applications. In Proceedings of the 11th European Conference on Information Systems.

Steinfield, C., Adelaar, T., & Lai, Y. (2002). Integrating brick and mortar locations with e-commerce: Understanding synergy opportunities. In *Proceedings of the 35th Annual Hawaii International Conference on System Sciences (HICSS'02)* (Vol. 8, p. 216.2).

Stone, M., Hobbs, M., & Khaleeli, M. (2002). Multichannel customer management: The benefits and challenges. *Journal of Database Marketing, 10*(1), 39–52. doi:10.1057/palgrave.jdm.3240093

Story, J., & Loroz, P. S. (2005). Technological congruence and perceived quality of brand extensions. *Journal of Product & Brand Management, 14*(7), 438-447.

Stowers, G. N. L. (2004). Measuring the performance of e-government. *E-government series* (pp.1-52).

Strang, T., & Linnhoff-Popien, C. (2004). A context modeling survey. In *Proceedings of the 1st International Workshop on Advanced Context Modelling, Reasoning And Management* (pp. 34-41).

Strauss, A., & Corbin, J. (1990). *Basics of Qualitative Research: Grounded Theory Procedures and Techniques.* Newbury Park: Sage Publications.

Street, C. T., & Meister, D. B. (2004). Small Business Growth and Internal Transparency, The Role of Information Systems. *MIS Quarterly, 28*, 473–506.

Subramaniam, C., Shaw, M. J., & Gardner, D. M. (2000). Product Marketing and Channel Management in Electronic Commerce. *Information Systems Frontiers, 1*(4), 363–378. doi:10.1023/A:1010061924822

Sukpanich & Chen (1999). Antecedents of desirable consumer behaviors in electronic commerce. In *Proceedings of 5th Americas Conference on Information Systems.* Milwaukee, WI: Omnipress.

Sun Microsystems Inc. (2003). *Enterprise JavaBeans 2.1 Specification* (Public Draft).

Sun. (2003). *Logistics sample application: Design.* Retrieved July 28, 2005, from http://docs.sun.com/source/817-1243-10/dvlpdsgn.html

Symbian Limited. (2009). *Fast Facts.* Retrieved March 10, 2009, from http://www.symbian.com/about/fastfacts/fastfacts.html

Szajna, B. (1994). Software Evaluation and Choice: Predictive Validation of the Technology Acceptance Instrument. *MIS Quarterly, 18*(3), 319–324. doi:10.2307/249621

Szajna, B. (1996). Empirical Evaluation of the Revised Technology Acceptance Model. *Management Science, 42*(1), 85–94. doi:10.1287/mnsc.42.1.85

Talukdar, D., Sudhir, K., & Ainslie, A. S. (2006). Investigating New Product Diffusion across Products and Countries. Unpublished Working paper. Yale SOM Working Paper No. MK-06.

Tarasewich, P. (2003). Designing mobile commerce applications. *Communications of the ACM, 46*(12), 57-60.

Tatnall, A., & Davey, B. (2007, 19-23 May 2007). *Researching the Portal.* Paper presented at the IRMA: Managing Worldwide Operations and Communications with Information Technology, Vancouver.

Tax, S. S., & Brown, S. W. (1998). Recovering and learning from service failure. *Sloan Management Review, 40*(1), 75-88.

Taylor, H. (2000). Does Internet research work? *Journal of the Market Research Society, Winter*, 13.

Taylor, S. A., & Todd, P. A. (1995). Understanding information technology usage: A test of competing models. *Information Systems Research, 6*(2), 144-176.

Taylor, S., & Todd, P. (1995a). Assessing IT Usage: The Role of Prior Experience. *MIS Quarterly, 19*(4), 561–570. doi:10.2307/249633

Taylor, S., & Todd, P. (1995b). Understanding Information Technology Usage: A Test of Competing Models. *Information Systems Research, 6*(2), 144–176. doi:10.1287/isre.6.2.144

Taylor, S., & Todd, P. (1997). Understanding the Determinants of Consumer Composting Behavior. *Journal of Applied Social Psychology, 27*(7), 602–628. doi:10.1111/j.1559-1816.1997.tb00651.x

Taylor, S.A. (1994). Distinguishing service quality from patient satisfaction in developing health care marketing strategies. *Hosp Health Serv Adm, 39*(2), 221-36.

The Editors Weblog. (2006a). Belgium*: E-paper test launch.* Retrieved November 15, 2007, from http://www.editorsweblog.org/news/2006/02/belgium_epaper_test_launch.php

The Editors Weblog. (2006b). *France: Les Echos publishes e-paper.* Retrieved November 15, 2007, from http://www.editorsweblog.org/print_newspapers/2006/03/france_les_echos_publishes_epaper.php

The Swedish Newspaper Publisher's Association. (2007). Retrieved November 15, 2007, from http://www.tu.se/

The Yankee Group. (2001). *Over 50% of large U.S. enterprises plan to implement a wireless/mobile solution by 2003.* Retrieved December 10, 2002 from http://www.yankeegroup.com/public/news_releases/news_release_detail.jsp?ID=PressReleases/news_09102002_wmec.htm

Thomson, K., Haziris, N., & Alekos, P. (1994). Attitudes and Food Consumption. *British Food Journal, 96*(11), 9–14. doi:10.1108/00070709410074632

Timmers, P. (1998). Business models for electronic markets. *Electronic Markets, 8*(2), 2-8.

TMWSuite Systems Inc. (2005). TMW Systems, Inc. Retrieved August 30, 2005, from http://www.tmwsystems.com/fileUploads/contentManagerDocuments/TMWSuiteBrochureFINALNOPics%5B0%5D.pdf

Towse, R. (Ed.). (2003). *A handbook of cultural economics.* Elgew Edward.

Trochim, W. M. K. (2001). *The research methods knowledge base* (2nd ed.). Retrieved May 2003, from at http://www.trochim.human.cornell.edu/kb/index.htm

Trochim, W. M. K. (2004). *The nomological network* (2nd ed.). Retrieved August 2004, from http://www.socialresearchmethods.net/kb/nomonet.htm

Tsalgatidou, A., & Pitoura, E. (2001). Business models and transactions in mobile electronic commerce: Requirements and properties. *Computer Networks, 37*(2), 221-236.

Tsang, M.M., Ho, S.-C., & Liang, T.-P, (2004). Consumer attitudes toward mobile advertising: An empirical study. *International Journal of Electronic Commerce, 8*(3), 65-78.

Tsetsos, V., Sekkas, O., Priggouris, I., & Hadjiefthymiades, S. (2006). A Scheduling Framework for Enterprise Services. [JSS]. *Journal of Systems and Software, 79*(2), 259–272. doi:10.1016/j.jss.2005.05.014

Turban, E. Lee, J. King, D. & Chung, H.M. (2006). *Electronic Commerce: A Managerial Perspective.* Prentice Hall Inc.

Turban, E., & Gehrke, D. (1999). Determinants of successful Web site design: Relative importance and recommendations for effectiveness. In *Proceedings of the 32nd Hawaii International Conference on System Sciences.*

Turban, E., King, D., Lee, J., & Viehland, D. (2004). *Electronic Commerce 2004: A Managerial Perspective.* Prentice Hall.

Turban, E., King, D., Lee, J., & Viehland, D. (2004). *Electronic commerce.* New Jersey: Prentice Hall.

Turban, E., King, D., Lee, J., Warkentin, M., Chung, H.M. (2002). *Electronic commerce 2002: A managerial perspective.* Prentice Hall.

Turban, E., Lee, J., King, D., & Chung H. M. (2000). *Electronic commerce—a managerial perspective.* Prentice Hall.

Urban, G. L., Sultan, F., & Qualls, W. J. (2000). Placing trust at the center of your Internet strategy. *Sloan Management Review, Fall.*

Valiente, P., & Heijden H. (2004). A method to identify opportunities for mobile business processes. In *Proceedings of E-Commerce and Web Technologies* (EC-Web 2004). Lecture Notes in Computer Science (pp. 238-247). Berlin, Heidelberg: Springer-Verlag.

Vallerand, R., Pelletier, L., Deshaies, P., Cuerrier, J., & Mongeau, C. (1992). Ajzen and Fishbein's Theory of Reasoned Action as Applied to Moral Behavior: A

Confirmatory Analysis. *Journal of Personality and Social Psychology, 62*(1), 98–109. doi:10.1037/0022-3514.62.1.98

Van De Kar, E. (2005). Guidelines for designing mobile information service systems in a value network. In *Proceedings of the 5th ACM International Conference on Electronic Commerce* (Vol. 50) (pp. 413-421). New York: ACM Press.

van der Beek, K., Swatman, P., & Krueger, C. (2005. January 3-6). Creating value from digital content: E-business model evolution in online news and music. In *Proceedings of the 38th Annual Hawaii International Conference on System Sciences* (CD-ROM). Computer Society Press.

van der Heijden, H. (2000a). *E-TAM: A revision of the technology acceptance model to explain Website revisits. Research Memorandum, 29.*

van der Heijden, H. (2000b). *Using the technology acceptance model to predict Web site usage: Extension and empirical test. Research Memorandum, 25.*

van der Heijden, H. (2001, June). Factors influencing the usage of Web sites: The case of a generic portal in the Netherlands. In *Proceedings of the 14th Bled Electronic Commerce Conference: e-everything: e-commerce, e-government, e-household, e-democracy.* Bled, Slovenia.

Van der Hejiden, H. (2000). Measuring IT core capabilities for electronic commerce: Results from a confirmatory analysis. In *Proceedings of the 21st International Conference on Information Systems*, Brisbane, Australia, (pp. 142-163).

van Riel, A. C. R., Liljander, V., & Jurriëns, P. (2001). Exploring consumer evaluations of e-services: a portal site. *International Journal of Service Industry Management, 12*(4), 359–377. doi:10.1108/09564230110405280

Varshney, U. (2003). Location management for mobile commerce applications in wireless Internet environment. *ACM Transactions on Internet Technology, 3*(3), 236-255.

Varshney, U., Vetter, R. J., & Kalakota, R. (2000). Mobile commerce: A new frontier. *IEEE Computer, 33*(10), 32–38.

Vatanasombut, B., Stylianou, A., & Igbaria, M. (2004). How to Retain Online Customers. *Communications of the ACM, 47*(6), 65–69. doi:10.1145/990680.990682

Vellido, A., Lisboa, P. J. G., & Meehan, K. (2000). Quantitative characterization and prediction of online purchasing behavior: A latent variable approach. *International Journal of Electronic Commerce, 4*(4).

Venetis, K. A., & Pervez, N. G. (2004). Service quality and customer retention: Building long-term relationships. *European Journal of Marketing, 38*(11/12), 1577-1598.

Venkatesh, V. (1999). Creation of Favorable User Perceptions: Exploring the Role of Intrinsic Motivation. *MIS Quarterly, 23*(2), 239–260. doi:10.2307/249753

Venkatesh, V. (2000). *Determinants of perceived ease of use: Integration control, intrinsic motivation and emotion into the technology acceptance model.* Retrieved May 2003, from http://isr.commerce.ubc.ca/Abstract/11-4-Venkatesh.html Wolfinbarger, M., & Gilly, M.. (2001). *Shopping online for freedom, control and fun.* California State University Long Beach: College of Business Administration.

Venkatesh, V., & Davis, F. (1996). A Model of the Antecedence of Perceived Ease of Use: Development and Test. *Decision Sciences, 27*(3), 451–481. doi:10.1111/j.1540-5915.1996.tb01822.x

Venkatesh, V., & Davis, F. (2000). A Theoretical Extension of the Technology Acceptance Model: Four Longitudinal Field Studies. *Management Science, 46*(2), 186–204. doi:10.1287/mnsc.46.2.186.11926

Vesanen, J., & Raulas, M. (2006). Building bridges for personalization: A process model for marketing. *Journal of Interactive Marketing, 20*(1), 5-20.

Voss, G. B., Parasuraman, A., & Grewal, G. (1998). The roles of price, performance, and expectations in determining satisfaction in service exchanges. *Journal of Marketing, 62*(October), 46-61.

Vrechopoulos, A., Constantiou, I., & Sideris, G. (2003). The critical role of consumer behavior research in mobile commerce. *International Journal of Mobile Communications, 1*(3), 329-340.

Wainwright, D., Green G., Mitchell, E., & Yarrow, D. (2005). Towards a framework for benchmarking ICT practice, competence and performance in small firms. *Performance Measurement and Metrics, 6*(1), 39-52.

Walker, J.L. (1995). Service encounter satisfaction: Conceptualized. *Journal of Services Marketing, 9*(1), 5-14.

Walker, R. H., Craig-Lees, M., Hecker, R., & Francis, H. (2002). Technology-enabled service delivery - An investigation of reasons affecting customer adoption and rejection. *International Journal of Service Industry Management, 13*(1), 91–106. doi:10.1108/09564230210421173

Wallace, B. (2006). *30 countries passed 100% mobile phone penetration in Q1.* Retrieved November 15, 2007, from http://www.telecommagazine.com/newsglobe/paper.asp?HH_ID=AR_2148

Walls, J. G., Widmeyer, G. R., & El Sawy, O. A. (1992). Building an information system design theory for Vigilant EIS. *Information Systems Research, 3*(1), 36-59.

Walter, Z., & Stolariva-Ornek, J. (2001). Strategies for local newspapers in the digital economy. In *Proceedings of the Seventh Americas Conference on Information Systems*, Boston, M.A.

Wang, F, & Head, M. (2001). Designing Web-based information systems for e-commerce. In *Proceedings of the 224 World Congress on the Management of Electronic Commerce*. Hamilton, Canada.

Wang, J., & Lin, J. (2003). Are personalization systems really personal? – Effects of conformity in reducing information overload. In *Proceedings of the 36th Hawaii International Conference on Systems Sciences* (HICSS'03) (p. 222c). IEEE Computer Society.

Wang, Y., Wang, Y., Lin, H., & Tang, T. (2003). Determinants of User Acceptance of Internet Banking: An Empirical Study. *International Journal of Service Industry Management, 14*(5), 501–519. doi:10.1108/09564230310500192

Warkentin, M.,& Bajaj, A. (2001). The On-demand delivery services model for e-commerce. In A. Gangopadhay (Ed.), *Managing business with electronic commerce: Issues and trends.* Hershey, PA: Idea Group.

Warshaw, P., & Davis, F. (1985). Disentangling Behavior Intention and Behavioral Expectation. *Journal of Experimental Social Psychology, 21,* 213–218. doi:10.1016/0022-1031(85)90017-4

Wernerfelt, B. (1984). A resource-based view of the firm. *Strategic Management Journal, 5*(2), 171-180.

Westlund, O. (2006, September 27-30). Beyond time and space Sweden and their mobile Internet news users (MINU).In *Proceedings of the AOIR 7.0 Conference "Internet and convergence,"* Brisbane.

Weston, F. C. (2003). ERP II: The extended enterprise system. *Business Horizons, 46*(6), 49–55. doi:10.1016/S0007-6813(03)00088-0

Whitman, M. (2004). In Defence of the Realm: Understanding the Threats to Information Security. *International Journal of Information Management, 24,* 43–57. doi:10.1016/j.ijinfomgt.2003.12.003

Who killed the newspaper?. (2006, August 24). *The Economist.*

Wigand, R. T., Picot, A., & Reichwald, R. (1997). *Information, organization and management: Expanding markets and corporate boundaries.* Chichester, England: Wiley.

Williamson, O. E. (1975). *Markets and hierarchies: Analysis and antitrust implications. A Study in the economics of internal organization.* New York: The Free Press.

Williamson, O. E. (1985). *The economic institutions of capitalism: Firms, markets, relational contracting.* New York: The Free Press.

Wilson, H., & Daniel, E. (2007). The multi-channel challenge: A dynamic capability approach. *Industrial Marketing Management, 36*(1), 10–20. doi:10.1016/j.indmarman.2006.06.015

Wind, J., & Rangaswamy, A. (2001). Customerization: The next revolution in mass customization. *Journal of Interactive Marketing, 15*(1), 13-32.

Wolfinbarger, M., & Gilly, M. C. (2001). Shopping online for freedom, control and fun. *California Management Review, 43*(2), 34-55.

Wonnacott, T., & Wonnacott, R. (1972). *Introductory statistics for business and economics* (2nd ed.). New York: John Wiley & Sons, Inc.

Woodcock, N., & Starkey, M. (2001). I wouldn't start from here' finding a way in CRM projects. *Journal of Database Marketing, 18*(1), 61-74.

Wright, A., & Ralston, D. (2002). The Lagging Development of Small Business Internet Banking in Australia. *Journal of Small Business Management*, *40*(1), 51–57. doi:10.1111/1540-627X.00038

Wright, W., Smith, R., Jesser, R., & Stupeck, M. (1999). Information technology, process reengineering and performance measurement: A balanced scorecard analysis of Compaq Computer Corporation. *Communications of the Association for Information Systems*, *8*(1), 1-47.

WTO. (1998, June 12). Postal *and courier services* (S/C/W/39). Retrieved April 12, 2008, from http://www.wto.org/english/tratop_e/serv_e/w39.doc

Wu, F., & Cavusgil, S. T. (2006). Organizational learning, commitment, and joint value creation in interfirm relationships. *Journal of Business Research, 59*, 81-89.

Wu, I. L., & Wu, K. W. (2005). A hybrid technology acceptance approach for exploring e-CRM adoption in organizations. *Behavior & Information Technology, 24*(4), 303-316.

Oman Ministry of Development. (1996). *Vision for Oman Economy* (Qaboos bin Said speech, June 2 1995). Muscat: Ministry of Development.

Oman Telecommunication Company. (2002). *Annual Book 2002*. Muscat: Oman Telecommunication Company.

Yahoo! (n.d.). *Yahoo! Mobile*. Retrieved February 21, 2009, from http://mobile.yahoo.com/

Yeung, W. L., & Lu, M.-t. (2004). Functional Characteristics of Commercial Web Sites: a Longitudinal Study in Hong Kong. *Information & Management*, *41*(4), 483–495. doi:10.1016/S0378-7206(03)00086-7

Yin, R. K. (2002). *Case Study Research. Design and Methods* (3rd ed., Vol. 5). London: Sage Publications.

Yin, R.K. (1989). *Case study design*. Newbury, CA.

Yukl, G.A. (1989). *Leadership in organizations* (2nd ed.). Englewood Cliffs, NJ: Prentice Hall.

Zahreddine, W., & Mahmoud, Q.H. (2005, March). An agent-based approach to composite mobile Web services. In *Proceedings of 19th International Conference on Advanced Information Networking and Applications* (AINA 2005) (pp. 189-192).

Zaltman, G., Duncan, R., & Holbeck, J. (1973). *Innovations and organizations*. New York: John Wiley and Sons.

Zeimpekis, V., Giaglis, G., & Lekakos, G. (2003). A taxonomy of indoor and outdoor positioning techniques for mobile location services. *ACM SIGECOM Exchanges, 3*(4), 19-27.

Zeithaml, V., (1988). Consumer perceptions of price, quality, and value: A means-end model and synthesis of evidence. *Journal of Marketing, 52*(3), 2-22.

Zeithaml, V., Bitner, M.J., & Gremler, D. (2006). *Services marketing: Integrating customer focus across the firm* (4th ed.). Singapore: McGraw-Hill.

Zeithaml, V., Parasuraman, A., & Malhotra, A. (2000). E-service quality: Definition, dimensions and conceptual model. Working Paper. Cambridge, MA: Marketing Science Institute.

Zhang, L-J., & Chang, H. (2002*). E-logistics processes integration framework*. Retrieved July 28, 2005, from http://www.Webservicesarchitect.com/content/articles/zhang01print.asp

Zhang, P., & von Dran, G. (2001). Expectations and rankings of Web site quality features: Results of two studies on user perceptions. In *Proceedings of the 34th Hawaii International Conference on System Sciences*.

Zhang, P., Small, R. V., von Dran, G. M., & Barcellos, S. (2000). A two factor theory for Web site design. In *Proceedings of the 33rd Hawaii International Conference on System Sciences*.

Zhang, X., & Prybutok, V. (2005). A consumer perspective of e-service quality. *IEEE Transactions on Engineering Management, 52*(4), 461-477.

Zimmermann, O., Doubrovski, V., Grundler, J., & Hogg, K. (2005). Service-Oriented Architecture and Business Process Choreography in an Order Management Scenario: Rationale, Concepts, Lessons Learned. In *Proceedings of Oopsla'05, San Diego, California*.

Zlotkin, G., & Rosenschein, J. S. (1991). Cooperation and conflict resolution via negotiation among autonomous agents in noncooperative domains'. *IEEE Transactions on Systems, Man and Cybernetics, 21*(6), 1317-1324.

Zucker, L. G. (1986). Production of trust: Institutional sources of economic structure: 1840-1920. In B.Staw & L.Cummings (Eds.), *Research in organizational behaviour*, 8, 53-111.

Zwass, V. (1999). Structure and macro-level impacts of electronic commerce: From technological infrastructure to electronic marketplaces. In K. E. Kendall (Ed.), *Emerging information technology* (pp. 517-526). Sage Publications.

About the Contributors

Mehdi Khosrow-Pour (DBA) received his Doctorate in Business Administration from the Nova Southeastern University (FL, USA). Dr. Khosrow-Pour taught undergraduate and graduate information system courses at the Pennsylvania State University–Harrisburg for 20 years where he was the chair of the Information Systems Department for 14 years. He is currently president and publisher of IGI Global, an international academic publishing house with headquarters in Hershey, PA and an editorial office in New York City (www.igi-global.com). He also serves as executive director of the Information Resources Management Association (IRMA) (www.irma-international.org), and executive director of the World Forgotten Children's Foundation (www.world-forgotten-children.org). He is the author/editor of over twenty books in information technology management. He is also the editor-in-chief of the *Information Resources Management Journal*, the *Journal of Cases on Information Technology*, the *Journal of Electronic Commerce in Organizations*, and the *Journal of Information Technology Research*, and has authored more than 50 articles published in various conference proceedings and journals.

* * *

Salim Al-Hajri is an Associate Professor in the Department of Business Studies at the Higher College of Technology in Muscat, Oman. His PhD involved a study in Internet Technology in the Banking Industry in which he investigated the inhibitors and enablers of Internet technology in the Omani banking industry compared with the Australian banking industry. His research interests include information technology, information systems in business organisations, technological innovation, electronic commerce, electronic business and electronic banking. He has published numerous, journal articles, conference papers, and book chapters and recently he is working on editing his new book the Electronic banking.

Lei Chen received a BE in Computer Science from the Nanjing University of Technology, China in 2000 and a PhD in Computer Science and Software Engineering from the Auburn University in 2007. He was a Vodafone Research Fellow from 2006 to 2007. He is currently an assistant professor in the Department of Computer Science at the Sam Houston State University. His research interests include computer networks, network security, and wireless & multimedia networking.

Stathes Hadjiefthymiades received his B.Sc., M.Sc. and Ph.D. degrees in Informatics from the Dept. of Informatics and Telecommunications, University of Athens (UoA). He also received a Joint Engineering-Economics M.Sc. from the National Technical University of Athens. In 1992 he joined the Greek consulting firm Advanced Services Group. In 1995 he joined the Communication Networks

Laboratory (CNL) of UoA. During the period 2001–2002, he served as a visiting assistant professor at the University of Aegean, Dept. of Information and Communication Systems Engineering. On the summer of 2002 he joined the faculty of the Hellenic Open University, Patras, Greece, as an assistant professor. Since December 2003, he is in the faculty of the Dept. of Informatics and Telecommunications, University of Athens, where he is presently an assistant professor. He has participated in numerous EU & National projects. His research interests are in the areas of web engineering, mobile/pervasive computing and networked multimedia. He has contributed to over 100 publications in these areas. Since 2004 he co-ordinates the Pervasive Computing Research Group of CNL.

Wen-Chen Hu received a BE, an ME, an MS, and a PhD all in Computer Science from the Tamkang University, Taiwan, National Central University, Taiwan, University of Iowa, Iowa City, and University of Florida, Gainesville in 1984, 1986, 1993, and 1998, respectively. He is currently an associate professor in the Department of Computer Science at the University of North Dakota. Dr. Hu has published over 70 articles on refereed journals, conferences, and books and advised more than 50 graduate students. He is the Editor-in-Chief of the International Journal of Handheld Computing Research. His current research interests are in handheld computing, electronic and mobile commerce systems, Web technologies, and databases.

Naima Kaabouch received a BS and an MS from the University of Paris 11, France in 1982 and 1985, respectively, and a PhD from the University of Paris 6, France in 1990. She is currently an assistant professor and graduate director in the Department of Electrical Engineering at the University of North Dakota. Her research interests include signal/image processing biomedical engineering, bioinformatics, embedded systems, digital communications, and nondestructive techniques.

Christine Legner is Deputy Chair of Enterprise Systems and Electronic Business at European Business School in Oestrich-Winkel (Germany). Before joining EBS, she was senior researcher and lecturer at University of St. Gallen (Switzerland). She has practical and academic experience in designing e-business solutions. Her research interests include business networking, collaborative business processes, inter- and intra-enterprise systems and service-oriented architectures. Her work has been published in leading academic journals, including the Journal of the Association of Information Systems, the International Journal of Technology Management, Electronic Markets and IEEE Pervasive Computing, and proceedings of international conferences, such as those of the European Conference on Information Systems and the Bled eConference. She holds a PhD in Information Management from the University of St. Gallen, a Master in Applied Economics from Université Paris-Dauphine (France) and a Master in Industrial Engineering from Technical University of Karlsruhe (Germany).

Ioannis Priggouris received his BSc in informatics from the Department of Informatics & Telecommunications at the University of Athens, Athens - Greece in 1997 and his MSc in communication systems and data networks from the same Department in 2000. Over the last years he has been a PhD candidate in the department. Since 1999, he has been a member of the Communication Networks Laboratory (CNL) of the University of Athens. He has participated in the EUROCITI (EUROpean CITIes platform for on-line transaction services) and the PoLoS (Integrated Platform for Location Based Services) project implemented in the context of IST. His research interests are in the areas of mobile computing, QoS and mobility support for IP networks. He is the author 20 publications in the above areas. He works for the Hellenic Telecommunications Organization since 2006.

Pauline Ratnasingam, is an Associate Professor of MIS, Department of Computer Information Systems, Harmon School of Business Administration, University of Central Missouri (UCM), Warrensburg, Missouri. Before joining UCM, she was an Assistant Professor at the School of Business Administration, University of Vermont, Burlington, Vermont. She received her Ph.D titled "Inter-organizational Trust in Business to Business Electronic Commerce" from Erasmus University, Rotterdam School of Management, The Netherlands in 2001. She lectured on topics such as; Project Management, Management of Information Systems, and Electronic Commerce in Australia, New Zealand, Europe and America. She is an associate member of the Association of Information Systems, and is a member of the Information Resources Management Association and Academy of Management. Her research interests include Business Risk Management; Internet based B2B E-Commerce, Organizational Behavior, Knowledge Management and Trust. She is a recipient of a National Science Foundation Grant and has published several articles related to this area in national and international conferences, refereed journals and chapters in books. Her biography is also published in the 58th Edition of Marquis Who's Who – Who's Who in America.

Dimitris Spanoudakis received his diploma in Electronic and Computer Engineering from the Department of Electronic and Computer Engineering, Technical University of Crete, Chania, Greece, in 2002, and his M.Sc. in Communication Systems & Networking, from the Department of Informatics and Telecommunications, University of Athens, Athens, Greece, in 2005. He has performed research on network traffic analysis and modelling, as well as architectures for mobile and wireless systems. He has been a member of the Information Systems Centre at the Technical University of Crete (ISC-TUC) from 1997 till 2000, and member of the Telecommunications Laboratory from 1999 till 2002. Since 2004 he has been a member of the Communication Networks Laboratory of the University of Athens (UoA-CNL). His current research interests are in the areas of network modelling, mobile computing, and autonomic communications. Since September 2006, he works as Informatics Teacher in Public Secondary Education.

Manos Spanoudakis received his B.Sc. in Computer Science and Telecommunications from the Department of Informatics and Telecommunication, University of Athens, Athens, Greece in 2000 and his M.Sc. (with honours) in Computer Systems Technology from the same Department in 2003. He is currently a Ph. D candidate at the same Department, specializing on Content Distribution Networks. He has performed research on WWW architectures for wireless/mobile systems and caching. Since 2000, he has been a member of the Communication Networks Laboratory of the University of Athens (UoA-CNL). He has been involved in the IST projects EURO-CITI (EUROpean CITIes platform for on-line transaction services) and PoLoS (integrated Platform for Location-based Services). Currently he is working as a software enginner for Credit Suisse, Switzerland. His research interests are in the area of wireless/mobile computing, caching technologies, CDNs.

Arthur Tatnall is an Associate Professor in the Graduate School of Business at Victoria University in Melbourne, Australia. His PhD, from Central Queensland University, involved a study in curriculum innovation in which he investigated the manner in which Visual Basic entered the curriculum of an Australian university. His research interests include technological innovation, history of technology, project management, information systems curriculum, information technology in educational management and electronic commerce. He is a Fellow of the Australian Computer Society, active in the International Federation for Information Processing (IFIP) and Chair of IFIP WG9.7 – 'History of Computing'. He is also Editor-in-Chief of the International Journal of Actor-Network Theory and Technological Innovation.

Yanjun Zuo received an MBA from the University of North Dakota in 2000, and an MS and a PhD in Computer Science from the University of Arkansas in 2003 and 2005, respectively. He is currently an assistant professor in the Department of Information Systems and Business Education at the University of North Dakota. His research interests include information and computer security, trustworthy computing, and database security.

Index

Q

R

S

T